Lecture Notes in Control and Information Sciences

Edited by A.V. Balakrishnan and M. Thoma

65

Yaakov Yavin

Numerical Studies in Nonlinear Filtering

Springer-Verlag
Berlin Heidelberg New York Tokyo

Author
Yaakov Yavin
c/o NRIMS
CSIR
P.O. Box 395
Pretoria 0001 – South Africa

ISBN 3-540-13958-3 Springer-Verlag Berlin Heidelberg New York Tokyo
ISBN 0-387-13958-3 Springer-Verlag New York Heidelberg Berlin Tokyo

Library of Congress Cataloging in Publication Data
Yavin, Yaakov
Numerical studies in nonlinear filtering.
(Lecture notes in control and information sciences; 65)
Includes bibliographies.
1. System analysis.
2. Filters (Mathematics).
3. Estimation theory.
I. Title.
II. Series.
QA402.Y3788 1985 003 84-23567

Offsetprinting: Color-Druck, G. Baucke, Berlin
Binding: Lüderitz und Bauer, Berlin
2061/3020-543210

PREFACE

State estimation techniques were developed for situations in engineering in which, based on nonlinear and noise-corrupted measurements of a process, and on a good model of the process, the process is estimated either on-line or off-line using the available measurements.

These techniques became known in the early sixties under the celebrated name of Kalman filtering, and were applied mainly to linear problems. Later developments by Kushner, Wonham and others led to solutions to nonlinear state estimation problems, in which, in general, infinite-dimensional filters are required. Practical algorithms, such as the linearized and extended Kalman filters, which involve only finite-dimen= sional filters, have been most frequently used as approximate solutions to these nonlinear state estimation problems.

The present work offers some new approaches to the construction of finite-dimensional filters as approximate solutions to nonlinear state estima= tion problems. Numerical procedures for the implementation of these filters are given, and the efficiency and applicability of these proce= dures is demonstrated by means of numerical experimentation.

It is my pleasant duty to record here my sincere thanks to the National Research Institute for Mathematical Sciences of the CSIR for encouraging this research. I gratefully acknowledge the contribution made by Mrs. H C Marais, Mrs. R de Villiers and Miss H M Tönsing, who wrote the com= puter programs for the examples presented here.

Finally, I should like to thank Mrs M Russouw for her excellent typing of the manuscript.

Yaakov Yavin

Pretoria, June 1984

CONTENTS

CHAPTER 1

PRELIMINARIES

1.1 INTRODUCTION

The problem of nonlinear filtering or state estimation can be described as follows. $\zeta_x = \{\zeta_x(t),\ t \geq 0\}$, called the signal or the state of the system, is an $\mathbb{R}^m$-valued stochastic process, direct observation of which is not possible. The data related to ζ_x are provided by an $\mathbb{R}^p$-valued measurement process $Y = \{y(t),\ t \geq 0\}$. The minimum variance estimate of $\zeta_x(t)$, based on the measurements $Y^t = \{y(s),\ 0 \leq s \leq t\}$, is given by the conditional expectation $E[\zeta_x(t)|Y^t]$. This work deals with the pro= blem of finding implementable approximations to $E[\zeta_x(t)|Y^t]$. The efforts here have been directed exclusively towards the derivation of finite-dimensional filters for computing approximations to $\{E[\zeta_x(t)|Y^t],\ 0 < t < \tau_T = \min(T,\tau)\}$, where τ is the first exit time of $\zeta_x(t)$ from a given open and bounded domain $D \subset \mathbb{R}^m$, and T is a given positive number.

The following nonlinear filtering problems have been considered here:

(a) Estimation of parameters via state observation.

The process ζ_x satisfies the following equation

$$\zeta_x(t) = x + \int_0^t [\eta(s)f(\zeta_x(s)) + g(\zeta_x(s))]ds + BW(t)\ ,\ t \geq 0,\ x \in \mathbb{R}^m \tag{1.1}$$

where $f : \mathbb{R}^m \to \mathbb{R}^m$ and $g : \mathbb{R}^m \to \mathbb{R}^m$ are given functions; $\{W(t),\ t \geq 0\}$ is an $\mathbb{R}^m$-valued standard Wiener process; and $\{\eta(t),\ t \geq 0\}$ is a continuous time Markov chain with a state space S which is at most countable. (The special case where $\eta(t) = \eta,\ t \geq 0$; i.e. η is a random element, is con=

sidered first). The measurement process Y is given by

$$y(t) = \zeta_x(t) \quad , \; t \geq 0 \; . \tag{1.2}$$

The problem is to find $\{E[\eta(t)|\zeta_x(s), 0 \leq s \leq t] \; , \; t \in (0,T]\}$.

(b) The 'standard' nonlinear filtering problem.

The process ζ_x satisfies the following equation

$$\zeta_x(t) = x + \int_0^t f(\zeta_x(s))ds + BW(t) \; , \; t \geq 0 \quad , \quad x \in \mathbb{R}^m \tag{1.3}$$

and the measurement process Y is given by

$$y(t) = \int_0^t g(\zeta_x(s))ds + \Gamma v(t) \quad , \quad t \geq 0 \tag{1.4}$$

where $f : \mathbb{R}^m \to \mathbb{R}^m$ and $g : \mathbb{R}^m \to \mathbb{R}^p$ are given functions; B and Γ are given m×m and p×p matrices respectively; $\{W(t), t \geq 0\}$ and $\{v(t), t \geq 0\}$ are $\mathbb{R}^m$-valued and $\mathbb{R}^p$-valued standard Wiener processes respectively. Let $D \subset \mathbb{R}^m$ be an open and bounded domain. The problem is to find approxima= tions to $\{\hat{\zeta}_x(t) \triangleq E[\zeta_x(t \wedge \tau_T -)|y(s), 0 \leq s \leq t \wedge \tau_T -], t \in [0,T]\}$ ($a \wedge b \triangleq \min(a,b)$, $\tau_T = \tau \wedge T$ where τ is the first exit time of $\zeta_x(t)$ from D).

(c) A modified Kalman filter.

The process ζ_x satisfies the following equation

$$\zeta_x(t) = x + \int_0^t A(\zeta_x(s))\zeta_x(s)ds + BW(t) \quad , \quad t \geq 0 \quad , \quad x \in \mathbb{R}^m \tag{1.5}$$

and the measurement process Y is given by

$$y(t_k) = \gamma(t_k)H(t_k)\zeta_x(t_k) + v(t_k) \quad , \quad t_k = k\Delta \; , \quad k=0,1,\ldots \tag{1.6}$$

where $A(x)$, $x \in \mathbb{R}^m$ and B are given m×m matrices; $\{H(t_k), k=0,1,\ldots\}$ are given p×m matrices; $\{W(t), t \geq 0\}$ is an $\mathbb{R}^m$-valued standard Wiener process and $\{v(t_k), k=0,1,\ldots\}$ is a sequence of independent R^p-valued random Gaussian elements. Two cases are considered:

(c-1) $\gamma(t_k) = 1 \quad , \quad k=0,1,\ldots$ (1.7)

(c-2) $\gamma(t_k) \in \{0,1\}$ according to

$$p(k) = P(\gamma(t_k) = 1) \quad , \quad k=0,1,2,\ldots \tag{1.8}$$

$$q(k) = 1 - p(k) = P(\gamma(t_k) = 0) \quad , \quad k=0,1,2,\ldots \tag{1.9}$$

where the sequence $\{p(k)\}$ is given.

Let $Y^k = \{y(t_o),y(t_1),\ldots,y(t_k)\}$. The problem is to find approximations to $\{E[\zeta_x(t_k)|Y^k], k=1,2,\ldots\}$ for cases (c-1) and (c-2) respectively.

(d) State estimation for systems with interrupted observations.

The process ζ_x satisfies the following equation

$$\zeta_x(t) = x + \int_o^t f(\zeta_x(s))ds + BW(t) \quad , \quad t \geq 0 \quad , \quad x \in \mathbb{R}^m \tag{1.3}$$

and the measurement process Y is given by

$$y(t) = \int_o^t \theta(s)g(\zeta_x(s))ds + \Gamma v(t) \quad , \quad t \geq 0 \tag{1.10}$$

where f,g,B,Γ, $\{W(t),\ t \geq 0\}$ and $\{v(t),\ t \geq 0\}$ are the same as described in Problem (b). $\{\theta(t),\ t \geq 0\}$ is a homogeneous jump Markov process with state space $\{0,1\}$. Let $D \subset \mathbb{R}^m$ be an open and bounded domain. The pro= blem is to find approximations to

$$\{E[(\zeta_x(t\wedge\tau_T-),\theta(t\wedge\tau_T-))|y(s) \quad , \quad 0 \leq s \leq t\wedge\tau_T-] \ , \ t \in [0,T]\}.$$

(e) Estimation in a multitarget environment.

Given L disjoint cells in the (x_1,x_2)-plane. In each of the cells there may be at most one target. Let $\theta = (\theta_1,\ldots,\theta_L)$ be a random element such that, for $j=1,\ldots,L$: $\theta_j=1$ if there is a target at the j-th cell, and $\theta_j=0$ otherwise.

The measurement process Y is given by

$$y_i(t) = \int_o^t c_{in(u)}\, \theta_{n(u)} du + \gamma_i v_i(t) \quad , \quad t \in [0,T], \quad i=1,2 \tag{1.11}$$

where (c_{1j},c_{2j}), j=1,...,L are the centres of the cells; $\{(v_1(t),v_2(t)), t \geq 0\}$ is an $\mathbb{R}^2$-valued standard Wiener process, and $\{n(t),t \geq 0\}$ is a continuous-time Markov chain with state space $\{1,\ldots,L\}$. Since, as tacitly assumed, the processes θ, $\{n(t),t \geq 0\}$ and $\{(v_1(t),v_2(t)), t \geq 0\}$ cannot be observed, it follows that the measurements cannot be associated with certainty with the various targets under consideration. The problem is to find $\{E[\theta|y(s), 0 \leq s \leq t], \ t \in [0,T]\}$ (i.e. to learn in which cells the targets are located).

(f) State and parameter estimation.

The process ζ_x satisfies the following equation

$$\zeta_x(t) = x + \int_0^t f(\zeta_x(s),\theta)ds + B(\theta)W(t), \quad t \geq 0, \quad x \in \mathbb{R}^m \tag{1.12}$$

and the measurement process is given by

$$y(t) = \int_0^t g(\zeta_x(s),\theta)ds + \Gamma v(t) \quad , \quad t \geq 0 \tag{1.13}$$

where θ is a random element with values in Ω_θ, $\Omega_\theta \subset \mathbb{R}^r$; $f : \mathbb{R}^m \times \Omega_\theta \to \mathbb{R}^m$ and $g : \mathbb{R}^m \times \Omega_\theta \to \mathbb{R}^p$ are given functions; $B(\theta)$, $\theta \in \Omega_\theta$, and Γ are given m×m and p×p matrices respectively, and $\{W(t), t \geq 0\}$ and $\{v(t),t \geq 0\}$ are $\mathbb{R}^m$-valued and $\mathbb{R}^p$-valued standard Wiener processes respectively. Let $D \subset \mathbb{R}^m$ be an open and bounded domain. The problem is to find approxima= tions to $\{E[(\zeta_x(t\wedge\tau_T-),\theta)|y(s), 0 \leq s \leq t\wedge\tau_T-], t \in [0,T]\}$.

(g) State estimation for jump diffusion processes.

The process ζ_x satisfies the following equation

$$\zeta_x(t) = x + \int_0^t f(\zeta_x(s-))ds + BW(t) + CN(t), \ t \geq 0, \ x \in \mathbb{R}^m \tag{1.14}$$

and the measurement process is given by

$$y(t) = \int_0^t g(\zeta_x(s))ds + \Gamma v(t) \quad , \quad t \geq 0 \tag{1.15}$$

where f,g,B,Γ, $\{W(t), t \geq 0\}$ and $\{v(t), t \geq 0\}$ are the same as described in Problem (b). $\{N(t) = (N_1(t),\dots,N_r(t)),\ t \ge 0\}$ is a vector of mutually independent Poisson processes with parameter $Q = (q_1,\dots,q_r)$, and C is a p×r given matrix. The problem is to find approximations to $\{E[\zeta_x(t\wedge\tau_T-)\,|\,y(s),\ 0 \le s \le t\wedge\tau_T-],\ \ t \in [0,T]\}$.

(h) The prediction problem.

The processes ζ_x and Y are given by equations (1.3) and (1.4) res= pectively. Let $D \subset \mathbb{R}^m$ be an open and bounded domain. The problem is to find approximations to $\{E[\zeta_x(t\wedge\tau_T-)\,|\,y(u),\ 0 \le u \le s\wedge\tau_T-],\ 0 \le s \le t \le T\}$.

Throughout this work, except for Problem (c) (Chapter 4), a unified approach towards the nonlinear filtering problem has been adopted. This approach is based on the implementation of the results obtained in Fuji= saki et al. [1.1]. These results are stated in Theorem 1.1, Section 1.2. In Section 1.3 the equations for the optimal least-squares estimator $\{E[\zeta_x(t)\,|\,y(s),\ 0 \le s \le t]\ ,\ t \in [0,T]\}$, in the case where the system is given by equations (1.14)-(1.15), are derived by means of the application of Theorem 1.1. The Kalman filter equations follow as a special case.

Chapter 2 is devoted to the treatment of Problem (a). Using the results of [1.1], the filter equations for computing $\{E[\eta(t)\,|\,\zeta_x(s),\ 0 \le s \le t]$, $t \in [0,T]\}$ are derived. An algorithm for the numerical solution of these equations is suggested and numerical results, from the numerical experi= mentation with the filter equations, are presented.

Chapter 3 is devoted to the treatment of Problem (b). First, the pro= cess $\{\zeta_x(t\wedge\tau_T),\ t \in [0,T]\}$ is approximated by a continuous-time Markov chain $\{\zeta_x^h(t\wedge\tau_T^h),\ t \in [0,T]\}$ with a finite state space S, $S \subset D$. Then, using the results of [1.1], an optimal least-squares filter is derived for the on-line computation of $\{E[\zeta_x^h(t\wedge\tau_T^h-)\,|\,y^h(s),\ 0 \le s \le t\wedge\tau_T^h-]$, $t \in [0,T]\}$(where $y^h(t) = \int_0^t g(\zeta_x^h(s))ds + \Gamma v(t)$). Based on this filter an estimator $\{\hat{\zeta}_x^{h,y}(k\Delta),\ k\Delta \in [0,T]\}$ is constructed as an approximation to $\{\hat{\zeta}_x(k\Delta) = E[\zeta_x(k\Delta \wedge \tau_T-) \,|\, y(\ell\Delta),\ 0 \le \ell\Delta \le k\Delta \wedge \tau_T-],\ \ k\Delta \in [0,T]\}$. Problems concerning the weak convergence of $\{\hat{\zeta}_x^{h,y}(k\Delta),\ k\Delta \in [0,T]\}$ to $\{\hat{\zeta}_x(k\Delta),\ k\Delta \in [0,T]\}$ as $h \downarrow 0$ are not discussed in this work. Instead, the role of $\hat{\zeta}_x^{h,y}$, as an approximation to $\hat{\zeta}_x$, is demonstrated by means of numerical experimentation with several examples.

Chapter 4 is devoted to the treatment of Problem (c). First, equation (1.5) is transformed into a sequence of integral equations from which a discrete in time representation for equation (1.5) is obtained. Based on this representation, a procedure for the discretization (in time) of the system given by equation (1.5) is suggested. Second, using the discretization procedure, a modified Kalman filter is proposed for the computation of a process $\{\hat{\zeta}(k|k),\ k\Delta \in [0,T]\}$. This process serves as an approximation to $\{\hat{\zeta}_x(t_k) = E[\zeta_x(t_k) \,|\, Y^k],\ t_k = k\Delta \in [0,T]\}$. The role of $\{\hat{\zeta}(k|k),\ k\Delta \in [0,T]\}$ as an approximation to $\{\hat{\zeta}_x(t_k),\ t_k \in [0,T]\}$, is illustrated throughout numerical experimentation with several examples.

Chapter 5 is devoted to the treatment of Problem (d). First, the process $\{(\zeta_x(t\wedge\tau_T),\theta(t\wedge\tau_T)),\ t \in [0,T]\}$ is approximated by a continuous-time Markov chain $\{(\zeta_x^h(t\wedge\tau_T^h),\tilde{\theta}(t\wedge\tau_T^h)),\ t \in [0,T]\}$ with a finite state space S, $S \subset D \times \{0,1\}$. Then, using the results of [1.1], an optimal least-squares filter is derived for the on-line computation of $(\hat{\zeta}_x^h(t),\hat{\theta}^h(t)) \triangleq E[(\zeta_x^h(t\wedge\tau_T^h-),\tilde{\theta}(t\wedge\tau_T^h-)) \,|\, y^h(s),\ 0 \le s \le t\wedge\tau_T^h-],\ t \in [0,T]$ (where $y^h(t) = \int_0^t \tilde{\theta}(s)g(\zeta_x^h(s))ds + \Gamma v(t))$. Based on this filter, an estimator $(\hat{\zeta}_x^{h,y},\ \hat{\theta}^{h,y}) = \{(\hat{\zeta}_x^{h,y}(k),\hat{\theta}^{h,y}(k)),\ k\Delta \in [0,T]\}$ is constructed as an approximation to $\{(\hat{\zeta}_x(k\Delta),\hat{\theta}(k\Delta)) = E[(\zeta_x(k\Delta\wedge\tau_T-),\theta(k\Delta\wedge\tau_T-)) \,|\, y(\ell\Delta),\ 0 \le \ell\Delta \le k\Delta\wedge\tau_T-],\ k\Delta \in [0,T]\}$. Problems concerning the weak convergence of $(\hat{\zeta}_x^{h,y},\ \hat{\theta}^{h,y})$ to $(\hat{\zeta}_x,\hat{\theta})$ as $h \downarrow 0$ are not discussed in this work. Numerical experimentation with several examples demonstrates the role of

$(\hat{\zeta}_x^{h,y},\hat{\theta}^{h,y})$ as an approximation to $(\hat{\zeta}_x,\hat{\theta})$.

Chapter 6 is devoted to the treatment of Problem (e). Using the results of [1.1], the filter equations for computing $\{\hat{\theta}(t) = E[\theta(t)|y(s),\ 0 \le s \le t],\ t \in [0,T]\}$ are derived. An algorithm for the numerical solution of these equations is suggested, and numerical experimentation is carried out.

Chapter 7 is devoted to the treatment of Problem (f). Following the procedures applied in Chapters 3 and 5, an estimator $\{(\hat{\zeta}_x^{h,y}(k),\hat{\theta}^{h_1,y}(k)),\ k\Delta \in [0,T]\}$ is constructed as an approximation to $\{E[(\zeta_x(k\Delta\wedge\tau_T-),\theta)|y(\ell\Delta),\ 0 \le \ell\Delta \le k\Delta\wedge\tau_T-],\ k\Delta \in [0,T]\}$. The role of $(\hat{\zeta}_x^{h,y},\hat{\theta}^{h_1,y})$ as an estimator of (ζ_x,θ) is illustrated by means of numerical experimentation.

Chapter 8 is devoted to the treatment of Problem (g). Here, again, the procedures applied in Chapters 3 and 5 are used and an estimator $\{\hat{\zeta}_x^{h,y}(k),\ k\Delta \in [0,T]\}$, which serves as an approximation to $\{E[\zeta_x(k\Delta\wedge\tau_T-)|y(\ell\Delta),\ 0 \le \ell\Delta \le k\Delta\wedge\tau_T-],\ k\Delta \in [0,T]\}$, is constructed. The role of $\hat{\zeta}_x^{h,y}$ as an estimator of ζ_x is illustrated by means of numerical experimentation.

Chapter 9 is devoted to the treatment of Problem (h). First, the process $\{\zeta_x(t\wedge\tau_T),\ t \in [0,T]\}$ is approximated by a continuous-time Markov chain $\{\zeta_x^h(t\wedge\tau_T^h),\ t \in [0,T]\}$ with a finite state space S, $S \subset D$. Then, an optimal least-squares filter is derived for the on-line computation of $\{\hat{\zeta}_x^h(t,s) = E[\zeta_x^h(t\wedge\tau_T^h-)|y^h(u),\ 0 \le u \le s\wedge\tau_T^h-],\ t \in [s,T]\}$ (where $y^h(t) = \int_0^t g(\zeta_x^h(u))du + \Gamma v(t))$. Based on this filter, an estimator $\{\hat{\zeta}_x^{h,y}(k,s),\ k\Delta \in [s,T]\}$ is constructed as an approximation to $\{\hat{\zeta}_x(k\Delta,s) = E[\zeta_x(k\Delta\wedge\tau_T-)|y(\ell\Delta),\ 0 \le \ell\Delta \le s\wedge\tau_T-],\ k\Delta \in [s,T]\}$. The role of $\hat{\zeta}_x^{h,y}$ as an estimator (predictor) is illustrated by means of numerical experimentation. Finally, some extensions of linear filtering are dealt with in Chapter 10.

Extensive work on nonlinear filtering on stochastic continuous-time sys=

tems has been done, various approaches having being used. For more details see, for example, Stratonovich [1.2], Kushner [1.3], Wonham [1.4], Bucy [1.5], Bucy and Joseph [1.6], Zakai [1.7], Jazwinski [1.8], Sage and Melsa [1.9], Fujisaki et al. [1.1], Frost and Kailath [1.10], McGarty [1.11], Liptser and Shiryayev [1.12], Anderson and Moore [1.13], Kallianpur [1.14], Davis and Wellings [1.15], Björk [1.16-1.17] and Takeuchi and Akashi [1.18]. A fairly comprehensive survey, which reflects the state of art in nonlinear filtering theory, is given in Hazewinkel and Willems [1.19], and in Kallianpur and Karandikar [1.32].

It was found, however, that for Problems (b), (c), (d) (see Sawaragi et al. [1.20]), (f),(g) and (h), the optimal least-squares filters are infinite dimensional and as a result are nonimplementable. Exceptions are the L.Q.G. case for which the Kalman filter is optimal, and a few other cases summarized in Van Schuppen [1.21]. Consequently, practical algorithms for approximating $\{E[\zeta_x(t)|y(s),\ 0 \le s \le t],\ t \in [0,T]\}$, where ζ_x satisfies an equation of form

$$\zeta_x(t) = x + \int_0^t f(\zeta_x(s))ds + \int_0^t B(\zeta_x(s))dW(s),\ t \ge 0,\ x \in \mathbb{R}^m \tag{1.16}$$

and Y is determined by equation (1.4), were developed. The best-known practical algorithms are linearized and extended Kalman filters (see [1.8], [1.9]) and these have been used most frequently.

The present work offers some new approaches to the construction of finite-dimensional filters as approximate solutions to the corresponding nonlinear state estimation problems.

1.2 THE FUJISAKI-KALLIANPUR-KUNITA FILTERING FORMULA

Let (Ω,F,P) be a complete probability space equipped with a nondecreasing family of sub-σ-fields $\{F_t\}_{t \ge 0}$ (i.e., $F_s \subset F_t \subset F$, $s \le t$), where F_0 is completed with P-null sets and $\{F_t\}_{t \geq 0}$ is right-continuous (i.e., $F_t = F_{t+} = \bigcap_{s > t} F_s$).

Let $\{M_t, t \geq 0\}$ be a family of $\mathbb{R}^d$-valued random elements on (Ω, F, P) such that for each $t \geq 0$, M_t is F_t-measurable. The process $\{M_t, t \geq 0\}$ is a (F_t, P)-martingale if:

$$E|M_t| < \infty \ , \quad t \geq 0 \qquad (|M_t|^2 \triangleq \sum_{i=1}^{d} M_{ti}^2) \tag{1.17}$$

$$E[M_t \mid F_s] = M_s \quad , \quad s \leq t. \tag{1.18}$$

For more details on martingales see, for example, Meyer [1.22], [1.12] and Shiryayev [1.23].

Let $\zeta = \{\zeta(t), t \in [0,T]\}$ be an S-valued Markov process on (Ω, F, P) where S is a complete separable metric space. Suppose that the measurements concerning ζ are provided by an observation process $Z = \{z(t), t \in [0,T]\}$ of the form

$$z(t) = \int_0^t h_s ds + v(t) \quad , \quad t \in [0,T] \quad , \tag{1.19}$$

where $\mathbf{V} = \{v(t), t \in [0,T]\}$ is an $\mathbb{R}^p$-valued standard Wiener process and $h_t(\omega)$ is a (t,ω)-measurable $\mathbb{R}^p$-valued process such that $\int_0^T E|h_t|^2 dt < \infty$. Define

$$F_t^z \triangleq \sigma(z(s) \ ; \ 0 \leq s \leq t), \ t \in [0,T] \tag{1.20}$$

$$G_t \triangleq \sigma(\zeta(s), v(s) \ ; \ 0 \leq s \leq t) \quad , \quad t \in [0,T] \tag{1.21}$$

i.e., F_t^z and G_t are the smallest σ-fields generated by $\{z(s); 0 \leq s \leq t\}$ and $\{\zeta(s), v(s) \ ; \ 0 \leq s \leq t\}$ respectively.

Let D(A) denote the class of all functions F, $F : S \to \mathbb{R}$, such that F is measurable on S,

$$E|F(\zeta(t))|^2 < \infty \quad \text{for all } t \in [0,T], \tag{1.22}$$

and there exists a jointly (t,ω)-measurable real function A_tF satisfying:

(i) For each $t \in [0,T]$, $(A_tF)(\zeta(t))$ is $\sigma(\zeta(s),z(s);\ 0 \le s \le t)$-measurable.

(ii) $\int_0^T E|(A_tF)(\zeta(t))|^2 dt < \infty$ (1.23)

(iii) $M_t(F) \triangleq F(\zeta(t)) - E[F(\zeta(0))|F_0^z] - \int_0^t (A_sF)(\zeta(s))ds$ is a (G_t,P)-martingale.

The following theorem is a straightforward conclusion of the results obtained in [1.1].

Theorem 1.1

Suppose that

(i) for each $t \in [0,T]$, the σ-fields G_t and $\sigma(v(t_2)-v(t_1);\ t < t_1 < t_2 \le T)$ are independent;

(ii) for each $t \in [0,T)$, h_t is G_t-measurable;

(iii) ζ and $\mathbf{V}$ are mutually independent.

Denote

$$\nu(t) \triangleq z(t) - \int_0^t E[h_s \mid F_s^z]ds \ , \ t \in [0,T] \ . \tag{1.24}$$

If $F \in D(A)$ satisfies $\int_0^T E \mid F(\zeta(t))h_t|^2 dt < \infty$, then

$$E[F(\zeta(t))|F_t^z] = EF(\zeta(0)) + \int_0^t E[(A_sF)(\zeta(s))|F_s^z]ds$$

$$+ \int_0^t (E[F(\zeta(s))h_s|F_s^z] - E[F(\zeta(s))|F_s^z]E[h_s|F_s^z],\ d\nu(s)) \tag{1.25}$$

$t \in [0,T]$,

where $(a,b) = \sum_{i=1}^{p} a_ib_i$.

Equation (1.25) is here called the *Fujisaki-Kallianpur-Kunita filtering formula.*

In this work we are interested only in cases where conditions (i) and (iii) of Theorem 1.1 are satisfied. Thus it is tacitly assumed here that the Markov chains $\{\zeta_x^h(t),\ t \in [0,T]\}$, introduced in Chapters 3, 8 and 9, and the Markov chains $\{(\zeta_x^h(t),\tilde{\theta}(t)),\ t \in [0,T]\}$ and $\{(\zeta_x^h(t),\ \theta^{h_1}(t)),\ t \in [0,T]\}$, introduced in Chapters 5 and 7 respectively, are constructed in such a manner that in each of the cases they are in= dependent of $\mathbf{V}$. Furthermore, we restrict the class of observation pro= cesses with which we deal in this work so that condition (i) of Theorem 1.1 is always satisfied.

1.3 THE 'STANDARD' NONLINEAR FILTERING PROBLEM

Consider an $\mathbb{R}^m$-valued Markov process $\zeta_x = \{\zeta_x(t),\ t \geq 0\}$ satisfying the equation

$$\zeta_x(t) = \zeta_x(0) + \int_0^t f(\zeta_x(s-))ds + \int_0^t B(\zeta_x(s-))dW(s) + \int_0^t \int_{\mathbb{R}^m} C(\zeta_x(s-),u)\mu(ds,du) \tag{1.26}$$

$$t \geq 0, \quad \zeta_x(0) \in \mathbb{R}^m \ , \quad E|\zeta_x(0)|^2 < \infty \ ,$$

with the noisy observation of ζ_x given by

$$y(t) = \int_0^t g(\zeta_x(s))ds + \Gamma v(t), \ t \geq 0 \tag{1.27}$$

where $\mathbf{W} = \{W(t),\ t \geq 0\}$ and $\mathbf{V} = \{v(t),\ t \geq 0\}$ are an $\mathbb{R}^m$-valued and an $\mathbb{R}^p$-valued standard Wiener processes respectively; μ is a zero-mean Poisson random measure on $[0,\infty) \times \mathbb{R}^m$, i.e.,

$$\mu(t,A) = \nu(t,A) - t\pi(A),\ t \geq 0,\ A \in \mathcal{B}(\mathbb{R}^m) \tag{1.28}$$

where $\mathcal{B}(\mathbb{R}^m)$ denotes the m-dimensional Borel σ-algebra, and $\{\nu(t,A), t \geq 0\}$, $A \in \mathcal{B}(\mathbb{R}^m)$, is a Poisson process with

$$E\nu(t,A) = t\pi(A), \; t \geq 0, \; A \in \mathcal{B}(\mathbb{R}^m). \tag{1.29}$$

For more details on $\{\nu(t,A);\ t \geq 0,\ A \in \mathcal{B}(\mathbb{R}^m)\}$ see Gihman and Skorohod [1.24]. It is assumed that $\pi(\mathbb{R}^m) < \infty$ and $\int_{\mathbb{R}^m} |u|^2\pi(du) < \infty$, and it is further assumed that $\zeta_x(0)$, $\mathbf{W}$, $\mathbf{V}$ and μ are mutually independent. (Note that equation (1.14) can be obtained as a special case of equation (1.26).) $f : \mathbb{R}^m \to \mathbb{R}^m$, $B : \mathbb{R}^m \to \mathbb{R}^{m\times m}$, $C : \mathbb{R}^m \times \mathbb{R}^m \to \mathbb{R}^m$ and $g : \mathbb{R}^m \to \mathbb{R}^p$ are given functions such that:

(i) there is an L for which

$$|f(x)|^2 + |B(x)|^2 + \int_{\mathbb{R}^m} |C(x,u)|^2\pi(du) + |g(x)|^2 \leq L(1 + |x|^2), \tag{1.30}$$

$$x \in \mathbb{R}^m$$

(ii) for any $R > 0$ there is a constant C_R, such that when $|x| < R$, $|x'| < R$, $x,x' \in \mathbb{R}^m$

$$|f(x) - f(x')|^2 + |B(x) - B(x')|^2 + \int_{\mathbb{R}^m} |C(x,u) - C(x',u)|^2\pi(du) + |g(x) - g(x')|^2 \leq C_R|x-x'|^2, \tag{1.31}$$

where $|f(x)|^2 = \sum_{i=1}^{m} f_i^2(x)$, $|B(x)|^2 = \sum_{i,j=1}^{m} B_{ij}^2(x)$, $|g(x)|^2 = \sum_{i=1}^{p} g_i^2(x)$ and $\mathbb{R}^{m\times m}$ denotes the space of all real $m\times m$ matrices. $\Gamma \in \mathbb{R}^{p\times p}$ is a given matrix such that Γ^{-1} exists.

Let $F : \mathbb{R}^m \to \mathbb{R}$ be bounded and twice continuously differentiable on $\mathbb{R}^m$ and define the following operator

$$(\mathcal{L}F)(x) \triangleq \sum_{i=1}^{m} f_i(x)\partial F(x)/\partial x_i + (\tfrac{1}{2}) \sum_{i,j=1}^{m} (B(x)B'(x))_{ij}\, \partial^2 F(x)/\partial x_i \partial x_j + \int_{\mathbb{R}^m} [F(x + C(x,u)) - F(x) - \sum_{i=1}^{m} C_i(x,u)\partial F(x)/\partial x_i]\pi(du). \tag{1.32}$$

It can be shown (Stroock [1.25]) that

$$m_t(F) \triangleq F(\zeta_x(t)) - E[\zeta_x(0)|\mathcal{F}_0^y] - \int_0^t (\mathcal{L}F)(\zeta_x(s))ds,\ t \in [0,T] \tag{1.33}$$

is a (G_t,P)-martingale on $[0,T]$, where $G_t \triangleq \sigma(\zeta_x(s),v(s);\ 0 \le s \le t)$ and $\mathcal{F}_t^y \triangleq \sigma(y(s);\ 0 \le s \le t)$, $t \in [0,T]$. Define

$$h_t \triangleq \Gamma^{-1}g(\zeta_x(t)),\ t \in [0,T], \tag{1.34}$$

$$z(t) \triangleq \int_0^t h_s ds + v(t),\ t \in [0,T] \tag{1.35}$$

$$\nu(t) = z(t) - \int_0^t E[h_s|\mathcal{F}_s^y]ds\ ,\ t \in [0,T]. \tag{1.36}$$

Note that $z(t) = \Gamma^{-1}y(t)$, $t \in [0,T]$. Consequently $\mathcal{F}_t^z = \mathcal{F}_t^y$, $t \in [0,T]$.

In order to derive the filter equations for $E[\zeta_x(t)|\mathcal{F}_t^y]$ take $F(x) = x_i$ and $A_tF = \mathcal{L}x_i$, $i \in \{1,\dots,m\}$. Then $F \in D(A)$ and the assumptions of Theorem 1.1 are easily seen to be satisfied. Hence equations (1.25) and (1.32) yield

$$\begin{aligned} E[\zeta_{xi}(t)|\mathcal{F}_t^y] &= E\zeta_{xi}(0) + \int_0^t E[f_i(\zeta_x(s))|\mathcal{F}_s^y]ds \\ &+ \int_0^t (E[\zeta_{xi}(s)h_s|\mathcal{F}_s^y] - E[\zeta_{xi}(s)|\mathcal{F}_s^y]E[h_s|\mathcal{F}_s^y], d\nu(s)) \end{aligned} \tag{1.37}$$

$i=1,\dots,m$, $t \in [0,T]$,

or, by using the notation $\pi_t(F) \triangleq E[F(\zeta_x(t))|\mathcal{F}_t^y]$,

$$\begin{aligned} \pi_t(\zeta_{xi}) &= E\zeta_{xi}(0) + \int_0^t \pi_s(f_i)ds \\ &+ \int_0^t (\Gamma^{-1}\{\pi_s(\zeta_{xi}g) - \pi_s(\zeta_{xi})\pi_s(g)\}, \Gamma^{-1}\{dy(s) - \pi_s(g)ds\}) \end{aligned} \tag{1.38}$$

$i=1,\dots,m$, $t \in [0,T]$.

Equations (1.38) constitute the filter equations for computing $\{E[\zeta_x(t)|\mathcal{F}_t^y],\ t \in [0,T]\}$, for the system given by equations (1.26)-(1.27).

In the case where $C(x,u) \equiv 0$, equations (1.38) constitute the filter equations for the 'standard' nonlinear filtering problem (see Problem (a)). Denote

$$\hat{\varepsilon}(t) \triangleq E[(\zeta_x(t) - \pi_t(\zeta_x))(\zeta_x(t) - \pi_t(\zeta_x))' | F_t^y], \ t \in [0,T] \ . \tag{1.39}$$

In order to derive the differential equations satisfied by $\hat{\varepsilon}$ take $F(x) = x_i x_j$ and $A_t F = \mathcal{L} x_i x_j$, $i,j \in \{1,\dots,m\}$. Then $F \in D(A)$ and the assumptions of Theorem 1.1 are easily seen to be satisfied. Hence, equations (1.25) and (1.32) yield

$$\begin{aligned}\pi_t(\zeta_{xi}\zeta_{xj}) = E[\zeta_{xi}(0)\zeta_{xj}(0)] + \int_0^t (&\pi_s(\zeta_{xj} f_i) + \pi_s(\zeta_{xi} f_j) + \pi_s((BB')_{ij}) \\ &+ \pi_s(\int_{\mathbb{R}^m} C_i(\cdot,u) C_j(\cdot,u)\pi(du)))ds \qquad (1.40) \\ + \int_0^t (\Gamma^{-1}\{\pi_s(\zeta_{xi}\zeta_{xj} g) &- \pi_s(\zeta_{xi}\zeta_{xj})\pi_s(g)\}, \Gamma^{-1}\{dy(s) - \pi_s(g)ds\})\end{aligned}$$

$i,j=1,\dots,m \quad , \quad t \in [0,T]$.

Equations (1.40), together with equation (1.41)

$$\hat{\varepsilon}(t) = \pi_t(\zeta_x \zeta_x') - \pi_t(\zeta_x)\pi_t(\zeta_x'), \ t \in [0,T], \tag{1.41}$$

constitute the equations for computing $\hat{\varepsilon}$.

In general, the solution of equations (1.38) involves all the conditional moments of ζ_x. This is illustrated by the following simple example. Consider the following system

$$\zeta_x(t) = \zeta_x(0) + \int_0^t a\, \zeta_x^2(s)ds + \sigma W(t), \ t \geq 0, \ \zeta_x(0) \in \mathbb{R}, \ E\zeta_x^2(0) < \infty, \tag{1.42}$$

$$y(t) = \int_0^t \zeta_x(s)ds + \gamma v(t) \quad , \quad t \geq 0 \tag{1.43}$$

where $\mathbf{W} = \{W(t), t \geq 0\}$ and $\mathbf{V} = \{v(t), t \geq 0\}$ are independent $\mathbb{R}$-valued standard Wiener processes; a, σ and γ are given positive numbers; and $\zeta_x(0)$ is independent of $\mathbf{W}$ and $\mathbf{V}$.

In this case equations (1.38) and (1.40) reduce to

$$\pi_t(\zeta_x) = E\zeta_x(0) + \int_0^t \pi_s(a\,\zeta_x^2)ds$$
$$+ \gamma^{-2}\int_0^t (\pi_s(\zeta_x^2) - \pi_s^2(\zeta_x))(dy(s) - \pi_s(\zeta_x)ds),\ t \in [0,T] \tag{1.44}$$

$$\pi_t(\zeta_x^2) = E\zeta_x^2(0) + \int_0^t (2\pi_s(a\,\zeta_x^3) + \sigma^2)ds$$
$$+ \gamma^{-2}\int_0^t (\pi_s(\zeta_x^3) - \pi_s(\zeta_x^2)\pi_s(\zeta_x))(dy(s) - \pi_s(\zeta_x)ds),\ t \in [0,T] \tag{1.45}$$

and, for $n \geq 3$,

$$\pi_t(\zeta_x^n) = E\zeta_x^n(0) + \int_0^t (na\pi_s(\zeta_x^{n+1}) + \tfrac{1}{2}\sigma^2\, n(n-1)\pi_s(\zeta_x^{n-2}))ds$$
$$+ \gamma^{-2}\int_0^t (\pi_s(\zeta_x^{n+1}) - \pi_s(\zeta_x^n)\pi_s(\zeta_x))(dy(s) - \pi_s(\zeta_x)ds),\ t \in [0,T]. \tag{1.46}$$

Hence, in order to compute $\{\pi_t(\zeta_x) = E[\zeta_x(t)|F_t^y],\ t \in [0,T]\}$ all the conditional moments $\{\pi_t(\zeta_x^n),\ t \in [0,T]\}$, $n=1,2,\ldots$, have to be computed simultaneously. In the case where $C(x,u) \neq 0$, even for a linear system, an infinite dimensional filter still emerges for the computation of $\{\pi_t(\zeta_x),\ t \in [0,T]\}$. This property is best illustrated by the following simple example.

Consider the following system

$$\zeta_x(t) = \zeta_x(0) + \int_0^t a\,\zeta_x(s)ds + cN(t),\ t \geq 0,\ \zeta_x(0) \in \mathbb{R},\ E\zeta_x^2(0) < \infty, \tag{1.47}$$

$$y(t) = \int_0^t \zeta_x(s)ds + \gamma v(t) \quad ,\quad t \geq 0, \tag{1.48}$$

where $\{N(t),\ t \geq 0\}$ is a Poisson process with parameter q; $\{v(t),\ t \geq 0\}$ is an $\mathbb{R}$-valued standard Wiener process; and a,c, γ and q are given positive numbers. It is assumed that $\zeta_x(0)$, $\{N(t),\ t \geq 0\}$ and $\{v(t), t \geq 0\}$

are mutually independent.

In this case, for $F \in D(A)$ we take

$$(A_t F)(x) = ax\partial F(x)/\partial x + q[F(x+c) - F(x)], \; x \in \mathbb{R}, \quad t \geq 0. \tag{1.49}$$

By using Theorem 1.1, the following set of equations is obtained

$$\begin{aligned}\pi_t(\zeta_x) &= E\zeta_x(0) + \int_0^t (a\pi_s(\zeta_x) + qc)ds \\ &\quad + \gamma^{-2} \int_0^t (\pi_s(\zeta_x^2) - \pi_s^2(\zeta_x))(dy(s) - \pi_s(\zeta_x)ds), \; t \in [0,T]\end{aligned} \tag{1.50}$$

and for $n \geq 1$ we obtain

$$\begin{aligned}\pi_t(\zeta_x^n) &= E\zeta_x^n(0) + \int_0^t (an\pi_s(\zeta_x^n) + q \sum_{k=1}^n \binom{n}{k} c^k \pi_s(\zeta_x^{n-k}))ds \\ &\quad + \gamma^{-2} \int_0^t (\pi_s(\zeta_x^{n+1}) - \pi_s(\zeta_x^n)\pi_s(\zeta_x))(dy(s) - \pi_s(\zeta_x)ds), \; t \in [0,T].\end{aligned} \tag{1.51}$$

A class of systems of interest in engineering is the class of bilinear systems (see for example Frick and Valavi [1.26] and the references cited there). Consider the following bilinear system given by

$$\begin{aligned}\zeta_x(t) &= \zeta_x(0) + \int_0^t A(s)\zeta_x(s)ds + \sum_{i=1}^q \int_0^t G_i(s)\zeta_x(s)dW_i(s), t \in [0,T], \; \zeta_x(0) \in \mathbb{R}^m, \\ &E|\zeta_x(0)|^2 < \infty\end{aligned} \tag{1.52}$$

with the observation

$$y(t) = \int_0^t H(s)\zeta_x(s)ds + \int_0^t \Gamma(s)dv(s) \quad , \quad t \in [0,T], \tag{1.53}$$

where $A(t) \in \mathbb{R}^{m\times m}$, $G_i(t) \in \mathbb{R}^{m\times m}$, $i=1,\dots,q$; $H(t) \in \mathbb{R}^{p\times m}$ and $\Gamma(t) \in \mathbb{R}^{p\times p}$, are given matrices with continuously differentiable components on $[0,T]$, Γ^{-1} exists and has continuous components on $[0,T]$.
$\mathbf{W} = \{W(t) = (W_1(t),\dots,W_q(t)), \; t \geq 0\}$ and $\mathbf{V} = \{v(t), \; t \geq 0\}$ are an $\mathbb{R}^q$-valued and an $\mathbb{R}^p$-valued standard Wiener processes respectively. It is

assumed that $\zeta_x(0)$, $\mathbf{W}$ and $\mathbf{V}$ are mutually independent. Define the following operator

$$\mathcal{L}V(t,x) \triangleq \partial V(t,x)/\partial t + \sum_{i=1}^{m} (A(t)x)_i \, \partial V(t,x)/\partial x_i$$

$$+ (\tfrac{1}{2}) \sum_{i,j=1}^{m} \sum_{\ell=1}^{q} (G_\ell(t)x)_i (G_\ell(t)x)_j \partial^2 V(t,x)/\partial x_i \partial x_j$$

$(t,x) \in [0,T] \times \mathbb{R}^m$, $V \in C_b^{1,2}([0,T] \times \mathbb{R}^m)$; the space of bounded functions on $[0,T] \times \mathbb{R}^m$ having one bounded continuous derivative in t and two bounded continuous derivatives in x.

Let $V \in C_b^{1,2}([0,T] \times \mathbb{R}^m)$. Then, by using Itô's formula (Gihman and Skorohod [1.27]) it can be shown that

$$V(t,\zeta_x(t)) - V(0,\zeta_x(0)) - \int_0^t \mathcal{L}V(s,\zeta_x(s))ds \qquad (1.54)$$

is a (G_t,P)-martingale on $[0,T]$. Note that since $z(t) = \int_0^t \Gamma^{-1}(u)dy(u)$, it follows that $F_t^y = F_t^z$, $t \in [0,T]$. Thus, by using Theorem 1.1 the following equations are obtained:

$$\pi_t(\zeta_{xi}) = E\zeta_{xi}(0) + \int_0^t (A(s)\pi_s(\zeta_x))_i ds$$

$$+ \int_0^t (\Gamma^{-1}(s)\{H(s)\pi_s(\zeta_{xi}\zeta_x) - \pi_s(\zeta_{xi})H(s)\pi_s(\zeta_x)\}, \Gamma^{-1}(s)\{dy(s)$$

$$- H(s)\pi_s(\zeta_x)ds\}) \qquad (1.55)$$

$$t \in [0,T] \text{ , } i=1,\dots m$$

and

$$\pi_t(\zeta_{xi}\zeta_{xj}) = E[\zeta_{xi}(0)\zeta_{xj}(0)] + \int_0^t [\sum_{\ell=1}^{m} (A_{i\ell}(s)\pi_s(\zeta_{x\ell}\zeta_{xj}) + A_{j\ell}(s)\pi_s(\zeta_{x\ell}\zeta_{xi}))$$

$$+ \sum_{\ell=1}^{q} \sum_{\nu_1,\nu_2=1}^{m} (G_\ell(t))_{i\nu_1}(G_\ell(t))_{j\nu_2}\pi_s(\zeta_{x\nu_1}\zeta_{x\nu_2})]ds$$

$$+ \int_0^t (\Gamma^{-1}(s)\{H(s)\pi_s(\zeta_{xi}\zeta_{xj}\zeta_x) - \pi_s(\zeta_{xi}\zeta_{xj})H(s)\pi_s(\zeta_x)\}, \quad (1.56)$$

$$\Gamma^{-1}(s)\{dy(s) - H(s)\pi_s(\zeta_x)ds\})$$

$t \in [0,T]$, $i,j=1,\dots,m$.

Hence, also for the bilinear system, the term $E[F(\zeta(s))h_s|F_s^z]$ in equation (1.25) leads to an infinite-dimensional filter, in the computation of $\{\pi_t(\zeta_x), t \in [0,T]\}$.

A vast number of papers have dealt with linear estimation theory, which is the natural basis for nonlinear estimation. A fairly comprehensive survey of linear filtering theory is that of Kailath [1.28]. The con= cept of the state and its use in estimation was introduced by Kalman. Kalman [1.29], and Kalman and Bucy [1.30] presented linear recursive equations with the related nonlinear Riccati equations for the least squares estimate of $\zeta_x(t)$. The continuous-time filter derived below is called the Kalman or Kalman-Bucy filter.

Consider the following linear system given by

$$\zeta_x(t) = \zeta_x(0) + \int_0^t A(s)\zeta_x(s)ds + \int_0^t B(s)dW(s),\ t \in [0,T],\ \zeta_x(0) \in \mathbb{R}^m, \quad (1.57)$$

$$E|\zeta_x(0)|^2 < \infty,$$

with the observation process given by equation (1.53), where $A(t) \in \mathbb{R}^{m\times m}$ $B(t) \in \mathbb{R}^{m\times m}$, $H(t) \in \mathbb{R}^{p\times m}$ and $\Gamma(t) \in \mathbb{R}^{p\times p}$ are given matrices with continuously differentiable components on $[0,T]$; Γ^{-1} exists and has continuous compo=

nents on [0,T]. $\mathbf{W} = \{W(t),\ t \geq 0\}$ and $\mathbf{V} = \{v(t),\ t \geq 0\}$ are an $\mathbb{R}^m$-valued and an $\mathbb{R}^p$-valued standard Wiener processes respectively. It is assumed that $\zeta_x(0)$, $\mathbf{W}$ and $\mathbf{V}$ are mutually independent.

Let $V \in C_b^{1,2}([0,T] \times \mathbb{R}^m)$, and take

$$(A_t V)(t,x) \triangleq \partial V(t,x)/\partial t + \sum_{i=1}^{m} (A(t)x)_i \partial V(t,x)/\partial x_i$$

$$+ (\tfrac{1}{2}) \sum_{i,j=1}^{m} (B(t)B'(t))_{ij} \partial^2 V(t,x)/\partial x_i \partial x_j \tag{1.58}$$

$(t,x) \in [0,T] \times \mathbb{R}^m$.

Then $V \in D(A)$, and by using Theorem 1.1, we obtain

$$\pi_t(\zeta_{xi}) = E\zeta_{xi}(0) + \int_0^t (A(s)\pi_s(\zeta_x))_i ds$$

$$+ \int_0^t (\Gamma^{-1}(s)\{H(s)\pi_s(\zeta_{xi}\zeta_x) - \pi_s(\zeta_{xi})H(s)\pi_s(\zeta_x)\}, \Gamma^{-1}(s)\{dy(s)$$

$$- H(s)\pi_s(\zeta_x)ds\}) \tag{1.59}$$

$t \in [0,T]$, $i=1,\ldots,m$.

Or

$$\pi_t(\zeta_x) = E\zeta_x(0) + \int_0^t A(s)\pi_s(\zeta_x)ds$$

$$+ \int_0^t [\pi_s(\zeta_x\zeta_x') - \pi_s(\zeta_x)\pi_s(\zeta_x')]H'(s)(\Gamma^{-1}(s))'\Gamma^{-1}(s)[dy(s)$$

$$- H(s)\pi_s(\zeta_x)ds], \quad t \in [0,T]. \tag{1.60}$$

We assume that $\zeta_x(0)$ is a Gaussian random element independent of $\mathbf{W}$ and $\mathbf{V}$. Thus, it follows that ζ_x is a Gaussian process.

Denote

$$P(t) \triangleq E[(\zeta_x(t) - \pi_t(\zeta_x))(\zeta_x(t) - \pi_t(\zeta_x))' | \mathcal{F}_t^y], \ t \in [0,T]. \tag{1.61}$$

Using the Gaussian property of ζ_x it can be shown that (see for example [1.14, pp.255-256])

$$P(t) = E[(\zeta_x(t) - \pi_t(\zeta_x))(\zeta_x(t) - \pi_t(\zeta_x))']$$

$$= E[\zeta_x(t)\zeta_x'(t) - \zeta_x(t)\pi_t(\zeta_x') - \pi_t(\zeta_x)\zeta_x'(t) + \pi_t(\zeta_x)\pi_t(\zeta_x')]$$

$$= E[\zeta_x(t)\zeta_x'(t)] - E\, E[\zeta_x(t)\pi_t(\zeta_x') + \pi_t(\zeta_x)\zeta_x'(t)|F_t^y]$$

$$+ E[\pi_t(\zeta_x)\pi_t(\zeta_x')] \tag{1.62}$$

$$= E[\zeta_x(t)\zeta_x'(t)] - 2\, E[\pi_t(\zeta_x)\pi_t(\zeta_x')] + E[\pi_t(\zeta_x)\pi_t(\zeta_x')]$$

$$= E[\zeta_x(t)\zeta_x'(t)] - E[\pi_t(\zeta_x)\pi_t(\zeta_x')], \; t \in [0,T].$$

Define

$$K(t) \triangleq P(t)H'(t)(\Gamma^{-1}(t))'\Gamma^{-1}(t) \quad , \quad t \in [0,T] , \tag{1.63}$$

and denote

$$\hat{x}(t) \triangleq \pi_t(\zeta_x) \quad , \quad t \in [0,T]. \tag{1.64}$$

Then, equation (1.60) can be written as

$$d\hat{x}(t) = A(t)\hat{x}(t)dt + K(t)(dy(t) - H(t)\hat{x}(t)dt), \; t \in (0,T), \; \hat{x}(0) = E\zeta_x(0) \tag{1.65}$$

It now remains to derive the differential equation satisfied by P(t). Let $V \in C_b^{1,2}([0,T] \times \mathbb{R}^m)$. Since

$$V(t,\zeta_x(t)) - V(0,\zeta_x(0)) - \int_0^t (A_s V)(s,\zeta_x(s))ds,$$

where $A_t V$ is given by equation (1.58) , is a (G_t,P)-martingale on $[0,T]$, it follows that

$$EV(t,\zeta_x(t)) = EV(0,\zeta_x(0)) + E\int_0^t (A_sV)(s,\zeta_x(s))ds. \tag{1.66}$$

Take $V(t,x) = x_i x_j$. Then equations (1.58) and (1.66) yield

$$\begin{aligned} E[\zeta_{xi}(t)\zeta_{xj}(t)] = E[\zeta_{xi}(0)\zeta_{xj}(0)] + E\int_0^t [(A(s)\zeta_x(s))_i\zeta_{xj}(s) \\ + (A(s)\zeta_x(s))_j\,\zeta_{xi}(s) + (B(s)B'(s))_{ij}]ds \end{aligned} \tag{1.67}$$

$t \in [0,T]$, $i,j=1,\ldots,m$,

from which it follows that

$$\begin{aligned} E[\zeta_x(t)\zeta_x'(t)] = E[\zeta_x(0)\zeta_x'(0)] + \int_0^t [A(s)E[\zeta_x(s)\zeta_x'(s)] \\ + E[\zeta_x(s)\zeta_x'(s)]A'(s) + B(s)B'(s)]ds,\ t \in [0,T]. \end{aligned} \tag{1.68}$$

By using the fact that equation (1.60) can be written as

$$d\hat{x}(t) = A(t)\hat{x}(t)dt + P(t)H'(t)(\Gamma^{-1}(t))'d\nu(t),\ t \in [0,T], \tag{1.69}$$

where $\{\nu(t) = \int_0^t \Gamma^{-1}(u)[dy(u) - H(u)\hat{x}(u)du],\ t \in [0,T]\}$ $(=\{z(t)-\int_0^t \Gamma^{-1}(u)H(u)\hat{x}(u)du,\ t \in [0,T]\})$ is a (F_t^y,P)-standard Wiener process (see [1.1] for example), it can be shown that, for $V \in C_b^{1,2}([0,T] \times \mathbb{R}^m)$,

$$V(t,\hat{x}(t)) - V(0,\hat{x}(0)) - \int_0^t (\tilde{A}_sV)(s,\hat{x}(s))ds,\ t \in [0,T], \tag{1.70}$$

is a (F_t^y,P)-martingale, where

$$\begin{aligned} (\tilde{A}_tV)(t,x) \triangleq \partial V(t,x)/\partial t + \sum_{\ell=1}^m (A(t)x)_\ell \partial V(t,x)/\partial x_\ell \\ + (\tfrac{1}{2}) \sum_{k,\ell=1}^m (P(t)H'(t)(\Gamma^{-1}(t))'\Gamma^{-1}(t)H(t)P(t))_{k\ell}\partial^2 V(t,x)/\partial x_k\partial x_\ell \end{aligned} \tag{1.71}$$

$(t,x) \in [0,T] \times \mathbb{R}^m$.

Then, by taking $V(t,x) = x_i x_j$, equations (1.70)-(1.71) yield

$$E[\hat{x}_i(t)\hat{x}_j(t)] = E[\hat{x}_i(0)\hat{x}_j(0)] + E\int_0^t [(A(s)\hat{x}(s))_i\hat{x}_j(s) + (A(s)\hat{x}(s))_j\hat{x}_i(s)$$

$$+ (P(s)H'(s)(\Gamma^{-1}(s))'\Gamma^{-1}(s)H(s)P(s))_{ij}]ds, \qquad (1.72)$$

$t \in [0,T]$, $i,j=1,\ldots,m$,

from which it follows that

$$E[\hat{x}(t)\hat{x}'(t)] = E[\hat{x}(0)\hat{x}'(0)] + \int_0^t (A(s)E[\hat{x}(s)\hat{x}'(s)] + E[\hat{x}(s)\hat{x}'(s)]A'(s)$$

$$+ P(s)H'(s)(\Gamma^{-1}(s))'\Gamma^{-1}(s)H(s)P(s)]ds, \quad t \in [0,T]. \qquad (1.73)$$

Now, by using equations (1.62), (1.68) and (1.73), we obtain

$$dP(t)/dt = A(t)P(t) + P(t)A'(t) + B(t)B'(t) - P(t)H'(t)(\Gamma^{-1}(t))'\Gamma^{-1}(t)H(t)P(t)$$

$$t \in (0,T) \qquad (1.74)$$

$$P(0) = E[\zeta_x(0) - E\zeta_x(0))(\zeta_x(0) - E\zeta_x(0))'].$$

In conclusion, the optimal (minimum variance) filter for the system given by equations (1.57) and (1.53), where $\zeta_x(0)$ is a Gaussian random element, is given by

$$d\hat{x}(t) = A(t)\hat{x}(t)dt + K(t)(dy(t) - H(t)\hat{x}(t)dt) \qquad t \in (0,T) \qquad (1.75)$$

$$K(t) = P(t)H'(t)R^{-1}(t) \quad , \quad R^{-1}(t) = (\Gamma^{-1}(t))'\Gamma^{-1}(t), t \in [0,T] \qquad (1.76)$$

$$dP(t)/dt = A(t)P(t) + P(t)A'(t) + B(t)B'(t) - P(t)H'(t)R^{-1}(t)H(t)P(t) \qquad (1.77)$$

$$t \in (0,T)$$

$$\hat{x}(0) = E\zeta_x(0) \quad , \quad P(0) = E[(\zeta_x(0) - \hat{x}(0))(\zeta_x(0) - \hat{x}(0))']. \qquad (1.78)$$

Equations (1.75)-(1.78) constitute the well known continuous-time Kalman-Bucy filter. Equations (1.76)-(1.78) are independent of the observed process and can be solved off-line. The nonlinear matrix equation (1.77) is the well-known matrix *Riccati* equation, and K(t) is the *Kalman gain* (*matrix*).

The nonlinear filtering problem for linear systems (equations (1.57) and (1.53)), where $\zeta_x(0)$ is a non-Gaussian random element, is treated in Beneš and Karatzas [1.31]. There, a finite-dimensional filter with a generalized random Kalman matrix gain is constructed for the computation of $\{\hat{x}(t), t \in [0,T]\}$.

An Important Note

Throughout this monograph (except in Chapters 4 and 10) $\zeta_x(0)$ and y(0) (in cases where y(0) $\neq$ 0) are assumed to be deterministic vectors. Also, throughout Chapter 4, y(0) is assumed to be a deterministic vector.

1.4 REFERENCES

1.1 M. Fujisaki, G. Kallianpur and H. Kunita, Stochastic differential equations for the nonlinear filtering problem, *Osaka J. Math.*, 9, pp 19-40, 1972.

1.2 R. Stratonovich, On the theory of optimal nonlinear filtration of random functions, *Theory of Probability and its* App., 4, pp 223-225, 1959.

1.3 H.J. Kushner, On the differential equations satisfied by conditional probability densities of Markov processes, with applications, *SIAM J. Control*, 2, pp 106-119, 1964.

1.4 W.M. Wonham, Some applications of stochastic differential equations to optimal nonlinear filtering, *J. SIAM Control*, 2, pp 347-369, 1965.

1.5 R.S.Bucy, Nonlinear filtering theory, *IEEE Trans. on Automatic Control*, AC-10, p 198, 1965.

1.6 R.S.Bucy and P.D.Joseph, *Filtering for Stochastic Processes with Applications to Guidance*, Interscience, New York, 1968.

1.7 M. Zakai, On the optimal filtering of diffusion processes, *Z. Wahrs. verw. Geb.*, 11, pp 230-243, 1969.

1.8 A.H. Jazwinski, *Stochastic Processes and Filtering Theory*, Academic Press, New York, 1970.

1.9 A.P.Sage and J.L.Melsa, *Estimation Theory with Applications to Communications and Control*, McGraw-Hill, New York, 1971.

1.10 P.A.Frost and T.K.Kailath, An innovation approach to least-squares estimation - Part III : Nonlinear estimation in white Gaussian noise, *IEEE Trans. Automatic Control*, AC-16, pp 217-226, 1971.

1.11 T.P.McGarty, *Stochastic Systems and State Estimation*, John Wiley & Sons, New York, 1974.

1.12 R.S.Liptser and A.N.Shiryayev, *Statistics of Random Processes*, Springer-Verlag, Berlin, Vol.I : 1977, Vol.II : 1978.

1.13 B.D.O. Anderson and J.B. Moore, *Optimal Filtering*, Prentice-Hall, Englewood Cliffs, 1979.

1.14 G.Kallianpur, *Stochastic Filtering Theory*, Springer-Verlag, New York, 1980.

1.15 M.H.A.Davis and P.H.Wellings, Computational problems in Nonlinear filtering, in *Analysis and Optimization of Systems*, Ed. A. Bensoussan and J.L. Lions, Lecture Notes in Control and Information Sciences, 28, Springer-Verlag, Berlin, pp 253-261, 1980.

1.16 T. Björk, Finite dimensional optimal filters for a class of Itô-processes with jumping parameters, *Stochastics*, 4, pp 167-183, 1980.

1.17 T. Björk, Finite optimal filters for a class of nonlinear diffusions with jumping parameters, *Stochastics*, 6, pp 121-138, 1982.

1.18 Y. Takeuchi and H. Akashi, Nonlinear filtering formulas for discrete-time observations, *SIAM J. Control and Optimization*, 19, pp 244-261, 1981.

1.19 M. Hazewinkel and J.C.Willems, Editors, *Stochastic Systems: The Mathematics of Filtering and Identification and Applications*, D.Reidel Publishing Comp., Dordracht, 1981.

1.20 Y. Sawaragi, T. Katayama and S. Fujishige, State estimation for continuous-time system with interrupted observation, *IEEE Trans. on Automatic Control*, AC-19, pp 307-314, 1974.

1.21 J.H.Van Schuppen, Stochastic filtering theory: A discussion of concepts, methods and results, in *Stochastic Control Theory and Stochastic Differential Systems*, M. Kohlmann and W. Vogel, Eds., Springer-Verlag, New York, pp 209-226, 1979.

1.22 P.A.Meyer, *Probability and Potentials*, Blaisdell, Waltham, Mass., 1966.

1.23 A.N. Shiryayev, Martingales: recent developments, results and appli= cations, *Int. Statistical Review*, 49, pp 199-233, 1981.

1.24 I.I.Gihman and A.V.Skorohod, *Stochastic Differential Equations*, Springer-Verlag, Berlin, 1972.

1.25 D.W. Stroock, Diffusion processes associated with Levy generators, *Z. Wahrs. verw. Gebiet*, 32, pp 209-244, 1975.

1.26 P.A.Frick and A.S.Valavi, Estimation and identification of bilinear systems, *Automatic Control Theory and App.*, 6, pp 1-7, 1978.

1.27 I.I.Gihman and A.V.Skorohod, *The Theory of Stochastic Processes III*, Springer-Verlag, Berlin, 1979.

1.28 T.Kailath, A view of three decades of linear filtering theory, *IEEE Trans. on Information Theory*, IT-20, pp 146-181, 1974.

1.29 R.Kalman, A new approach to linear filtering and prediction problems, *J. Basic Eng. (Trans. ASME, Series D)*, 82, pp 35-45, 1960.

1.30 R.Kalman and R.Bucy, New results in linear filtering and predic= tion theory, *J. Basic Eng.* (*Trans.* ASME, *Series D*), 83, pp 95-108, 1961.

1.31 V.E.Benes and I. Karatzas, Estimation and control for linear par= tially observable systems with non-Gaussian initial distribution, *Stochastic Processes and their Applications*, 14, pp 233-248, 1983.

1.32 G. Kallianpur and R.L. Karandikar, Some recent developments in non= linear filtering theory, *Acta Applicandae Mathematicae*, 1 pp 399-434, 1983.

CHAPTER 2

ESTIMATION OF PARAMETERS VIA STATE OBSERVATION

2.1 INTRODUCTION

Let a nonlinear stochastic system be given by

$$dx = [\eta f(x) + g(x)]dt + BdW, \quad t > 0, \quad x \in \mathbb{R}^m \tag{2.1}$$

where $f : \mathbb{R}^m \to \mathbb{R}^m$ and $g : \mathbb{R}^m \to \mathbb{R}^m$ are given continuously differentiable functions on $\mathbb{R}^m$, which are bounded on any bounded domain in $\mathbb{R}^m$; $B \in \mathbb{R}^{m\times m}$ ($\mathbb{R}^{m\times m}$ denotes the space of $m\times m$ matrices) is a given symmetric positive definite matrix. $\mathbf{W} = \{W(t) = (W_1(t),\ldots,W_m(t)),\ t \geq 0\}$ is an $\mathbb{R}^m$-valued standard Wiener process. η is a random variable, $\eta \in S = \{\eta_1,\ldots,\eta_L\}$ with $\pi_j = \text{Prob.}\{\eta = \eta_j\}$, $j=1,\ldots,L$. It is assumed here that $\mathbf{W}$ and η are independent. The methods used here can be applied to the case where η is a vector or matrix valued random element. For the sake of simplicity η is taken here as a random variable.

By writing equation (2.1) in the form

$$\left\{\begin{aligned} dx &= [\eta f(x) + g(x)]dt + BdW \\ d\eta &= 0 \end{aligned}\right. \qquad t > 0, \ (x,\eta) \in \mathbb{R}^{m+1} \tag{2.2}$$

it can be shown (Gihman and Skorohod [2.1]) that for each $(x,\eta(\omega)) \in \mathbb{R}^{m+1}$, with $E\eta^2(\omega) < \infty$, equations (2.2) have a unique solution $(\zeta_x,\eta) = \{(\zeta_x(t), \eta(\omega)) = (\zeta_{x,1}(t),\ldots,\zeta_{x,m}(t),\eta(\omega)),\ t \geq 0\}$ with continuous sample paths and such that $(\zeta_x(0),\eta(\omega)) = (x,\eta(\omega))$. Furthermore (ζ_x,η) is a Markov process on a probability space $(\Omega,\mathcal{B},P)$. In the sequel we confine ourselves to the case where $(\zeta_x(0),\eta(\omega)) = (x,\eta(\omega)) \in \mathbb{R}^m \times S$. We assume that for each $t \geq 0$, $\zeta_x(t)$ is completely observable.

The problem dealt with in the first five sections of this chapter is to find an estimate $\hat{\eta}_x(t)$ to η such that

$$\hat{\eta}_x(t) = E\,[\eta \mid \zeta_x(s),\ 0 \leq s \leq t]\ ,\quad t \geq 0 \tag{2.3}$$

In the next section a filter is constructed for the computation of $\hat{\eta}_x = \{\hat{\eta}_x(t),\ t \geq 0\}$.

2.2 DERIVATION OF THE FILTER

In this section the random variable η is treated as a Markov process on $(\Omega,\mathcal{B},P)$.

Denote $y = B^{-1}x$ and $y(t) = B^{-1}\,\zeta_x(t)$, $t \geq 0$. Then, equation (2.1) yields

$$dy = B^{-1}[\eta f(By) + g(By)]dt + dW,\ t > 0,\ y \in \mathbb{R}^m. \tag{2.4}$$

In the sequel the following notations will be used:

$$\mathcal{F}_t^y \triangleq \sigma(y(s)\ ;\ 0 \leq s \leq t),\quad t \geq 0\ , \tag{2.5}$$

$$\mathcal{F}_t^{\eta,W} \triangleq \sigma(\eta,W(s);\ 0 \leq s \leq t)\ ,\quad t \geq 0 \tag{2.6}$$

$$h_t \triangleq B^{-1}[\eta f(By(t)) + g(By(t))]\ ,\quad t \geq 0 \tag{2.7}$$

$$\nu(t) \triangleq y(t) - \int_0^t E[h_s \mid \mathcal{F}_s^y]ds\ ,\quad t \geq 0, \tag{2.8}$$

$\mathcal{F}_t^y$ is the smallest σ-field generated by the family of random elements $\{y(s);\ 0 \leq s \leq t\}$, and $\mathcal{F}_t^{\eta,W}$ is the smallest σ-field generated by $\{\eta,W(s);\ 0 \leq s \leq t\}$. We further assume that $\int_0^T E\ |h_t|^2 dt < \infty$ for any $0 < T < \infty$ ($|h_t|^2 = \sum_{i=1}^m h_{ti}^2$).

For each $t \geq 0$, the σ-fields $\mathcal{F}_t^{\eta,W}$ and $\sigma(W(v) - W(u);\ t < u < v)$ are inde=

pendent, and h_t is $F_t^{\eta,w}$-measurable. Thus, the results of Theorem 1.1(see also [2.2]) can be applied to our problem. Let $F : \mathbb{R} \to \mathbb{R}$ be a bounded and measurable function. Then, equation (1.25)

$$dE[F(\eta)|F_t^y] = (E[F(\eta)h_t|F_t^y] - E[F(\eta)|F_t^y]E[h_t|F_t^y], d\nu(t)),\ t > 0 \tag{2.9}$$

(where for each $a,b \in \mathbb{R}^m$, $(a,b) = \sum_{i=1}^{m} a_i b_i$).

Let

$$\chi_i(\eta) \triangleq \begin{cases} 1 & \text{if } \eta = \eta_i \\ 0 & \text{if } \eta \neq \eta_i \end{cases} \qquad i=1,\dots,L \tag{2.10}$$

and

$$P_i(t) \triangleq E[\chi_i(\eta)|F_t^y] = P(\eta=\eta_i|F_t^y) = P(\eta=\eta_i|\zeta_x(s),\ 0 \le s \le t),\ i=1,\dots,L. \tag{2.11}$$

By inserting $F(\eta) = \chi_i(\eta)$ into (2.9) and using (2.11) we obtain

$$dP_i(t) = (E[\chi_i(\eta)h_t|F_t^y] - P_i(t)E[h_t|F_t^y], d\nu(t)),\ t > 0. \tag{2.12}$$

Using that

$$E[\chi_i(\eta)h_t|F_t^y] = B^{-1}[\eta_i f(By(t)) + g(By(t))]P_i(t),\ t \ge 0,\ i=1,\dots,L \tag{2.13}$$

and

$$E[h_t|F_t^y] = B^{-1}\sum_{j=1}^{L}[\eta_j f(By(t)) + g(B\,y(t))]P_j(t) = B^{-1}(\hat{\eta}_x(t)f(By(t)) + g(By(t)))\ ,\ t \ge 0 \tag{2.14}$$

where

$$\hat{\eta}_x(t) = \sum_{i=1}^{L} \eta_i P_i(t)\ ,\ t \ge 0, \tag{2.15}$$

equation (2.12) reduces to

$$dP_i(t)=P_i(t)(\eta_i-\hat{\eta}_x(t))(B^{-1}f(\zeta_x(t)),B^{-1}[d\zeta_x(t)-(\hat{\eta}_x(t)f(\zeta_x(t))+g(\zeta_x(t)))dt]) \tag{2.12'}$$

$t \geq 0, \quad i=1,\ldots,L.$

Equations (2.16) can be written in the following form

$$dP_i(t)=P_i(t)(\eta_i-\hat{\eta}_x(t)) \sum_{q,r=1}^{m} (B^{-1})^2_{qr} f_q(\zeta_x(t))[d\zeta_{xr}(t)-(\hat{\eta}_x(t)f_r(\zeta_x(t)) + g_r(\zeta_x(t)))dt], \quad t > 0,\ i=1,\ldots,L. \tag{2.16}$$

Equations (2.15)-(2.16) constitute the filter equations for computing $\hat{\eta}_x(t)$ (eq. (2.3) or (2.15)). In general, the numbers $\pi_i = P(\eta=\eta_i)$, $i=1,\ldots,L$, are unknown. In the next section an algorithm for computing $\hat{\eta}_x$ is suggested.

2.3 AN ALGORITHM FOR COMPUTING $\hat{\eta}_x$

In the sequel the following notation is used:

$$P_i(k) \triangleq P_i(k\Delta), \quad i=1,\ldots,L \ , \quad k=0,1,\ldots$$
$$\zeta_x(k) \triangleq \zeta_x(k\Delta), \quad \hat{\eta}_x(k) \triangleq \hat{\eta}_x(k\Delta) \ , \ k=0,1,\ldots \quad . \tag{2.17}$$

Let $\varepsilon > 0$, $\Delta > 0$, $L > 0$, $N > 0$ (L,N are integers) and $P_i(0) = 1/L$, $i=1,\ldots,L$.

Algorithm 2.3

1. $k=0$, $\hat{\eta}_x(0) = \sum_{i=1}^{L} \eta_i/L$
2. For $i=1,\ldots,L$ calculate

$$P_i(k+1):=P_i(k)+P_i(k)(\eta_i-\hat{\eta}_x(k)) \sum_{j,\ell=1}^{m} (B^{-1})^2_{j\ell} f_j(\zeta_x(k))[\zeta_{x\ell}(k+1) - \zeta_{x\ell}(k) - (\hat{\eta}_x(k)f_\ell(\zeta_x(k)) + g_\ell(\zeta_x(k)))\Delta] \tag{2.18}$$

$$P_i(k+1):=\max(0,P_i(k+1)) \tag{2.19}$$

3. $$Z(k+1):= \sum_{i=1}^{L} P_i(k+1) \tag{2.20}$$

4. If $Z(k+1) \geq \varepsilon$ then : for $i=1,\dots,L$, $P_i(k+1):=P_i(k+1)/Z(k+1)$. (2.21)

   Otherwise: stop.

5. $$\hat{\eta}_x(k+1):= \sum_{i=1}^{L} \eta_i P_i(k+1) \tag{2.22}$$

6. If $k=N$ stop. Otherwise $k:=k+1$ and go to 2.

Remark 2.3.1: Note that if for some i and k, say $i=i_o$ and $k=k_o$, we have $P_{i_o}(k_o) = 0$, then $P_{i_o}(k) = 0$ for all $k \geq k_o$. On the other hand if for some $i=i_o$ and $k=k_o$, $P_{i_o}(k_o) = 1$ (and then $P_i(k_o) = 0$ for all $i \neq i_o$), then $P_{i_o}(k) = 1$ and $P_i(k) = 0$ for $i \neq i_o$, for all $k \geq k_o$.

In the sequel a numerical study of Algorithm 2.3 will be carried out via numerical experimentation. Each experiment is called a *run*. Here a run always consists of two stages. In the first stage the following proce= dure for simulating (2.1) is applied:

For $k = 0,1,\dots,N$

1. For $i=1,\dots,m$ calculate

$$X_i(k+1) = \zeta_{xi}(k) + [\eta_\ell\, f_i(\zeta_x(k)) + g_i(\zeta_x(k))]\Delta + \sqrt{\Delta} \sum_{j=1}^{m} B_{ij}\, W_j(k) \tag{2.23}$$

2. For $i=1,\dots,m$ calculate

$$\zeta_{xi}(k+1) = \zeta_{xi}(k) + [\eta_\ell(f_i(\zeta_x(k)) + f_i(X(k+1))) + g_i(\zeta_x(k)) + g_i(X(k+1))]\Delta/2 + \sqrt{\Delta} \sum_{j=1}^{m} B_{ij}\, W_j(k) \tag{2.24}$$

where $\eta_\ell \in \{\eta_1,\dots,\eta_L\}$ and $\{W(k)\}_{k=0}^{N}$ is a sequence of independent $\mathbb{R}^m$-valued Gaussian elements with

$$EW(k) = 0 \text{ and } E[W(k)W'(\ell)] = \delta_{k\ell}\, I_m,\ k,\ell=0,1,\dots,N \tag{2.25}$$

(I_m denotes the unit m×m matrix); and the sequences $\{\zeta_x(k)\}_{k=0}^N$ and $\{\zeta_x(k+1) - \zeta_x(k)\}_{k=0}^N$ are stored. In the second stage Algorithm 2.3 is applied, where the sequences $\{\zeta_x(k)\}_{k=0}^N$ and $\{\zeta_x(k+1) - \zeta_x(k)\}_{k=0}^N$ act as the input to the filter, and $\{\hat{\eta}_x(k)\}_{k=0}^N$ constitute the output of the fil= ter. If for some k, say $k=k_o$, $\hat{\eta}_x(k) = \hat{\eta}_x(k_o)$ for all $k \geq k_o$ then we take $\hat{\eta}_{xo} \triangleq \hat{\eta}_x(k_o)$ as the estimator of η in the sense $\hat{\eta}_{xo} = E[\eta|\ \zeta_x(k),\ 0 \leq k \leq N]$.

In many practical situations η is a random variable with values in (α,β) where $\alpha,\beta \in \mathbb{R}$ are given numbers. In these cases Algorithm 2.3 can be adapted in order to find approximations to $E[\eta|\ \zeta_x(k),\ 0 \leq k \leq \ell]$, $\ell=0,\dots,N$. Such cases are illustrated in the following examples.

2.4 EXAMPLE 2.1: SINE-WAVE OSCILLATOR

2.4.1 Introduction

Consider the noise-driven nonlinear sine-wave oscillator given by

$$\begin{cases} dx_1 = [-a_o\, x_2 + bx_1(\rho^2 - x_1^2 - x_2^2)]dt + \sigma_1 dW_1 \\ \qquad\qquad\qquad\qquad\qquad\qquad\qquad\qquad t > 0 \\ dx_2 = [a_o\, x_1 + bx_2(\rho^2 - x_1^2 - x_2^2)]dt + \sigma_2 dW_2 \end{cases} \tag{2.26}$$

where b,ρ,σ_1 and σ_2 are given positive numbers; $\mathbf{W} = \{W(t) = (W_1(t),W_2(t)),\ t \geq 0\}$ is an $\mathbb{R}^2$-valued standard Wiener process; and a_o is an unknown number, $a_o \in (\alpha,\beta)$, where $\alpha > 0$ and $\beta > 0$ are given. The (deterministic) case where a_o is given and $\sigma_1 = \sigma_2 = 0$ is considered in Kaplan [2.3-2.4]. It is shown in [2.3] that $\{\zeta_x^o(t),\ t \geq 0\}$ (the solution to (2.26) where $\sigma_1 = \sigma_2 = 0$) is the state of a sine-wave oscillator with a stable limit cycle $\{x : x_1^2 + x_2^2 = \rho^2\}$ and frequency a_o.

The problem considered in this example is to find approximations to $E[\eta|\zeta_x(s), 0 \le s \le t], t \ge 0$, where $\zeta_x = \{\zeta_x(t), t \ge 0\}$ is the solution to (2.26), and η stands for a_o.

In order to perform numerical experimentation it was assumed here that $a_o = 50$, $b = 50$, $\rho = 1$ and $\sigma_1 = \sigma_2 = 0.01$. All the computations were carried out with $\Delta = 10^{-3}$ and $N = 100$.

2.4.2 Results

Several sets of runs, each consisting of ten runs were carried out. In each of the runs first, equations (2.26) (with $a_o = 50$) were simulated by using the procedure given by (2.23) - (2.24). Second, Algorithm 2.3 was applied. In all the cases computed, $\hat{\eta}_x(k)$ converged to $\hat{\eta}_{xo}$ in less than 10 time steps (i.e. $\hat{\eta}_{xo} = \hat{\eta}_x(10)$ in all the cases computed). Typical extracts from the sets of runs are presented below:

(a) $\eta_i = 45 + i/4$, $i=1,2,...,40$ (L=40)

(b) $\eta_i = 45 + i/2$, $i=1,2,...,20$ (L=20)

(c) $\eta_i = 47 + i/5$, $i=1,2,...,20$ (L=20)

(d) $\eta_i = 48 + i/10$, $i=1,2,...,20$ (L=20)

(e) $\eta_i = 50 + i/10$, $i=1,2,...,20$ (L=20)

The results obtained in these runs are

	Run 1	Run 2	Run 3	Run 4	Run 5	Run 6	Run 7	Run 8	Run 9	Run 10
(a) $\hat{\eta}_{xo}$:	49.25	48.75	49.25	49.00	49.25	48.75	49.25	49.25	49.25	49.00
(b) $\hat{\eta}_{xo}$:	48.50	49.00	48.50	49.00	49.50	49.00	49.50	48.50	48.50	48.50
(c) $\hat{\eta}_{xo}$:	48.60	49.00	49.00	49.00	49.00	48.60	49.00	48.80	49.00	48.80
(d) $\hat{\eta}_{xo}$:	48.80	48.50	48.50	48.50	48.70	48.40	48.50	48.90	48.50	48.50
(e) $\hat{\eta}_{xo}$:	50.10	50.10	50.10	50.10	50.10	50.10	50.10	50.10	50.10	50.10

2.5 EXAMPLE 2.2 : TRIANGULAR-WAVES GENERATOR

2.5.1 Introduction

Consider the noise-driven triangular wave generator given by

$$\begin{cases} dx_1 = [a_o \operatorname{sign}(x_2) + b(\rho - |x_1| - |x_2|)x_1]dt + \sigma_1 dW_1 \\ \qquad\qquad\qquad\qquad\qquad\qquad\qquad\qquad t > 0, \\ dx_2 = [-a_o \operatorname{sign}(x_1) + b(\rho - |x_1| - |x_2|)x_2]dt + \sigma_2 dW_2 \end{cases} \qquad (2.27)$$

where b,ρ,σ_1 and σ_2 are given positive numbers; $\mathbf{W} = \{W(t) = (W_1(t),W_2(t)),$ $t \geq 0\}$ is an $\mathbb{R}^2$-valued standard Wiener process; and a_o is an unknown number, $a_o \in (\alpha,\beta)$, where $\alpha > 0$ and $\beta > 0$ are given. The (deterministic) case where a_o is given and $\sigma_1 = \sigma_2 = 0$ is considered in Kaplan and Tatrash [2.5] and Kaplan [2.6], while a controlled stochastic version of (2.27) is dealt with in Huisman and Yavin [2.7]. An oscillator of the type given by (2.27), where $\sigma_1 = \sigma_2 = 0$ generates precise triangular waves and square waves. The solution to (2.27) (with $\sigma_1 = \sigma_2 = 0$) has a stable limit cycle $\{x : |x_1| + |x_2| = \rho\}$ and the frequency of the tri= angular waves depends on the value of a_o and on the initial conditions.

The problem considered in this example is to find approximations to $E[\eta|\ \zeta_x(s),\ 0 \leq s \leq t],\ t \geq 0$, where $\zeta_x \doteq \{\zeta_x(t),\ t \geq 0\}$ is the solution to (2.27), and η stands for a_o. (It is tacitly assumed here that $|x_i|$ and $\operatorname{sign}(x_i)$ are expressions used for $x_i \cdot \tanh(ax_i)$ and $\tanh(ax_i)$, res= pectively, for some $a >> 1$.. Under these assumptions, equations (2.27) have an unique solution ζ_x).

In order to perform numerical experimentation it was assumed here that $a_o = 70$, $b = 50$, $\rho = 1$ and $\sigma_1 = \sigma_2 = 0.01$. Also, the values of $\Delta = 10^{-3}$ and $N = 100$ have been used throughout all the computations.

2.5.2 Results

Several sets of runs, each consisting of ten runs were carried out. In each of the runs, first, equations (2.27) (with $a_o = 70$) were simulated, using the procedure given by (2.23)-(2.24). Next, Algorithm 2.3 was applied. In all the cases computed, $\hat{\eta}_x(k)$ converged to $\hat{\eta}_{xo}$ in less than forty time steps (i.e. $\hat{\eta}_{xo} = \hat{\eta}_x(40)$ in all the cases computed). Some extracts from the sets of runs are presented below:

(a) $\eta_i = 68 + i/10$, $i=1,\ldots,20$ $(L=20)$

(b) $\eta_i = 69.9 + i/10$, $i=1,\ldots,20$ $(L=20)$

(c) $\eta_i = 59.5 + i/2$, $i=1,\ldots,20$ $(L=20)$

(d) $\eta_i = 70 + i/2$, $i=1,\ldots,20$ $(L=20)$

The results obtained in these runs are

	Run 1	Run 2	Run 3	Run 4	Run 5	Run 6	Run 7	Run 8	Run 9	Run 10
(a) $\hat{\eta}_{xo}$:	69.70	69.80	69.70	69.70	69.60	69.90	69.80	69.80	69.90	69.60

(b) $\hat{\eta}_{xo}$: for all the runs $\hat{\eta}_{xo} = 69.99991642$

(c) $\hat{\eta}_{xo}$: for all the runs $\hat{\eta}_{xo} = 69.49998807$

(d) $\hat{\eta}_{xo}$: for all the runs $\hat{\eta}_{xo} = 70.50$

2.6 ESTIMATION OF A MARKOV CHAIN

A natural extension of the previous discussion would be to treat the case where η is a continuous-time Markov chain. Thus in the rest of this chapter the following system is considered

$$x(t) = x + \int_0^t [\eta(s)f(x(s)) + g(x(s))]ds + \int_0^t BdW(s),\ t \geq 0,\ x \in \mathbb{R}^m \tag{2.28}$$

where $f : \mathbb{R}^m \to \mathbb{R}^m$ and $g : \mathbb{R}^m \to \mathbb{R}^m$ are given continuously differentiable functions on $\mathbb{R}^m$, which are bounded on any bounded domain in $\mathbb{R}^m$; and $B \in \mathbb{R}^{m\times m}$ is a given symmetric positive definite matrix. On a probability space (Ω,F,P) $\mathbf{W}$ is an $\mathbb{R}^m$-valued standard Wiener process and $\{\eta(t),\ t \geq 0\}$ is a continuous-time Markov chain with a state space S and transition probabilities

$$P(\eta(t+\Delta) = \beta|\eta(t) = \alpha) = \lambda_{\alpha\beta}\Delta + 0(\Delta^2),\quad \alpha \neq \beta,\quad \alpha,\beta \in S \tag{2.29}$$

$$P(\eta(t+\Delta) = \alpha|\eta(t) = \alpha) = 1 + \lambda_{\alpha\alpha}\Delta + 0(\Delta^2),\ \alpha \in S \tag{2.30}$$

$$\lambda_{\alpha\beta} \geq 0,\ \alpha \neq \beta\ ,\ \sum_{\beta \in S} \lambda_{\alpha\beta} = 0;\ P(\eta(0)=\alpha) = \pi_\alpha,\ \alpha \in S \tag{2.31}$$

It is assumed here that S is at most countable, $\sup_{t \in [0,T]} E\eta^2(t) < \infty$ for any $T < \infty$, and that $\mathbf{W}$ and $\eta = \{\eta(t),\ t \geq 0\}$ are mutually independent. Also, for the sake of simplicity, we assume that $S \subset \mathbb{R}$.

Let $x \in \mathbb{R}^m$, then following Sergeeva and Teterina [2.8] and Sergeeva [2.9] it can be shown that equation (2.28) has an unique solution $\zeta_x = \{\zeta_x(t),\ t \geq 0\}$ which is such that $\zeta_x(0) = x$. Also, in the same manner as in [2.9] it can be shown that (ζ_x,η) is a Markov process on (Ω,F,P).

The problem dealt with in the rest of this chapter is to find an estimate $\hat{\eta}_x$ to η such that

$$\hat{\eta}_x(t) = E[\eta(t)|\ \zeta_x(s),\ \ 0 \leq s \leq t]\ ,\quad 0 \leq t \leq T \tag{2.32}$$

where $T < \infty$ is a given number.

In the next section a filter is constructed for the computation of $\hat{\eta}_x = \{\hat{\eta}_x(t),\ t \in [0,T]\}$.

2.7 THE EQUATIONS OF OPTIMAL FILTERING

As in Section 2.2, denote $y = B^{-1}x$ and $y(t) = B^{-1}\zeta_x(t)$, $t \geq 0$. Then, equation (2.28) yields

$$y(t) = y + \int_0^t h_s ds + W(t) \quad , \quad t \geq 0 \tag{2.33}$$

where

$$h_t \triangleq B^{-1}[\eta(t)f(By(t)) + g(By(t))] \quad , \quad t \geq 0 \tag{2.34}$$

We further assume that $\int_0^T E|h_t|^2 dt < \infty$ for any $0 < T < \infty$.

Let F_t^y and $\{\nu(t), t \geq 0\}$ be defined as in (2.5) and (2.8) respectively and let

$$F_t^{\eta,W} \triangleq \sigma(\eta(s), W(s); 0 \leq s \leq t), \quad t \geq 0 \quad . \tag{2.35}$$

For each $t \geq 0$, the σ-fields $F_t^{\eta,W}$ and $\sigma(W(v) - W(u); t < u < v)$ are inde= pendent, and h_t is $F_t^{\eta,W}$-measurable. Thus, the results of Theorem 1.1 can be applied to our problem. Let $F : S \to \mathbb{R}$ be a bounded and measurable function. Then, equation (1.25)

$$\begin{aligned} dE[F(\eta(t))|F_t^y] &= E[A_t F(\eta(t))|F_t^y]dt \\ &+ (E[F(\eta(t))h_t|F_t^y] - E[F(\eta(t))|F_t^y]E[h_t|F_t^y], d\nu(t)) \end{aligned} \tag{2.36}$$

$t \in (0,T)$

(where for $a,b \in \mathbb{R}^m$, $(a,b) = \sum_{i=1}^m a_i b_i$) where $A_t F$ is such that $\int_0^T E|A_t F(\eta(t))|^2 dt < \infty$ and

$$M_t(F) \triangleq F(\eta(t)) - E[F(\eta(0))|F_0^y] - \int_0^t A_s F(\eta(s))ds \tag{2.37}$$

is a $(F_t^{\eta,W}, P)$-martingale.

Denote

$$P_{\alpha\beta}(s,t) \triangleq P(\eta(t) = \beta | \eta(s) = \alpha), \quad s \le t, \quad \alpha,\beta \in S, \tag{2.38}$$

then, Liptser and Shiryayev [2.10, Vol.I, p.331]

$$P_{\alpha\beta}(s,t) = \delta_{\alpha\beta} + \int_s^t \sum_{\gamma \in S} \lambda_{\gamma\beta} P_{\alpha\gamma}(s,u)du. \tag{2.39}$$

Let

$$\chi_\alpha(t) \triangleq \begin{cases} 1 & \text{if } \eta(t) = \alpha \\ & \qquad\qquad \alpha \in S \\ 0 & \text{if } \eta(t) \neq \alpha \quad . \end{cases} \tag{2.40}$$

Suppose that

$$A_t \chi_\alpha(t) \triangleq \sum_{\gamma \in S} \lambda_{\gamma\alpha} \chi_\gamma(t) \quad , \alpha \in S, \quad t \in [0,T] \tag{2.41}$$

and assume that for any $\alpha \in S$

$$\int_0^T E| \sum_{\gamma \in S} \lambda_{\gamma\alpha} \chi_\gamma(t)|^2 dt < \infty \quad .$$

Then, using (2.37) and (2.41) we obtain

$$M_t(\chi_\alpha) - M_s(\chi_\alpha) = \chi_\alpha(t) - \chi_\alpha(s) - \int_s^t \sum_{\gamma \in S} \lambda_{\gamma\alpha} \chi_\gamma(u)du$$

Hence

$$E[M_t(\chi_\alpha) - M_s(\chi_\alpha) | F_s^{\eta,w}]$$

$$= E[\chi_\alpha(t) - \chi_\alpha(s) - \int_s^t \sum_{\gamma \in S} \lambda_{\gamma\alpha} \chi_\gamma(u)du | F_s^{\eta,w}] \tag{2.42}$$

and by using the Markov property of $\{\eta(t), t \ge 0\}$

$$= E[\chi_\alpha(t) - \chi_\alpha(s) - \int_s^t \sum_{\gamma \in S} \lambda_{\gamma\alpha} \chi_\gamma(u)du | \eta(s)]$$

$$= P_{\eta(s),\alpha}(s,t) - \delta_{\eta(s),\alpha} - \int_s^t \sum_{\gamma \in S} \lambda_{\gamma\alpha} P_{\eta(s),\gamma}(s,u)du \tag{2.43}$$

Thus, equations (2.42)-(2.43) and (2.39) yield

$$E[M_t(\chi_\alpha) - M_s(\chi_\alpha)|F_s^{\eta,w}]$$

$$= P_{\eta(s),\alpha}(s,t) - \delta_{\eta(s),\alpha} - \int_s^t \sum_{\gamma \in S} \lambda_{\gamma\alpha} P_{\eta(s),\gamma}(s,u)du = 0, \tag{2.44}$$

from which it follows that $M_t(\chi_\alpha)$ is a $(F_t^{\eta,w},P)$-martingale. This result verifies the assumption given by equation (2.41) (see [2.2] for further details on the operator A_t).

Denote

$$P_\alpha(t) \triangleq P(\eta(t) = \alpha|F_t^y) = P(\eta(t) = \alpha|\zeta_x(s),\ 0 \le s \le t),\ \alpha \in S \tag{2.45}$$

then, using the property

$$E[\chi_\alpha(t)|F_t^y] = P(\eta(t) = \alpha|F_t^y) \tag{2.46}$$

and inserting $F(\eta(t)) = \chi_\alpha(t)$; $\alpha \in S$, equation (2.36) yields

$$dP_\alpha(t) = \sum_{\gamma \in S} \lambda_{\gamma\alpha} P_\gamma(t)dt$$

$$+ (E[\chi_\alpha(t)h_t|F_t^y] - P_\alpha(t)E[h_t|F_t^y],\ d\nu(t)) \tag{2.47}$$

$t \in (0,T)$, $\alpha \in S$.

Using that

$$E[\chi_\alpha(t)h_t|F_t^y] = B^{-1}[\alpha f(By(t)) + g(By(t))]P_\alpha(t)$$

$$= B^{-1}[\alpha f(\zeta_x(t)) + g(\zeta_x(t))]P_\alpha(t),\ t \ge 0,\ \alpha \in S \tag{2.48}$$

and

$$E[h_t|F_t^y] = B^{-1}[\hat{\eta}_x(t)f(By(t)) + g(By(t))]$$

$$= B^{-1}[\hat{\eta}_x(t)f(\zeta_x(t)) + g(\zeta_x(t))]\quad,\ t \ge 0, \tag{2.49}$$

where

$$\hat{\eta}_x(t) = \sum_{\alpha \in S} \alpha P_\alpha(t) = E[\eta(t) | \zeta_x(s), 0 \le s \le t] , t \ge 0 \qquad (2.50)$$

equations (2.47) reduce to

$$dP_\alpha(t) = \sum_{\gamma \in S} \lambda_{\gamma\alpha} P_\gamma(t)dt$$
$$+ P_\alpha(t)(\alpha-\hat{\eta}_x(t)) \sum_{i,j=1}^{m} (B^{-1})^2_{ij} f_i(\zeta_x(t))[d\zeta_{xj}(t) - (\hat{\eta}_x(t)f_j(\zeta_x(t))$$
$$+ g_j(\zeta_x(t)))dt], \qquad (2.51)$$

$t \in (0,T)$, $\alpha \in S$.

Remark 2.7.1: Equations (2.51), for the case m=1, are derived in [2.10, Vol.I] by using a different method.

Also, the problem of parameter estimation via state observation is trea= ted, by using approaches other than the one used here, in [2.10, Vol.II].

Equations (2.51) and (2.50) constitute the filter equations for computing $\hat{\eta}_x(t)$. We assume that the numbers $\pi_\alpha = P(\eta(0) = \alpha)$, $\alpha \in S$, are unknown. The problem of computing $\hat{\eta}_x(t)$ is discussed in the following examples.

2.8 EXAMPLE 2.3: POISSON PROCESS

2.8.1 Introduction

In this section we consider the case where the process η appearing in equation (2.28) is a Poisson process with parameter λ. More explicitly, $\eta = N = \{N(t), t \ge 0\}$ is a Markov process with state space $S = \{0,1,2,...\}$ and where

$$\lambda_{ij} = \begin{cases} \lambda & \text{if } j = i+1 \\ 0 & \text{if } j \ne i+1 \end{cases} \qquad i = 0,1,2,... \qquad (2.52)$$

where λ is a given positive number. Thus, equations (2.50)-(2.51) reduce here to

$$\begin{cases} dP_i(t) = \lambda[P_{i-1}(t) - P_i(t)]dt \\ \qquad + P_i(t)(i - \hat{N}_x(t)) \sum_{q,s=1}^{m} (B^{-1})^2_{qs} f_q(\zeta_x(t))[d\zeta_{xs}(t) - (\hat{N}_x(t)f_s(\zeta_x(t)) \\ \qquad\qquad + g_s(\zeta_x(t)))dt] \\ i=0,1,2,\ldots,\ t \in (0,T)\ ;\ P_{-1}(t) = 0\ ,\ t \in [0,T] \\ P_o(0) = 1;\ P_i(0) = 0,\ i \geq 1\ ;\ \hat{N}_x(0) = 0 \\ \hat{N}_x(t) = \sum_{i=1}^{\infty} i\,P_i(t)\ ,\ t \in [0,T]. \end{cases} \tag{2.53}$$

$$(\hat{N}_x(t) \stackrel{\Delta}{=} E[N(t) \mid \zeta_x(s)\ ,\ 0 \leq s \leq t],\ t \in [0,T])$$

2.8.2 Algorithm 2.8

In this subsection an algorithm for computing $\{\hat{N}_x(t),\ t \in [0,T]\}$ is suggested.

Choose $\varepsilon > 0$, $\Delta > 0$. Let $N\Delta = T$.

1. $k=0$; $P_o(0) = 1$, $P_1(0) = 0$, $\hat{N}_x(0) = 0$

2. $L = k+1$

3. For $i=0,\ldots,L$ calculate

$$P_{-1}(k) := 0 \tag{2.54}$$

$$P_i(k+1) := \lambda[P_{i-1}(k) - P_i(k)]\Delta + P_i(k)(i - \hat{N}_x(k)) \sum_{q,s=1}^{m} (B^{-1})^2_{qs} f_q(\zeta_x(k)) \tag{2.55}$$

$$[\zeta_{xs}(k+1) - \zeta_{xs}(k) - (\hat{N}_x(k)f_s(\zeta_x(k)) + g_s(\zeta_x(k)))\Delta]$$

$$P_i(k+1) := \max(0, P_i(k+1)) \tag{2.56}$$

4. $P_{L+1}(k+1) := 0$ (2.57)

5. $Z(k+1) := \sum_{i=0}^{L} P_i(k+1)$ (2.58)

6. If $Z(k+1) \geq \varepsilon$ then: for $i=1,\ldots,L$ $P_i(k+1) := P_i(k+1)/Z(k+1)$ (2.59)

   Otherwise: stop.

7. $\hat{N}_x(k+1) = \sum_{i=1}^{L} i\, P_i(k+1)$ (2.60)

8. If $k=N$ stop. Otherwise $k:=k+1$ and go to 2.

Algorithm 2.8 uses the property that, in computing $\{P_i(k)\}$ by applying Euler's method on (2.53), $P_j(\ell) = 0$ for all $j \geq \ell+1$, $\ell=0,1,2,\ldots$.

2.8.3 The simulation procedure

In the sequel a numerical study of Algorithm 2.8 will be carried out via numerical experimentation. Each experiment is called a *run*. Here a run always consists of three stages:

(i) A sample path of $\{N(t),\ t \in [0,T]\}$ is constructed by applying the following procedure:

1. Read the numbers $R_i, i=1,\ldots,M+1$ from a random number gene= rator with a uniform probability density on (0,1).

2. Calculate

$$z_i = (-1/\lambda)\ell n R_i, \quad i=1,\ldots,M+1$$

$$T_i = \sum_{\ell=1}^{i} z_\ell \quad , \quad i=1,\ldots,M+1$$

where M is determined by $T_M \leq T < T_{M+1}$

3. Define the following function

$$N(k) = 0 \quad , \quad 0 \leq k\Delta < T_1$$

$$N(k) = i, \quad T_i \leq k\Delta < T_{i+1}, \quad i=1,\ldots,M$$

(ii) Simulate equations (2.28) by applying the procedure described by equations (2.23)-(2.24), where $\{N(k),\ k\Delta \in [0,T]\}$ and $\{W(k)\}$ are as described in Section 2.3. Store the sequences $\{\zeta_x(k)\}$ and $\{\zeta_x(k+1) - \zeta_x(k)\}$.

(iii) Apply Algorithm 2.8.

2.8.4 A numerical example

Consider the system dealt with in Section 2.4 but with the frequency hop= ping as a Poisson process, i.e.,

$$\begin{cases} dx_1 = [-a_o(1 + N(t))x_2 + bx_1(\rho^2 - x_1^2 - x_2^2)]dt + \sigma_1 dW_1 \\ \qquad\qquad\qquad\qquad\qquad\qquad\qquad\qquad\qquad t > 0 \\ dx_2 = [a_o(1 + N(t))x_1 + bx_2(\rho^2 - x_1^2 - x_2^2)]dt + \sigma_2 dW_2 \end{cases} \tag{2.61}$$

where $\{N(t),\ t \geq 0\}$ is a Poisson process with parameter λ. It is assumed that a_o, b, ρ, σ_1, σ_2 and λ are given positive numbers, and that $\mathbf{W}$ and $\{N(t),\ t \geq 0\}$ are mutually independent. Using the notation

$$\begin{cases} f_1(x) = -a_o\, x_2\ ,\ g_1(x) = -a_o\, x_2 + b\, x_1(\rho^2 - x_1^2 - x_2^2) \\ f_2(x) = a_o\, x_1\ ,\ g_2(x) = a_o\, x_1 + b\, x_2(\rho^2 - x_1^2 - x_2^2) \end{cases} \tag{2.62}$$

equations (2.61) can be written in the form of equations (2.28).

Numerical experimentation has been carried out for the following set of parameters: $a_o = 50$, $b = 50$, $\rho = 1$, $\sigma_1 = \sigma_2 = 0.01$, $\Delta = 10^{-3}$, and $\lambda = 1$, 2.5,3,4,5.

Typical extracts from the sets of runs are presented below.

TABLE 2.1: $N(k)$, $\hat{N}_x(k)$, $|\zeta_x(k)|$ and $Z(k)$ as functions of k for $\lambda = 2.5$. Here $\{T_1,\ldots,T_9\} = \{.2378, .4071, .5824, .6451, .9883, 1.796, 1.900, 1.941, 2.223\}$

k	N(k)	$\hat{N}_x(k)$	$\lvert\zeta_x(k)\rvert$	Z(k)
1	0	.2500E-02	.1351E+01	.1000E+01
51	0	.1411E-10	.1000E+01	.3321E+04
101	0	.1079E-04	.9992E+00	.1158E+01
151	0	.2496E-02	.1000E+01	.1001E+01
201	0	.4642E-02	.9987E+00	.1000E+01
251	1	.1000E+01	.9979E+00	.1441E+01
301	1	.1000E+01	.9980E+00	.1842E+01
351	1	.1003E+01	.9977E+00	.1000E+01
401	1	.2000E+01	.9986E+00	.6286E+02
451	2	.2000E+01	.9943E+00	.9637E+03
501	2	.2000E+01	.9941E+00	.2092E+01
551	2	.2025E+01	.9939E+00	.1000E+01
601	3	.3000E+01	.9911E+00	.2524E+03
651	4	.4005E+01	.9913E+00	.1000E+01
701	4	.4002E+01	.9886E+00	.1000E+01
751	4	.4000E+01	.9881E+00	.1246E+01
801	4	.4002E+01	.9880E+00	.1000E+01
851	4	.4000E+01	.9880E+00	.1708E+01
901	4	.4000E+01	.9881E+00	.1477E+01
951	4	.4002E+01	.9895E+00	.1000E+01
1001	5	.5000E+01	.9873E+00	.1287E+01
1051	5	.5000E+01	.9884E+00	.2218E+01
1101	5	.5002E+01	.9878E+00	.1000E+01
1151	5	.5000E+01	.9872E+00	.1815E+01
1201	5	.5002E+01	.9861E+00	.1000E+01
1251	5	.5002E+01	.9869E+00	.1001E+01
1301	5	.5000E+01	.9869E+00	.1528E+01
1351	5	.5000E+01	.9882E+00	.1349E+01
1401	5	.5002E+01	.9869E+00	.1001E+01
1451	5	.5002E+01	.9875E+00	.1000E+01
1501	5	.5002E+01	.9880E+00	.1001E+01
1551	5	.5000E+01	.9866E+00	.1939E+01
1601	5	.5000E+01	.9868E+00	.1239E+01
1651	5	.5003E+01	.9864E+00	.1000E+01
1701	5	.5000E+01	.9862E+00	.1595E+01
1751	5	.5002E+01	.9888E+00	.1000E+01
1801	6	.6000E+01	.9859E+00	.1695E+01
1851	6	.6000E+01	.9867E+00	.1699E+01
1901	7	.6000E+01	.9873E+00	.1980E+01
1951	8	.8000E+01	.9947E+00	.3330E+01
2001	8	.8000E+01	.9958E+00	.3885E+01
2051	8	.8000E+01	.9961E+00	.4162E+01
2101	8	.8000E+01	.9956E+00	.3178E+01
2151	8	.8000E+01	.9959E+00	.4002E+01
2201	8	.8000E+01	.9950E+00	.3219E+01

(Here $|\zeta_x(k)| \stackrel{\Delta}{=} [\sum_{i=1}^{2} \zeta_{xi}^2(k)]^{\frac{1}{2}}$).

TABLE 2.2: $N(k)$, $\hat{N}_x(k)$, $|\zeta_x(k)|$ and $Z(k)$ as functions of k for $\lambda = 3$. Here $\{T_1,\ldots,T_{10}\}$ = {1.159, 1.270, 1.275, 1.771, 2.855, 2.933, 3.082, 3.641, 3.978, 3.987}

k	N(k)	$\hat{N}_x(k)$	$\lvert\zeta_x(k)\rvert$	Z(k)
1	0	.3000E-02	.1351E+01	.1000E+01
51	0	.5772E+00	.1000E+01	.1000E+01
101	0	.8356E-10	.9981E+00	.2503E+01
151	0	.1199E+00	.1000E+01	.1000E+01
201	0	.1015E-01	.9983E+00	.1000E+01
251	0	.2999E-02	.9988E+00	.1000E+01
301	0	.1000E+01	.9995E+00	.7525E+02
351	0	.1000E+01	.9982E+00	.7506E+02
401	0	.1000E+01	.9988E+00	.7492E+02
451	0	.1000E+01	.1000E+01	.7600E+02
501	0	.1000E+01	.9998E+00	.7534E+02
551	0	.1000E+01	.9994E+00	.7526E+02
601	0	.1000E+01	.9978E+00	.7534E+02
651	0	.1000E+01	.9986E+00	.7626E+02
701	0	.1000E+01	.9995E+00	.7638E+02
751	0	.1000E+01	.9983E+00	.7498E+02
801	0	.1000E+01	.9989E+00	.7569E+02
851	0	.1000E+01	.9990E+00	.7523E+02
901	0	.1000E+01	.1000E+01	.7561E+02
951	0	.1000E+01	.9994E+00	.7560E+02
1001	0	.1000E+01	.9993E+00	.7592E+02
1051	0	.1000E+01	.9997E+00	.7466E+02
1101	0	.1000E+01	.9994E+00	.7523E+02
1151	0	.1000E+01	.9986E+00	.7555E+02
1201	1	.1000E+01	.9974E+00	.1044E+01
1251	1	.1000E+01	.9971E+00	.2230E+01
1301	3	.3000E+01	.9939E+00	.1639E+01
1351	3	.3000E+01	.9924E+00	.7284E+02
1401	3	.3000E+01	.9916E+00	.3614E+04
1451	3	.3002E+01	.9912E+00	.1001E+01
1501	3	.3002E+01	.9913E+00	.1002E+01
1551	3	.3000E+01	.9926E+00	.1088E+01
1601	3	.3000E+01	.9920E+00	.1246E+01
1651	3	.3000E+01	.9923E+00	.1370E+01
1701	3	.3000E+01	.9915E+00	.1730E+01
1751	3	.3002E+01	.9916E+00	.1001E+01
1801	4	.4000E+01	.9885E+00	.1659E+01
1851	4	.4002E+01	.9902E+00	.1001E+01
1901	4	.4788E+01	.9891E+00	.1000E+01
1951	4	.5002E+01	.9886E+00	.1005E+01
2001	4	.5002E+01	.9894E+00	.1005E+01
2051	4	.5002E+01	.9894E+00	.1006E+01
2101	4	.5002E+01	.9897E+00	.1005E+01
2151	4	.5002E+01	.9885E+00	.1006E+01
2201	4	.5002E+01	.9880E+00	.1005E+01
2251	4	.5002E+01	.9892E+00	.1005E+01

TABLE 2.2 (CONTINUED)

k	N(k)	$\hat{N}_x(k)$	$\lvert\zeta_x(k)\rvert$	Z(k)
2301	4	.5002E+01	.9888E+00	.1005E+01
2351	4	.5002E+01	.9888E+00	.1005E+01
2401	4	.5002E+01	.9904E+00	.1005E+01
2451	4	.5002E+01	.9889E+00	.1006E+01
2501	4	.5002E+01	.9891E+00	.1005E+01
2551	4	.5002E+01	.9879E+00	.1005E+01
2601	4	.5002E+01	.9896E+00	.1005E+01
2651	4	.5002E+01	.9891E+00	.1005E+01
2701	4	.5002E+01	.9883E+00	.1005E+01
2751	4	.5002E+01	.9897E+00	.1005E+01
2801	4	.5002E+01	.9879E+00	.1005E+01
2851	4	.5002E+01	.9883E+00	.1005E+01
2901	5	.5004E+01	.9867E+00	.1000E+01
2951	6	.6000E+01	.9876E+00	.1412E+01
3001	6	.6000E+01	.9875E+00	.1909E+01
3051	6	.6000E+01	.9853E+00	.2582E+01
3101	7	.7002E+01	.9892E+00	.1003E+01
3151	7	.7002E+01	.9896E+00	.1001E+01
3201	7	.7002E+01	.9897E+00	.1002E+01
3251	7	.7002E+01	.9903E+00	.1003E+01
3301	7	.7002E+01	.9901E+00	.1002E+01
3351	7	.7002E+01	.9905E+00	.1004E+01
3401	7	.7002E+01	.9900E+00	.1002E+01
3451	7	.7002E+01	.9895E+00	.1003E+01
3501	7	.7002E+01	.9893E+00	.1002E+01
3551	7	.7002E+01	.9887E+00	.1004E+01
3601	7	.7002E+01	.9906E+00	.1005E+01
3651	8	.8000E+01	.9935E+00	.4634E+01
3701	8	.8000E+01	.9949E+00	.4452E+01
3751	8	.8000E+01	.9954E+00	.3901E+01
3801	8	.8000E+01	.9963E+00	.3972E+01
3851	8	.8000E+01	.9948E+00	.3713E+01
3901	8	.8000E+01	.9957E+00	.3808E+01
3951	8	.8000E+01	.9969E+00	.4423E+01

TABLE 2.3: $N(k)$, $\hat{N}_x(k)$, $|\zeta_x(k)|$ and $Z(k)$ as functions of k for $\lambda = 4$. Here $\{T_1,\ldots,T_{12}\}$ = {.1487, .2545, .3640, .4032, .6177, 1.123, 1.188, 1.213, 1.389, 1.752, 1.786, 2.103}

k	N(k)	$\hat{N}_x(k)$	$\lvert\zeta_x(k)\rvert$	Z(k)
1	0	.4000E-02	.1351E+01	.1000E+01
51	0	.4000E-02	.1001E+01	.1000E+01
101	0	.6618E-01	.9993E+00	.1000E+01
151	1	.1999E+01	.9985E+00	.1428E+04
201	1	.1004E+01	.9983E+00	.1000E+01
251	1	.1483E+01	.9981E+00	.1000E+01
301	2	.2000E+01	.9943E+00	.3288E+01
351	2	.3003E+01	.9948E+00	.1007E+01
401	3	.3004E+01	.9923E+00	.1000E+01
451	4	.5003E+01	.9881E+00	.1008E+01
501	4	.5003E+01	.9892E+00	.1007E+01
551	4	.5003E+01	.9886E+00	.1007E+01
601	4	.5003E+01	.9893E+00	.1007E+01
651	5	.5000E+01	.9877E+00	.3269E+04
701	5	.5000E+01	.9876E+00	.1330E+01
751	5	.5003E+01	.9864E+00	.1005E+01
801	5	.6000E+01	.9874E+00	.9877E+02
851	5	.6000E+01	.9882E+00	.9903E+02
901	5	.6000E+01	.9879E+00	.9829E+02
951	5	.6000E+01	.9868E+00	.9890E+02
1001	5	.6000E+01	.9861E+00	.9753E+02
1051	5	.6000E+01	.9866E+00	.9758E+02
1101	5	.6000E+01	.9867E+00	.9866E+02
1151	6	.6000E+01	.9871E+00	.2340E+01
1201	7	.7000E+01	.9889E+00	.4584E+01
1251	8	.8000E+01	.9957E+00	.5767E+01
1301	8	.8000E+01	.9967E+00	.6063E+01
1351	8	.8000E+01	.9954E+00	.4799E+01
1401	9	.9000E+01	.1004E+01	.8368E+01
1451	9	.9000E+01	.1006E+01	.8289E+01
1501	9	.9000E+01	.1005E+01	.7568E+01
1551	9	.9000E+01	.1007E+01	.7828E+01
1601	9	.9000E+01	.1006E+01	.9903E+01
1651	9	.9000E+01	.1005E+01	.7797E+01
1701	9	.9000E+01	.1006E+01	.9076E+01
1751	9	.9000E+01	.1006E+01	.7563E+01
1801	11	.1100E+02	.1039E+01	.1007E+01
1851	11	.1100E+02	.1043E+01	.1007E+01
1901	11	.1100E+02	.1042E+01	.1007E+01
1951	11	.1100E+02	.1042E+01	.1007E+01
2001	11	.1100E+02	.1042E+01	.1007E+01
2051	11	.1100E+02	.1043E+01	.1008E+01
2101	11	.1100E+02	.1043E+01	.1007E+01

TABLE 2.4: $N(k)$, $\hat{N}_x(k)$, $|\zeta_x(k)|$ and $Z(k)$ as functions of k for $\lambda = 5$. Here $\{T_1,\dots,T_{19}\}$ = {.4369, .5542, .8807, 1.150, 1.195, 1.587, 1.665, 2.172, 2.189, 2.214, 2.217, 2.472, 2.504, 2.572, 2.660, 3.039, 3.065, 3.398, 3.510}

k	N(k)	$\hat{N}_x(k)$	$\|\zeta_x(k)\|$	Z(k)
1	0	.5000E-02	.1351E+01	.1000E+01
51	0	.4963E-02	.1001E+01	.1007E+01
101	0	.2700E-04	.9994E+00	.1851E+01
151	0	.1228E-01	.9977E+00	.1000E+01
201	0	.1000E+01	.9994E+00	.1254E+03
251	0	.1000E+01	.9988E+00	.1267E+03
301	0	.1000E+01	.9990E+00	.1250E+03
351	0	.1000E+01	.9986E+00	.1251E+03
401	0	.1000E+01	.1000E+01	.1253E+03
451	1	.2005E+01	.9978E+00	.1000E+01
501	1	.2004E+01	.9994E+00	.1009E+01
551	1	.2004E+01	.9973E+00	.1009E+01
601	2	.2005E+01	.9948E+00	.1000E+01
651	2	.2004E+01	.9940E+00	.1003E+01
701	2	.2000E+01	.9955E+00	.7503E+03
751	2	.2004E+01	.9953E+00	.1001E+01
801	2	.2983E+01	.9947E+00	.1000E+01
851	2	.2000E+01	.9953E+00	.4202E+01
901	3	.3007E+01	.9910E+00	.1000E+01
951	3	.3000E+01	.9912E+00	.3000E+01
1001	3	.3005E+01	.9892E+00	.1000E+01
1051	3	.3000E+01	.9912E+00	.2072E+01
1101	3	.3000E+01	.9908E+00	.2718E+01
1151	4	.3005E+01	.9900E+00	.1000E+01
1201	5	.5004E+01	.9882E+00	.1003E+01
1251	5	.5000E+01	.9882E+00	.2448E+01
1301	5	.5004E+01	.9882E+00	.1003E+01
1351	5	.5009E+01	.9869E+00	.1000E+01
1401	5	.5004E+01	.9863E+00	.1009E+01
1451	5	.5004E+01	.9879E+00	.1004E+01
1501	5	.6000E+01	.9873E+00	.1213E+03
1551	5	.6000E+01	.9872E+00	.1212E+03
1601	6	.6000E+01	.9873E+00	.3316E+01
1651	6	.6000E+01	.9874E+00	.2967E+01
1701	7	.7000E+01	.9904E+00	.5498E+01
1751	7	.7000E+01	.9902E+00	.3912E+01
1801	7	.7000E+01	.9905E+00	.5079E+01
1851	7	.7000E+01	.9904E+00	.3985E+01
1901	7	.7000E+01	.9902E+00	.5940E+01
1951	7	.7000E+01	.9904E+00	.5530E+01
2001	7	.7000E+01	.9904E+00	.2915E+01
2051	7	.7000E+01	.9894E+00	.5678E+01
2101	7	.7000E+01	.9902E+00	.4700E+01

TABLE 2.4 : (CONTINUED)

k	N(k)	$\hat{N}_x(k)$	$\lvert\zeta_x(k)\rvert$	Z(k)
2151	7	.7000E+01	.9890E+00	.4822E+01
2201	9	.9004E+01	.1004E+01	.1007E+01
2251	11	.1100E+02	.1041E+01	.1010E+01
2301	11	.1100E+02	.1041E+01	.1009E+01
2351	11	.1100E+02	.1041E+01	.1009E+01
2401	11	.1100E+02	.1042E+01	.1009E+01
2451	11	.1100E+02	.1042E+01	.1010E+01
2501	12	.1200E+02	.1067E+01	.1009E+01
2551	13	.1300E+02	.1099E+01	.5409E+02
2601	14	.1400E+02	.1135E+01	.1009E+01
2651	14	.1400E+02	.1135E+01	.1010E+01
2701	15	.1500E+02	.1175E+01	.1095E+03
2751	15	.1500E+02	.1176E+01	.1118E+03
2801	15	.1500E+02	.1176E+01	.1124E+03
2851	15	.1500E+02	.1176E+01	.1117E+03
2901	15	.1500E+02	.1176E+01	.1108E+03
2951	15	.1500E+02	.1176E+01	.1102E+03
3001	15	.1500E+02	.1176E+01	.1125E+03
3051	16	.1600E+02	.1220E+01	.1009E+01
3101	17	.1700E+02	.1272E+01	.2236E+03
3151	17	.1700E+02	.1272E+01	.2235E+03
3201	17	.1700E+02	.1272E+01	.2230E+03
3251	17	.1700E+02	.1272E+01	.2206E+03
3301	17	.1700E+02	.1272E+01	.2244E+03
3351	17	.1700E+02	.1272E+01	.2226E+03
3401	18	.1700E+02	.1299E+01	.6154E+02
3451	18	.1700E+02	.1327E+01	.9217E+02
3501	18	.1700E+02	.1326E+01	.9170E+02

2.9 EXAMPLE 2.4: RANDOM TELEGRAPH SIGNAL

2.9.1 Introduction

In this section we consider the case where the process η appearing in equation (2.28) is the *random telegraph signal*. More explicitly, $\eta = \{\eta(t),\ t \geq 0\}$ is a Markov process with state space $S = \{-1,1\}$ and transition probabilities

$$P(\eta(t+\Delta) = j \mid \eta(t) = i) = \begin{cases} \lambda\Delta + O(\Delta^2) & \text{if } j \neq i \\ 1 - \lambda\Delta + O(\Delta^2) & \text{if } j = i \end{cases} \tag{2.63}$$

$i,j = -1,1$, where λ is a given positive number. It is assumed here that $\pi_i = P(\eta(0) = i)$, $i=-1,1$ are given numbers. Thus equations (2.50)-(2.51) reduce here to

$$dP_i(t) = \sum_{j=-1,1} \lambda_{ji} P_j(t)dt + P_i(t)(i - \hat{\eta}_x(t)) \sum_{q,s=1}^{m} (B^{-1})^2_{qs} f_q(\zeta_x(t))$$

$$[d\zeta_{xs}(t) - (\hat{\eta}_x(t)f_s(\zeta_x(t)) + g_s(\zeta_x(t)))dt], \qquad (2.64)$$

$i=-1,1$, $t \in (0,T)$,

$$\hat{\eta}_x(t) = P_1(t) - P_{-1}(t) \quad , \quad t \in [0,T] \qquad (2.65)$$

where

$$\lambda_{ji} = \begin{cases} \lambda & \text{if } j \neq i \\ & \qquad\qquad i,j=-1,1 \\ -\lambda & \text{if } j = i \quad . \end{cases} \qquad (2.66)$$

2.9.2 Algorithm 2.9

In this subsection an algorithm for computing $\{\hat{\eta}_x(t), t \in [0,T]\}$ is sug= gested.

Choose $\varepsilon > 0$ and $\Delta > 0$. Let $N\Delta = T$.

1. $k=0$, $P_1(0) = P_{-1}(0) = \frac{1}{2}$, $\hat{\eta}_x(0) = 0$

2. For $i=-1,1$, calculate

$$P_i(k+1):=P_i(k) + \sum_{j=-1,1} \lambda_{ji} P_j(k)\Delta + P_i(k)(i - \hat{\eta}_x(k)) \sum_{q,s=1}^{m} (B^{-1})^2_{qs} \qquad (2.67)$$

$$f_q(\zeta_x(k))[\zeta_{xs}(k+1) - \zeta_{xs}(k) - (\hat{\eta}_x(k)f_s(\zeta_x(k)) + g_s(\zeta_x(k)))\Delta]$$

$$P_i(k+1):= \max(0,P_i(k+1)) \qquad (2.68)$$

3. $Z(k+1) := P_1(k+1) + P_{-1}(k+1)$ (2.69)

4. If $Z(k+1) \geq \varepsilon$ then: for $i=1,-1$, $P_i(k+1) := P_i(k+1)/Z(k+1)$
Otherwise: stop.

5. $\hat{\eta}_x(k+1) = P_1(k+1) - P_{-1}(k+1)$ (2.70)

6. If $k=N$ stop. Otherwise: $k:=k+1$ and go to 2.

2.9.3 <u>The simulation procedure</u>

(i) A sample path of $\{\eta(t),\ t \in [0,T]\}$ is constructed by applying the following procedure:

1. Read the numbers R_i, $i=1,\ldots,M+1$, from a random number generator with a uniform probability density on (0,1).

2. Calculate

$$Z_i = (-1/\lambda)\ell n R_i \quad , \quad i=1,\ldots,M+1$$

$$T_o = 0 \quad T_i = \sum_{\ell=1}^{i} Z_\ell \quad , \quad i=1,\ldots,M+1,$$

where M is determined by $T_M \leq T < T_{M+1}$

3. Define the following function

(a) $\eta(k) = (-1)^i \qquad T_i \leq k\Delta < T_{i+1} \quad i=0,1,\ldots,M,$

or

(b) $\eta(k) = (-1)^{i+1} \qquad T_i \leq k\Delta < T_{i+1} \quad i=0,1,\ldots,M$

(ii) Follow paragraph (ii) of subsection 2.8.3.

(iii) Apply Algorithm 2.9.

2.9.4 A numerical example

Consider the system dealt with in Section 2.5 but with the frequency hopping according to a random telegraph signal $\{\eta(t),\ t \geq 0\}$, i.e.

$$\begin{cases} dx_1 = [(a_o + \alpha\eta(t))\mathrm{sign}(x_2) + b(\rho - |x_1| - |x_2|)x_1]dt + \sigma_1 dW_1 \\ \qquad\qquad\qquad\qquad\qquad\qquad\qquad\qquad\qquad\qquad\qquad t > 0, \\ dx_2 = [-(a_o + \alpha\eta(t))\mathrm{sign}(x_1) + b(\rho - |x_1| - |x_2|)x_2]dt + \sigma_2 dW_2 \ . \end{cases} \tag{2.71}$$

It is assumed that a_o, α, b, ρ, σ_1, σ_2 and λ are given positive numbers and that $\mathbf{W}$ and $\{\eta(t),\ t \geq 0\}$ are mutually independent. Using the nota= tions

$$\begin{cases} f_1(x) = \alpha\ \mathrm{sign}(x_2)\ ,\quad g_1(x) = a_o\ \mathrm{sign}(x_2) + b(\rho - |x_1| - |x_2|)x_1 \\ f_2(x) = -\ \alpha\ \mathrm{sign}(x_1)\ ,\quad g_2(x) = -a_o\ \mathrm{sign}(x_1) + b(\rho - |x_1| - |x_2|)x_2 \end{cases} \tag{2.72}$$

equations (2.71) can be written in the form of equations (2.28).

Numerical experimentation has been carried out for the following set of parameters: $a_o = 100$, $\alpha = 10$, $b = 50$, $\rho = 1$, $\sigma_1 = \sigma_2 = 0.01$, $\lambda = 2,3,4,5$. Some extracts of the results are presented below. Here $|\zeta_x(k)|_1 \triangleq |\zeta_{x1}(k)| + |\zeta_{x2}(k)|$.

TABLE 2.5: $\eta(k)$ and $\hat{\eta}_x(k)$ as functions of k for $\lambda = 2$. Here $\{T_1,\ldots,T_9\} = \{.6289,\ 1.440,\ 3.094,\ 4.109,\ 4.197,\ 5.642,\ 8.181,\ 8.434,\ 8.689\}$

k	$\eta(k)$	$\hat{\eta}_x(k)$	k	$\eta(k)$	$\hat{\eta}_x(k)$
1	1	.1000E+01	2301	1	.9960E+00
51	1	.1000E+01	2351	1	.9960E+00
101	1	.1000E+01	2401	1	.1000E+01
151	1	.9960E+00	2451	1	.1000E+01
201	1	.9960E+00	2501	1	-.1000E+01
251	1	.1000E+01	2551	1	.9960E+00
301	1	.9960E+00	2601	1	.9396E+00
351	1	.1000E+01	2651	1	.3179E+00
401	1	-.9960E+00	2701	1	.1000E+01
451	1	.9960E+00	2751	1	.9960E+00
501	1	.1000E+01	2801	1	.9960E+00
551	1	.1000E+01	2851	1	.1000E+01
601	1	.9960E+00	2901	1	.8721E+00
651	-1	-.9960E+00	2951	1	-.1000E+01
701	-1	-.1000E+01	3001	1	.9960E+00
751	-1	-.9960E+00	3051	1	.1000E+01
801	-1	-.1000E+01	3101	-1	-.1000E+01
851	-1	-.1000E+01	3151	-1	-.1000E+01
901	-1	-.1000E+01	3201	-1	-.9960E+00
951	-1	-.4154E+00	3251	-1	-.1000E+01
1001	-1	-.9960E+00	3301	-1	-.9960E+00
1051	-1	-.9960E+00	3351	-1	-.9960E+00
1101	-1	-.9960E+00	3401	-1	-.9966E+00
1151	-1	-.9960E+00	3451	-1	-.1000E+01
1201	-1	-.1000E+01	3501	-1	-.1000E+01
1251	-1	-.9960E+00	3551	-1	-.9495E+00
1301	-1	-.9960E+00	3601	-1	-.9960E+00
1351	-1	-.9960E+00	3651	-1	-.9960E+00
1401	-1	-.7294E+00	3701	-1	-.9960E+00
1451	1	.9960E+00	3751	-1	-.1000E+01
1501	1	.1000E+01	3801	-1	-.6992E+00
1551	1	.1000E+01	3851	-1	-.9960E+00
1601	1	.9960E+00	3901	-1	-.1000E+01
1651	1	.9960E+00	3951	-1	-.9960E+00
1701	1	-.9960E+00	4001	-1	-.1000E+01
1751	1	.9960E+00	4051	-1	-.1000E+01
1801	1	.9960E+00	4101	-1	-.1000E+01
1851	1	.9960E+00	4151	1	.6747E+00
1901	1	.9960E+00	4201	-1	-.1000E+01
1951	1	.1000E+01	4251	-1	-.1000E+01
2001	1	.1000E+01	4301	-1	-.9960E+00
2051	1	.9960E+00	4351	-1	-.1000E+01
2101	1	.1000E+01	4401	-1	-.9960E+00
2151	1	.1000E+01	4451	-1	-.9960E+00
2201	1	.9960E+00	4501	-1	-.9960E+00
2251	1	.1000E+01	4551	-1	-.9960E+00

TABLE 2.5 (CONTINUED)

k	$\eta(k)$	$\hat{\eta}_x(k)$
4601	-1	-.9960E+00
4651	-1	-.1000E+01
4701	-1	-.1000E+01
4751	-1	-.8255E+00
4801	-1	-.1000E+01
4851	-1	-.9960E+00
4901	-1	-.9960E+00
4951	-1	-.1000E+01
5001	-1	-.9960E+00
5051	-1	-.1000E+01
5101	-1	-.9960E+00
5151	-1	-.7099E+00
5201	-1	-.1000E+01
5251	-1	-.1000E+01
5301	-1	-.9960E+00
5351	-1	-.1000E+01
5401	-1	-.1000E+01
5451	-1	-.1000E+01
5501	-1	-.3283E+00
5551	-1	-.1000E+01
5601	-1	-.1000E+01
5651	1	.1000E+01
5701	1	.1000E+01
5751	1	.9960E+00
5801	1	.9960E+00
5851	1	.9960E+00
5901	1	.9960E+00
5951	1	.9960E+00
6001	1	.1000E+01
6051	1	.1000E+01
6101	1	.1000E+01
6151	1	.1000E+01
6201	1	.8945E+00
6251	1	.1000E+01
6301	1	.9960E+00
6351	1	-.1000E+01
6401	1	.1000E+01
6451	1	-.9960E+00
6501	1	.9960E+00
6551	1	.1000E+01
6601	1	.9960E+00
6651	1	.9960E+00
6701	1	.6405E+00
6751	1	.9553E+00
6801	1	-.1000E+01
6851	1	.9960E+00
6901	1	-.9960E+00
6951	1	.9960E+00
7001	1	.1000E+01
7051	1	.1000E+01
7101	1	.9960E+00
7151	1	.8562E+00
7201	1	.9684E+00
7251	1	.9960E+00
7301	1	.9960E+00
7351	1	-.9960E+00
7401	1	.9960E+00
7451	1	.1000E+01
7501	1	.9960E+00
7551	1	.1000E+01
7601	1	.9960E+00
7651	1	.1000E+01
7701	1	-.1000E+01
7751	1	.9960E+00
7801	1	-.9960E+00
7851	1	.1000E+01
7901	1	.1000E+01
7951	1	.1000E+01
8001	1	.9960E+00
8051	1	.5106E+00
8101	1	.1000E+01
8151	1	-.1000E+01
8201	-1	-.1000E+01
8251	-1	-.1000E+01
8301	-1	-.1000E+01
8351	-1	-.1000E+01
8401	-1	-.1000E+01
8451	1	.1000E+01
8501	1	.1000E+01
8551	1	.9960E+00
8601	1	.1000E+01
8651	1	.1000E+01

TABLE 2.6: $\eta(k)$ and $\hat{\eta}_x(k)$ as functions of k for $\lambda = 3$. Here $\{T_1,\ldots,T_9\}$ = {.086, .1235, .3025, .4997, 1.029, 1.379, 1.494, 1.661, 1.876}

k	$\eta(k)$	$\hat{\eta}_x(k)$	k	$\eta(k)$	$\hat{\eta}_x(k)$
1	1	.1000E+01	951	1	.5019E+00
51	1	.1000E+01	1001	1	-.1000E+01
101	-1	-.9940E+00	1051	-1	-.1000E+01
151	1	.9940E+00	1101	-1	-.9940E+00
201	1	-.9940E+00	1151	-1	-.9940E+00
251	1	.9940E+00	1201	-1	-.1000E+01
301	1	.9940E+00	1251	-1	-.9940E+00
351	-1	-.6221E+00	1301	-1	-.9940E+00
401	-1	-.1000E+01	1351	-1	-.1000E+01
451	-1	-.9940E+00	1401	1	.1000E+01
501	1	-.9940E+00	1451	1	.9940E+00
551	1	-.1000E+01	1501	-1	-.1000E+01
601	1	.1000E+01	1551	-1	-.9940E+00
651	1	.1000E+01	1601	-1	-.1000E+01
701	1	.1000E+01	1651	-1	-.1000E+01
751	1	.9940E+00	1701	1	.1000E+01
801	1	.9940E+00	1751	1	.9940E+00
851	1	.9940E+00	1801	1	.1000E+01
901	1	.4687E+00	1851	1	.1000E+01

TABLE 2.7: $\eta(k)$ and $\hat{\eta}_x(k)$ as functions of k for $\lambda = 4$. Here $\{T_1,\ldots,T_9\}$ = {.2089, .9699, 1.059, 1.135, 1.179, 1.786, 1.877, 2.004, 2.005}

k	$\eta(k)$	$\hat{\eta}_x(k)$	k	$\eta(k)$	$\hat{\eta}_x(k)$
1	1	.1000E+01	1051	1	.9920E+00
51	1	.9920E+00	1101	-1	-.9920E+00
101	1	.1000E+01	1151	1	.9920E+00
151	1	.9920E+00	1201	-1	-.9920E+00
201	1	.9920E+00	1251	-1	-.9217E+00
251	-1	-.9920E+00	1301	-1	-.9920E+00
301	-1	-.9920E+00	1351	-1	-.3782E+00
351	-1	-.1000E+01	1401	-1	-.1000E+01
401	-1	-.1000E+01	1451	-1	-.9920E+00
451	-1	-.1000E+01	1501	-1	-.9920E+00
501	-1	-.9920E+00	1551	-1	-.9920E+00
551	-1	-.9920E+00	1601	-1	-.1000E+01
601	-1	-.1000E+01	1651	-1	-.1000E+01
651	-1	-.9920E+00	1701	-1	-.9920E+00
701	-1	-.9920E+00	1751	-1	-.9920E+00
751	-1	-.9920E+00	1801	1	.1000E+01
801	-1	.5461E+00	1851	1	.9920E+00
851	-1	-.9454E+00	1901	-1	-.9920E+00
901	-1	-.4321E-01	1951	-1	-.9920E+00
951	-1	-.1000E+01	2001	-1	-.1000E+01
1001	1	.1589E+00			

TABLE 2.8: $\eta(k)$, $\hat{\eta}_x(k)$ and $|\zeta_x(k)|_1$ as functions of k for $\lambda = 5$. Here $\{T_1,\ldots,T_9\}$ = {.0856, .3099, .4114, .8241, .9053, 1.062, 1.155, 1.213, 1.435}

k	$\eta(k)$	$\hat{\eta}_x(k)$	$\|\zeta_x(k)\|_1$
1	1	.1000E+01	.1907E+01
51	1	.1000E+01	.9830E+00
101	-1	-.9900E+00	.1015E+01
151	-1	-.1000E+01	.9878E+00
201	-1	-.9900E+00	.9924E+00
251	-1	-.9900E+00	.9887E+00
301	-1	-.9900E+00	.9914E+00
351	1	.9900E+00	.9852E+00
401	1	.1000E+01	.9913E+00
451	-1	-.1000E+01	.9911E+00
501	-1	-.9900E+00	.9952E+00
551	-1	-.1000E+01	.9913E+00
601	-1	-.9900E+00	.9864E+00
651	-1	-.1000E+01	.9906E+00
701	-1	-.1000E+01	.9898E+00
751	-1	-.1000E+01	.9912E+00
801	-1	-.1000E+01	.9851E+00
851	1	.1000E+01	.9666E+00
901	1	.9900E+00	.9895E+00
951	-1	-.9793E+00	.9943E+00
1001	-1	-.9900E+00	.9840E+00
1051	-1	-.1000E+01	.9913E+00
1101	1	.1000E+01	.9949E+00
1151	1	.8413E+00	.9880E+00
1201	-1	-.9900E+00	.9900E+00
1251	1	.9900E+00	.9903E+00
1301	1	.9943E+00	.9985E+00
1351	1	.9900E+00	.9858E+00
1401	1	-.1000E+01	.9904E+00

2.10 REMARKS

Equations (2.26) and (2.27) constitute models for a sine-wave oscillator [2.3-2.4] and a triangular-wave generator [2.5-2.6], both having an unknown frequency a_o, respectively. Equations (2.61) and (2.71) constitute models for a sine-wave oscillator with frequency $a_o(1 + N(t))$, and a triangular-wave generator with frequency $a_o(1 + \alpha\eta(t))$, respectively, where $\{N(t), t \geq 0\}$ is a Poisson process and $\{\eta(t), t \geq 0\}$ is a random telegraph signal (for more details on this process see, for example, [2.11]).

One basic technique used in spread-spectrum communication or in radar is that of frequency hopping. With this technique the transmitter rapidly changes frequency in a coded sequence, and the frequencies and sequence can be changed as often as needed [2.12].

Assume that an enemy receives the sine-wave or the triangular-wave signal, but does not know the specific sequence code in use at that moment. Then, using these measurements, the enemy has to estimate the frequency code sequence of the signal. The examples dealt with in this chapter constitute a two-stage approximate solution to this estimation problem. First, the transmitter is modelled by a sine-wave generator or a triangular-wave generator . Then, using the procedures derived in this chapter, minimum variance filter equations are obtained.

The results obtained here can be applied to cases where other forms of wave shapes are in use, such as square waves, trapezoidal waves, etc.

2.11 REFERENCES

2.1 I.I.Gihman and A.V. Skorohod, *Stochastic Differential Equations*, Springer-Verlag, Berlin, 1972.

2.2 M. Fujisaki, G. Kallianpur and H. Kunita, Stochastic differential equations for the non linear filtering problem, *Osaka J. Math.* 9, pp 19-40, 1972.

2.3 B.Z. Kaplan, On second-order non-linear systems with conservative limit-cycles, *Int. J. Non-Linear Mechanics*, 13, pp 43-51, 1978.

2.4 B.Z. Kaplan, On the simplified implementation of quadrature oscilla= tor models and the expected quality of their operation as VCO's, *Proceed. of the IEEE*, 68, pp 745-746, 1980.

2.5 B.Z. Kaplan and Y. Tatrash, New method for generating precise tri= angular waves and square waves, *Electronic Letters*, 13, pp 71-73, 1977.

2.6 B.Z. Kaplan, An implementation of a new method for generating triangu= lar and square waves, *Int. J. Electronics*, 46, pp 299-308, 1979.

2.7 W.C. Huisman and Y. Yavin, Numerical studies of the performance of an optimally controlled nonlinear stochastic oscillator, *Computer Methods in Applied Mechanic and Eng.*, 21, pp 171-191, 1980.

2.8 L.V. Sergeeva and N.I. Teterina, Investigation of the solution of a stochastic equation with random coefficients,*Theor. Probability and Math. Statist.*, pp 145-158, 1974.

2.9 L.V. Sergeeva, On a certain generalization of diffusion processes, *Theor. Probability and Math Statist.*, pp 161-169, 1976.

2.10 R.S. Liptser and A.N. Shiryayev, *Statistics of Random Processes*, Springer-Verlag, New York, Vol.I, 1977; Vol.II, 1978.

2.11 J.L. Melsa and A.P. Sage, *An Introduction to Probability and Stochas= tic Processes*, Prentice-Hall, Englewood Cliffs, 1973.

2.12 P.J. Klass, *Spread spectrum use likely to expand*, Aviation Week & Space Technology, January 3, pp 55-59, 1983.

CHAPTER 3

FILTERING VIA MARKOV CHAINS APPROXIMATION

1. INTRODUCTION

Consider an $\mathbb{R}^m$-valued Markov process $\zeta_x = \{\zeta_x(t),\ t \geq 0\}$ satisfying the equation

$$\zeta_x(t) = x + \int_0^t f(\zeta_x(s))ds + \int_0^t BdW(s),\ t \geq 0,\ x \in \mathbb{R}^m \tag{3.1}$$

with the noisy observations of ζ_x given by

$$y(t) = \int_0^t g(\zeta_x(s))ds + \int_0^t \Gamma dv(s)\ ,\quad t \geq 0,\quad y(t) \in \mathbb{R}^p\ , \tag{3.2}$$

where $f : \mathbb{R}^m \to \mathbb{R}^m$ and $g : \mathbb{R}^m \to \mathbb{R}^p$ are given continuously differentiable functions on $\mathbb{R}^m$ satisfying $|f(x)|_m^2 \leq M(1 + |x|_m^2)$ and $|g(x)|_p^2 \leq M(1 + |x|_m^2)$, $x \in \mathbb{R}^m$, for some $0 < M < \infty$(here $|f(x)|_m^2 = \sum_{i=1}^m f_i^2(x)$ and $|g(x)|_p^2 = \sum_{i=1}^p g_i^2(x)$); $B \in \mathbb{R}^{m\times m}$ and $\Gamma \in \mathbb{R}^{p\times p}$ are given symmetric positive definite matrices. $\mathbf{W} \triangleq \{W(t) = (W_1(t),\ldots,W_m(t)), t \geq 0\}$ and $\mathbf{V} \triangleq \{v(t) = (v_1(t),\ldots,v_p(t)),$ $t \geq 0\}$ are $\mathbb{R}^m$-valued and $\mathbb{R}^p$-valued standard Wiener processes respectively. It is assumed that $\mathbf{W}$ and $\mathbf{V}$ are mutually independent.

Denote by F_t^y the smallest σ-field generated by the family of random ele= ments $Y^t = \{y(s) : 0 \leq s \leq t\}$. The problem dealt with in this chapter is to find an approximation $\hat{\zeta}_x^{h,y}(k)$ to

$$\hat{\zeta}_x(t) \triangleq E[\zeta_x(t\wedge\tau_T-)|F^y_{t\wedge\tau_T-}]\ ,\qquad t \in [0,T] \tag{3.3}$$

at the instants $t_k = k\Delta$, $k\Delta \in [0,T]$, where $\tau_T = \tau_T(x)$ is the first exit time of $\zeta_x(t)$ from an open and bounded domain $D \subset \mathbb{R}^m$ and $t \wedge \tau_T = \min(t,\tau_T)$. The problem of finding $E[\zeta_x(t)|F_t^y]$ (which is the minimum least-squares estimate of $\zeta_x(t)$ given Y^t), is called the *nonlinear filtering problem.* Extensive work has been done on nonlinear filtering on stochastic con=tinuous time systems and various approaches have been used. For more details see, for example, Stratonovich [3.1], Kushner [3.2-3.3], Wonham [3.4], Bucy [3.5], Bucy and Joseph [3.6], Zakai [3.7], Jazwinski [3.8], Frost and Kailath [3.9], McGarty [3.10], Fujisaki et al. [3.11] and Kalli=anpur [3.12].

In most of the works mentioned, recursive formulae were obtained for up=dating the least-squares estimate $E[\zeta_x(t)|F_t^y]$. It was found, however, that in general, (the exceptions being the linear-Gaussian cases in which the Kalman filter is optimal, and a very few other known cases, see for example Van Schuppen [3.13]), the formulae involve all the conditional moments, so that an infinite set of coupled stochastic differential equa=tions is formed.

In order to find approximations to $E[\zeta_x(t)|F_t^y]$, several practical algo=rithms have been suggested of which the best known are linearized and extended Kalman filters (see Jazwinski [3.8]) and these have been used most frequently.

In this chapter, the nonlinear filtering problem is treated by using methods different from those used in [3.1] - [3.10]. Given an open and bounded domain D in $\mathbb{R}^m$. Let $\tau_T = \tau_T(x)$ be the first exit time of $\zeta_x(t)$ from D, during the time interval [0,T]. First, the process $\{\zeta_x(t\wedge\tau_T),$ $t \in [0,T]\}$ is approximated by a continuous-time Markov chain $\{\zeta_x^h(t\wedge\tau_T^h),\ t \in [0,T]\}$ with a finite state space S, $S \subset D$. Second, an optimal least-squares filter is derived for the on-line computation of $\hat{\zeta}_x^h(t) \triangleq E[\zeta_x^h(t\wedge\tau_T^h-)|F^{y,h}_{t\wedge\tau_T^h-}]$. ($\tau_T^h$ and $F_t^{y,h}$ are defined in Section 3.2). Third, an estimator $\{\hat{\zeta}_x^{h,y}(k),\ k\Delta \in [0,T]\}$ is constructed as an approximation to $\{\hat{\zeta}_x(k\Delta),\ k\Delta \in [0,T]\}$ (equation (3.3)) and this estimator is simulated for a variety of examples.

In Kushner [3.14] and Di Masi and Runggaldier [3.15] analogous approxima= tions of $\{\zeta_x(t\wedge\tau_T),\ t \in [0,T]\}$ by discrete-time Markov chains, [3.14], or by continuous-time Markov chains, [3.15], are obtained. There, in both cases, the nonlinear filtering problem is treated via the Kallianpur-Striebel formula [3.16].

3.2 CONSTRUCTION OF THE MARKOV CHAIN

Let $\mathbb{R}_h^m$ be a grid on $\mathbb{R}^m$ with a constant mesh size h along all axes, i.e.,

$$\mathbb{R}_h^m \triangleq \{x \in \mathbb{R}^m : x_i = n_i h \ ,\ \ i=1,\dots,m \ ,\ \ n_i = 0,\pm 1,\pm 2,\dots\} \ , \tag{3.4}$$

and denote by e^i the unit vector along the i-th axis, i=1,...,m.

Throughout this chapter it is assumed, for the sake of simplicity, that the matrices B and Γ are of the form

$$B_{ij} = \sigma_i\ \delta_{ij},\ i,j=1,\dots,m \text{ and } \Gamma_{ij} = \gamma_i\ \delta_{ij},\ i,j=1,\dots,p \tag{3.5}$$

where σ_i, i=1,...,m and γ_i, i=1,...,p are given positive numbers.
Define

$$\lambda(x,x) \triangleq -\sum_{i=1}^{m} (\sigma_i^2 + h|f_i(x)|)/h^2 \ ,\ \ x \in \mathbb{R}_h^m \tag{3.6}$$

$$\lambda(x,x+e^i h) \triangleq (\sigma_i^2/2 + hf_i^+(x))/h^2 \ ,\ \ x \in \mathbb{R}_h^m,\ \ i=1,\dots,m \tag{3.7}$$

$$\lambda(x,x-e^i h) \triangleq (\sigma_i^2/2 + hf_i^-(x))/h^2,\ \ x \in \mathbb{R}_h^m,\ \ i=1,\dots,m \tag{3.8}$$

$$\lambda(x,y) \triangleq 0 \ ,\ \ x \in \mathbb{R}_h^m \text{ and } y \in U_x \tag{3.9}$$

where for each $\alpha \in \mathbb{R}$, $\alpha^+ = \max(0,\alpha)$, $\alpha^- = -\min(0,\alpha)$ and

$$U_x \triangleq \{y \in \mathbb{R}_h^m : y \neq x \text{ and } y \neq x \pm e^i h \ ,\ \ i=1,\dots,m\} \ . \tag{3.10}$$

Note that $\lambda(x,y) \geq 0$ for each $x,y \in \mathbb{R}_h^m$, $x \neq y$ and

$\lambda(x,x) + \sum_{\substack{y \\ y \neq x}} \lambda(x,y) = 0$. Hence, given $x \in \mathbb{R}_h^m$, we can construct a continuous-time Markov chain $\{\zeta_x^h(t),\ t \geq 0\}$, with state space $\mathbb{R}_h^m$, by defining the following set of transition probabilities

$$\begin{cases} P(\zeta_x^h(t+\Delta) = z \pm e^i h \mid \zeta_x^h(t) = z) \stackrel{\Delta}{=} \lambda(z, z \pm e^i h)\Delta + 0(\Delta^2) \\ \\ i=1,\dots,m \quad , \quad z \in \mathbb{R}_h^m \end{cases} \tag{3.11}$$

$$P(\zeta_x^h(t+\Delta) = z \mid \zeta_x^h(t) = z) \stackrel{\Delta}{=} 1 + \lambda(z,z)\Delta + 0(\Delta^2), \quad z \in \mathbb{R}_h^m \tag{3.12}$$

$$\sum_{y \in U_z} P(\zeta_x^h(t+\Delta) = y \mid \zeta_x^h(t) = z) \stackrel{\Delta}{=} 0(\Delta^2), \quad z \in \mathbb{R}_h^m \quad , \tag{3.13}$$

and

$$P(\zeta_x^h(0) = x) = 1, \quad x \in \mathbb{R}_h^m. \tag{3.14}$$

Thus, using (3.11)-(3.13) it follows that, for $x,z \in \mathbb{R}_h^m$

$$E[\zeta_{xi}^h(t+\Delta) - \zeta_{xi}^h(t) \mid \zeta_x^h(t) = z] = f_i(z)\Delta + h0(\Delta^2), \ i=1,\dots,m \tag{3.15}$$

and

$$\begin{aligned} &E[(\zeta_{xi}^h(t+\Delta) - \zeta_{xi}^h(t))(\zeta_{xj}^h(t+\Delta) - \zeta_{xj}^h(t)) \mid \zeta_x^h(t) = z] \\ &= \delta_{ij}(\sigma_i^2 + h|f_i(z)|)\Delta + (1 + \delta_{ij})h^2 0(\Delta^2), \ i,j=1,\dots,m \ . \end{aligned} \tag{3.16}$$

Consequently, for $x,z \in \mathbb{R}_h^m$,

$$|E[\zeta_{xi}^h(t+\Delta) - \zeta_{xi}^h(t) \mid \zeta_x^h(t) = z] - f_i(z)\Delta| = h0(\Delta^2), \ i=1,\dots,m, \tag{3.17}$$

and

$$\begin{aligned} &|E[(\zeta_{xi}^h(t+\Delta) - \zeta_{xi}^h(t))(\zeta_{xj}^h(t+\Delta) - \zeta_{xj}^h(t)) \mid \zeta_x^h(t) = z] - \delta_{ij}\ \sigma_i^2\Delta| \\ &= \delta_{ij}\ h0(\Delta) + h^2\ 0(\Delta^2) \quad , \quad i,j=1,\dots,m. \end{aligned} \tag{3.18}$$

Equations (3.17)-(3.18) illustrate the relations between the Markov chain $\{\zeta_x^h(t),\ t \geq 0\}$ and the Markov process ζ_x. In [3.15] it is shown that $\{\zeta_x^h(t),\ t \geq 0\}$ converges weakly to $\{\zeta_x(t),\ t \geq 0\}$ as $h \downarrow 0$.

Assume, without loss of generality, that

$$D \triangleq \{x \in \mathbb{R}^m : |x_i| < a_i + \delta,\ i=1,\dots,m\} \tag{3.19}$$

where a_i, $i=1,\dots,m$ and δ are given positive numbers, and $\delta < h$. Let $D_h \triangleq \mathbb{R}_h^m \cap D$. Given $T > 0$, define the following stopping times

$$\tau_T(x) \triangleq \begin{cases} T & \text{if } \zeta_x(t) \in D \text{ for all } t \in [0,T] \\ \inf\{t : 0 \leq t \leq T,\ \zeta_x(t) \notin D\} & \text{otherwise} \end{cases} \tag{3.20}$$

$$\tau_T^h(x) \triangleq \begin{cases} T & \text{if } \zeta_x^h(t) \in D_h \text{ for all } t \in [0,T] \\ \inf\{t : 0 \leq t \leq T,\ \zeta_x^h(t) \notin D_h\} & \text{otherwise} \end{cases} \tag{3.21}$$

It is shown in [3.15], under a nonrestrictive additional condition, that $\{\zeta_x^h(t\wedge\tau_T^h),\ t \in [0,T]\}$ converges weakly to $\{\zeta_x(t\wedge\tau_T),\ t \in [0,T]\}$ as $h \downarrow 0$.

In the next section an optimal least-squares filter is constructed for the computation of $E[\zeta_x^h(t\wedge\tau_T^h -)\,|\,F_{t\wedge\tau_T^h -}^{y,h}]$, where

$$y^h(t) = \int_0^t g(\zeta_x^h(s))ds + \int_0^t \Gamma dv(s), \quad t \geq 0 \tag{3.22}$$

and $F_t^{y,h}$ is the σ-field generated by $\{y^h(s),\ 0 \leq s \leq t\}$.

3.3 THE EQUATIONS OF THE OPTIMAL FILTER

Assume that $\sup_{t \in [0,T]} E|\zeta_x^h(t)|^2 < \infty$, $x \in \mathbb{R}_h^m$, and denote

$$F_t^{\zeta,h} \triangleq \sigma(\zeta_x^h(s),\ v(s);\ \ 0 \leq s \leq t\ ),\quad x \in \mathbb{R}_h^m \tag{3.23}$$

$$h_t \triangleq \Gamma^{-1}\, g(\zeta_x^h(t)) \quad , \quad t \geq 0 \quad , \quad x \in \mathbb{R}_h^m$$

$$z^h(t) \triangleq \int_0^t \Gamma^{-1} g(\zeta_x^h(s))ds + v(t) \quad (= \Gamma^{-1} y^h(t)),\ t \geq 0,\ x \in \mathbb{R}_h^m \tag{3.24}$$

$$\nu^h(t) \triangleq z^h(t) - \int_0^t E[h_s | F_s^{y,h}]ds \quad , \quad t \geq 0. \tag{3.25}$$

$$\tilde{P}_\alpha(t) \triangleq P(\zeta_x^h(t) = \alpha \mid F_t^{y,h}) \quad , t \geq 0 \quad , \quad x,\ \alpha \in \mathbb{R}_h^m \tag{3.26}$$

$$P_\alpha(t) \triangleq P(\zeta_x^h(t \wedge \tau_T^h -) = \alpha \mid F_{t\wedge\tau_T^h-}^{y,h}),\quad t \geq 0,\quad x,\alpha \in D_h\ . \tag{3.27}$$

It is further assumed that $\int_0^T E|h_t|^2 dt < \infty$, $0 < T < \infty$.

For each $t \geq 0$, the σ-fields $F_t^{\zeta,h}$ and $\sigma(v(s_2) - v(s_1);\ \ t < s_1 < s_2)$ are independent and h_t is $F_t^{\zeta,h}$-measurable. Thus, by following the same develop= ment given in Section 2.7 we obtain

$$d\tilde{P}_\alpha(t) = \sum_{\gamma \in \mathbb{R}_h^m} \lambda(\gamma,\alpha)\tilde{P}_\gamma(t)dt + \tilde{P}_\alpha(t) \sum_{\ell=1}^{p} \gamma_\ell^{-2}(g_\ell(\alpha) - \hat{\hat{g}}_\ell(t))(dy_\ell^h(t) - \hat{\hat{g}}_\ell(t)dt), \tag{3.28}$$

$\alpha \in \mathbb{R}_h^m$, $t \in (0,T)$

$$\hat{\hat{g}}_\ell(t) = \sum_{\alpha \in \mathbb{R}_h^m} g_\ell(\alpha)\tilde{P}_\alpha(t) = E[g_\ell(\zeta_x^h(t)) | F_t^{y,h}],\ \ell=1,\ldots,p,\ t \in [0,T], \tag{3.29}$$

and

$$E[\zeta_x^h(t) | F_t^{y,h}] = \sum_{\alpha \in \mathbb{R}_h^m} \alpha\, \tilde{P}_\alpha(t),\quad t \in [0,T], \tag{3.30}$$

where $\lambda(\alpha,\beta)$, $\alpha,\ \beta \in \mathbb{R}_h^m$ are given by equations (3.6)-(3.10).

Remark 3.3.1: Equations similar to equations (3.28)-(3.30), for the case p=1 are derived in [3.4] and [3.17,Vol.I] by using approaches other than those used here.

Let $\lambda(x,z)$, $x,z \in \mathbb{R}_h^m$ be defined by equations (3.6)-(3.10) together with the condition

$$\lambda(x,z) = 0,\quad x \in \mathbb{R}_h^m - D_h \quad , \quad z \in \mathbb{R}_h^m \ . \tag{3.31}$$

Given $x \in D_h$, we consider a continuous-time Markov chain $Z_x^h = \{Z_x^h(t), t \in [0,T]\}$ with state space $S = \mathbb{R}_h^m$, by defining the following set of transition probabilities

$$P(Z_x^h(t+\Delta) = z \pm e^i h | Z_x^h(t) = z) \triangleq \begin{cases} \lambda(z,z\pm e^i h)\Delta + O(\Delta^2), & z \in D_h \\ 0 & z \in \mathbb{R}_h^m - D_h \end{cases} \tag{3.32}$$

$i=1,\ldots,m,$

$$P(Z_x^h(t+\Delta) = z | Z_x^h(t) = z) \triangleq \begin{cases} 1 + \lambda(z,z)\Delta + O(\Delta^2), & z \in D_h \\ 1 & z \in \mathbb{R}_h^m - D_h \end{cases} \tag{3.33}$$

$$\sum_{y \in U_z} P(Z_x^h(t+\Delta) = y | Z_x^h(t)=z) \triangleq \begin{cases} O(\Delta^2) & , \quad z \in D_h \\ 0 & , \quad z \in \mathbb{R}_h^m - D_h \end{cases} \tag{3.34}$$

and

$$P(Z_x^h(0) = x) = 1 \quad , \quad x \in D_h \tag{3.35}$$

Taking equations (3.6)-(3.10) and (3.31)-(3.35) into account we can choose the sample paths of Z_x^h to satisfy (with probability 1)

$$Z_x^h(t) = \begin{cases} \zeta_x^h(t) & , \quad 0 \le t < \tau_T^h(x) \\ \zeta_x^h(\tau_T^h) & , \quad \tau_T^h(x) \le t \le T \end{cases} \qquad x \in D_h$$

Keeping this choice in mind we can write

$$Z_x^h(t) = \zeta_x^h(t \wedge \tau_T^h) \quad , \quad \text{w.p.1} \quad , \quad t \in [0,T].$$

Thus, for $x \in D_h$, the equations for $\hat{\zeta}_x^h(t) = E[\zeta_x^h(t \wedge \tau_T^h -) | F^{y,h}_{t \wedge \tau_T^h -}]$ are given by

$$dP_\alpha(t) = \sum_{\gamma \in D_h} \lambda(\gamma,\alpha)P_\gamma(t)dt$$

$$+ P_\alpha(t) \sum_{\ell=1}^{p} \gamma_\ell^{-2}(g_\ell(\alpha) - \hat{g}_\ell(t))(dy_\ell^h(t) - \hat{g}_\ell(t)dt) \tag{3.36}$$

$$\alpha \in D_h\ , \quad t \in (0,T)$$

$$\hat{g}_\ell(t) = \sum_{\alpha \in D_h} g_\ell(\alpha)P_\alpha(t) = E[g_\ell(\zeta_x^h(t\wedge\tau_T^h-))|F_{t\wedge\tau_T^h-}^{y,h}] \tag{3.37}$$

$$\ell=1,\ldots,p$$

$$\hat{\zeta}_x^h(t) = \sum_{\alpha \in D_h} \alpha\, P_\alpha(t), \quad t \in [0,T], \tag{3.38}$$

where in equation (3.19), $a_i = L_i h$, $i=1,\ldots,m$; $\{L_i\}$ are given positive integers, and $\{\lambda(\gamma,\alpha)\}$ are given by (3.6)-(3.10) and (3.31).

Given $Y^t = \{y(s),\ 0 \le s \le t\}$, $t \in [0,T]$. Then, in order to compute an approximation to $\hat{\zeta}_x(t)$ (equation (3.3)), equations (3.36)-(3.37) are solved, where in equations (3.36), the increment dy^h is replaced by dy. Let $\{P_\alpha^y(t),\ \alpha \in D_h,\ t \in [0,T]\}$, denote the solution to equations (3.36)-(3.37) (where dy replaces dy^h in (3.36)). Then, a process $\hat{\zeta}_x^{h,y} = \{\hat{\zeta}_x^{h,y}(t),\ t \in [0,T]\}$ is defined by

$$\hat{\zeta}_x^{h,y}(t) \triangleq \sum_{\alpha \in D_h} \alpha\, P_\alpha^y(t) \quad , \quad t \in [0,T]. \tag{3.39}$$

$\hat{\zeta}_x^{h,y}$ serves here as an approximation to $\hat{\zeta}_x$. In the next section a proce= dure for computing $\{\hat{\zeta}_x^{h,y}(k\Delta),\ k\Delta \in [0,T]\}$ is suggested. We assume that $\pi_\alpha = P_\alpha^y(0)$, $\alpha \in D_h$, are unknown.

3.4 AN ALGORITHM FOR COMPUTING $\hat{\zeta}_x^{h,y}$

In the sequel the following notations are used:

$$P_\alpha^y(k) \triangleq P_\alpha^y(k\Delta),\ \alpha \in D_h,\ k=0,1,\ldots,N; \quad \zeta_x(k) \triangleq \zeta_x(k\Delta),\ k=0,1,\ldots,N,$$

$$y(k) \triangleq y(k\Delta),\ k=0,1,\ldots,N; \quad \hat{\zeta}_x^{h,y}(k) \triangleq \hat{\zeta}_x^{h,y}(k\Delta),\ k=0,1,\ldots,N. \tag{3.40}$$

Let $\varepsilon > 0$, $\Delta > 0$, $N\Delta = T$; $P^y_\alpha(0) = [\prod_{i=1}^{m} (2L_i+1)]^{-1}$, $\alpha \in D_h$; and

$\hat{g}_\ell(0) = [\prod_{i=1}^{m} (2L_i+1)]^{-1} \sum_{\alpha \in D_h} g_\ell(\alpha)$, $\ell=1,\dots,p$.

Algorithm 3.4

1. k=0

2. For $\alpha \in D_h$ (ζ^h_x has $\prod_{i=1}^{m} (2L_i+1)$ states in D_h) calculate

$$P^y_\alpha(k+1) := P^y_\alpha(k) + \sum_{\gamma \in D_h} \lambda(\gamma,\alpha) P^y_\gamma(k)\Delta + P^y_\alpha(k) \sum_{\ell=1}^{p} \gamma_\ell^{-2}(g_\ell(\alpha) - \hat{g}_\ell(k))(y_\ell(k+1) - y_\ell(k) - \hat{g}_\ell(k)\Delta) \tag{3.41}$$

$$P^y_\alpha(k+1) := \max(0, P^y_\alpha(k+1)) \tag{3.42}$$

3. $Z(k+1) := \sum_{\alpha \in D_h} P^y_\alpha(k+1)$ (3.43)

4. If $Z(k+1) \geq \varepsilon$ then: for $\alpha \in D_h$, $P^y_\alpha(k+1) := P^y_\alpha(k+1)/Z(k+1)$ (3.44)

   Otherwise : stop.

5. For $\ell=1,\dots,p$ calculate

$$\hat{g}_\ell(k+1) := \sum_{\alpha \in D_h} g_\ell(\alpha) P^y_\alpha(k+1) \tag{3.45}$$

$$\hat{\zeta}^{h,y}_x(k+1) := \sum_{\alpha \in D_h} \alpha\, P^y_\alpha(k+1) \text{ , } (\hat{\zeta}^{h,y}_x(0) = \prod_{i=1}^{m} (2L_i+1)^{-1} \sum_{\alpha \in D_h} \alpha), \tag{3.46}$$

6. If k=N or if $\hat{\zeta}^{h,y}_x(k+1) \notin D$ then stop. Otherwise : k:=k+1 and go to 2.

Note that the set D_h can be written as

$$D_h \triangleq \{\alpha = (i_1 h,\dots,i_m h): i_k = 0,\pm 1,\dots,\pm L_k;\ \ k=1,\dots,m\}. \tag{3.47}$$

The problem of establishing conditions for the weak convergence of $\{\hat{\zeta}^{h,y}_x(t),\ t \in [0,T]\}$ to $\{\hat{\zeta}_x(t),\ t \in [0,T]\}$ (equation 3.3) as $h \downarrow 0$ is out of the scope of this work. Instead, the role of $\hat{\zeta}^{h,y}_x$ as an approxima=

tion to $\hat{\zeta}_x$ is demonstrated in the sequel by numerical experimentation with several examples.

The numerical study of Algorithm 3.4 will be carried out via numerical experimentation. Each experiment is called a *run*. Here a run always consists of two stages:

(i) Simulate equations (3.1)-(3.2) by applying the following procedure: (using the notations $\zeta_x(k) = \zeta_x(k\Delta)$, k=0,1,...)

$\zeta_x(0) = x$ and $y(0) = 0$. For k=0,1,...,N

1. For i=1,...,m calculate

$$\bar{X}_i(k+1) = \zeta_{xi}(k) + f_i(\zeta_x(k))\Delta + \sqrt{\Delta} \sum_{j=1}^{m} B_{ij} W_j(k) \tag{3.48}$$

2. For i=1,...,m calculate

$$\zeta_{xi}(k+1) = \zeta_{xi}(k) + [f_i(\zeta_x(k)) + f_i(\bar{X}(k+1))]\Delta/2 + \sqrt{\Delta} \sum_{j=1}^{m} B_{ij} W_j(k) \tag{3.49}$$

3. $$y(k+1) = y(k) + g(\zeta_x(k))\Delta + \sqrt{\Delta}\,\Gamma\, v(k) \tag{3.50}$$

where $\{W(k)\}_{k=0}^N$ is a sequence of independent $\mathbb{R}^m$-valued Gaussian elements with

$$EW(k) = 0 \text{ and } E[W(k)W'(\ell)] = \delta_{k\ell}\, I_m, \quad k,\ell=0,1,\dots,N, \tag{3.51}$$

and $\{v(k)\}_{k=0}^N$ is a sequence of independent $\mathbb{R}^p$-valued Gaussian elements with

$$Ev(k) = 0 \text{ and } E[v(k)v'(\ell)] = \delta_{k\ell}\, I_p\,, \quad k,\ell=0,1,\dots,N. \tag{3.52}$$

It is assumed that $\{W(k)\}_{k=0}^N$ and $\{v(k)\}_{k=0}^N$ are mutually independent.

4. Store the sequence $\{y(k)\}_{k=0}^N$.

(ii) Apply Algorithm 3.4, where $\{y(k+1) - y(k)\}_{k=0}^{N-1}$ serves as an input to the filter and $\{\hat{\zeta}_x^{h,y}(k)\}_{k=1}^{N}$ serves as the filter's output.

3.5 EXAMPLES : THE CASE m=1

In this section cases where $\zeta_x = \{\zeta_x(t),\ t \geq 0\}$ (equations (3.1)) is an $\mathbb{R}$-valued Markov process are considered. In these cases

$$D_h \triangleq \{ih : i=0,\pm 1,\ldots,\pm L\} \tag{3.53}$$

and Algorithm 3.4 reduces to:

1. $k=0$, $P_i^y(0) = 1/(2L+1)$, $i=0,\pm 1,\ldots,\pm L$; $\hat{g}_\ell(0) = \sum_{j=-L}^{L} g_\ell(jh)/(2L+1)$ $\ell=1,\ldots,p$.

2. For $i=-L, -L+1,\ldots,L$ calculate

$$P_i^y(k+1) := P_i^y(k) + [\lambda(i-1,i)P_{i-1}^y(k) + \lambda(i,i)P_i^y(k) + \lambda(i+1,i)P_{i+1}^y(k)]\Delta$$

$$+ P_i^y(k) \sum_{\ell=1}^{p} \gamma_\ell^{-2}(g_\ell(ih) - \hat{g}_\ell(k))(y_\ell(k+1) - y_\ell(k) - \hat{g}_\ell(k)\Delta)$$

$$P_i^y(k+1) := \max(0, P_i^y(k+1))$$

3. $Z(k+1) := \sum_{i=-L}^{L} P_i^y(k+1)$

4. If $Z(k+1) \geq \varepsilon$ then: for $i=-L, -L+1,\ldots,L$, $P_i^y(k+1) := P_i^y(k+1)/Z(k+1)$ Otherwise : stop.

5. For $\ell=1,\ldots,p$

$$\hat{g}_\ell(k+1) := \sum_{j=-L}^{L} g_\ell(jh)P_j^y(k)$$

$$\hat{\zeta}_x^{h,y}(k+1) := \sum_{j=-L}^{L} jhP_j^y(k+1)$$

6. If $k=N$ or $|\hat{\zeta}_x^{h,y}(k+1)| > Lh$ then stop. Otherwise $k:=k+1$ and go to 2,

where

$$\lambda(i-1,i) = [\sigma^2/2 + hf^+(ih-h)]/h^2 \quad , -L+1 \le i \le L$$

$$\lambda(i,i) \quad = -[\sigma^2 \; + h|f(ih)|]/h^2 \quad , -L \le i \le L$$

$$\lambda(i+1,i) = [\sigma^2/2 + h\, f^-(ih+h)]/h^2, \; -L \le i \le L-1$$

$$\lambda(-L-1,L) = \lambda(L+1,L) = 0.$$

Two sets of systems were considered. The first set consisted of the following systems:

(a) $dx = 0 \quad ; \quad dy = xdt + \gamma_o dv \quad , \quad x,y \in \mathbb{R},\ t > 0$ (3.54)

(b) $dx = 0 \quad ; \quad dy = x^3 dt + \gamma_o dv \quad , \quad x,y \in \mathbb{R},\ t > 0$ (3.55)

(c) $dx = -0.2(x-o.5)dt + \sigma dW;\ dy = xdt + \gamma dv,\ \ x,y \in \mathbb{R},\ t > 0$ (3.56)

(d) $dx = -0.2(x-0.5)^3 dt + \sigma dW;\ dy = \arctan(x/2)dt + \gamma dv,\ x,y \in \mathbb{R}, t > 0$ (3.57)

(e) $dx = -0.2(x-0.5)dt + \sigma dW;\ dy = \arctan(x/2)dt + \gamma dv,\ x,y \in \mathbb{R},\ t > 0$ (3.58)

(f) $dx = -0.2(x-0.5)^3 dt + \sigma dW;\ dy = xdt + \gamma dv,\ x,y \in \mathbb{R},\ t > 0$ (3.59)

(g) $dx = 0.8\sin(0.5x)dt + \sigma dW;\ dy = xdt + \gamma dv,\ \ x,y \in \mathbb{R},\ t > 0$ (3.60)

(h) $dx = 0.5x^2 dt + \sigma dW;\ dy = xdt + \gamma dv,\ \ x,y \in \mathbb{R},\ \ t > 0$ (3.61)

where $\gamma_o = 0.01$, $\sigma = 0.01$, $\gamma = 0.02$, $h = 0.01$, $L = 200$ and $\Delta = 10^{-3}$.
Note that here

$$D = \{x \in \mathbb{R} : |x| < 2 + \delta\} \qquad \delta < 0.01 \tag{3.62}$$

Some of the results from the corresponding runs are given in Tables 3.1-3.2 (cases (a) and (b) respectively), and in Figs. 3.1-3.6 (cases (c)-(g)). All the graphs in this section were plotted using the set of points $\{t'_k = 100k\Delta:\ k=0,1,\ldots,200\}$

TABLE 3.1: $\zeta_x(k)$ and $\hat{\zeta}_x^{h,y}(k)$ as functions of k; for dx=0, $\zeta_x(0) = 0.5$; $dy = x dt + \gamma_o dv$

k	$\zeta_x(k)$	$\hat{\zeta}_x^{h,y}(k)$
99	.5000	.4204
199	.5000	.4778
299	.5000	.5029
399	.5000	.5086
499	.5000	.5037
599	.5000	.5065
699	.5000	.4992
799	.5000	.4968
899	.5000	.4959
999	.5000	.4947
1099	.5000	.5010
1199	.5000	.5019
1299	.5000	.5047
1399	.5000	.5064
1499	.5000	.5078
1599	.5000	.5083
1699	.5000	.5094
1799	.5000	.5069
1899	.5000	.5090
1999	.5000	.5071
2099	.5000	.5072
2199	.5000	.5082
2299	.5000	.5069
2399	.5000	.5076
2499	.5000	.5066
2599	.5000	.5051
2699	.5000	.5071
2799	.5000	.5050
2899	.5000	.5057
2999	.5000	.5060
3099	.5000	.5073
3199	.5000	.5060
3299	.5000	.5059
3399	.5000	.5060
3499	.5000	.5058
3599	.5000	.5057
3699	.5000	.5055
3799	.5000	.5043
3899	.5000	.5039
3999	.5000	.5026
4099	.5000	.5040
4199	.5000	.5038
4299	.5000	.5020
4399	.5000	.5015
4499	.5000	.5023
4599	.5000	.5024
4699	.5000	.5024

TABLE 3.2: $\zeta_x(k)$ and $\hat{\zeta}_x^{h,y}(k)$ as functions of k; for $dx = 0, \zeta_x(0) = 0.5$; $dy = x^3 dt + \gamma_o dv$

$\zeta_x(k)$	$\hat{\zeta}_x^{h,y}(k)$
.5000	.1986
.5000	.4657
.5000	.5039
.5000	.5114
.5000	.5050
.5000	.5087
.5000	.4990
.5000	.4956
.5000	.4944
.5000	.4927
.5000	.5014
.5000	.5026
.5000	.5062
.5000	.5084
.5000	.5103
.5000	.5110
.5000	.5123
.5000	.5091
.5000	.5118
.5000	.5093
.5000	.5095
.5000	.5107
.5000	.5091
.5000	.5100
.5000	.5087
.5000	.5068
.5000	.5092
.5000	.5066
.5000	.5074
.5000	.5079
.5000	.5095
.5000	.5078
.5000	.5077
.5000	.5079
.5000	.5076
.5000	.5074
.5000	.5071
.5000	.5058
.5000	.5054
.5000	.5038
.5000	.5054
.5000	.5053
.5000	.5030
.5000	.5024
.5000	.5035
.5000	.5037
.5000	.5037

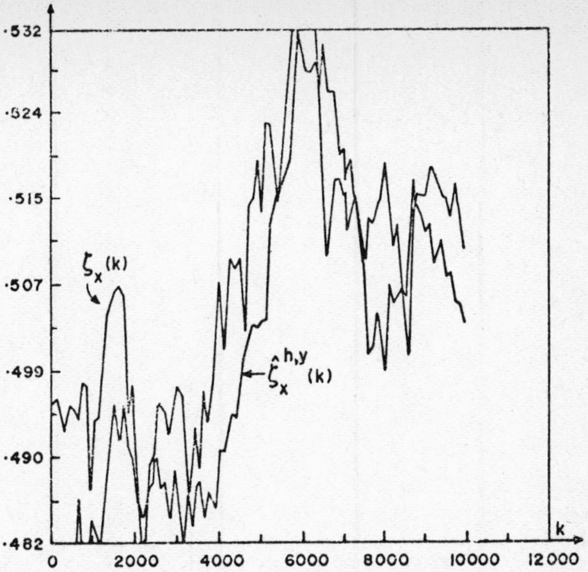

Fig.3.1: $\zeta_x(k)$ and $\hat{\zeta}_x^{h,y}(k)$ as functions of k for the system given by equations (3.56).

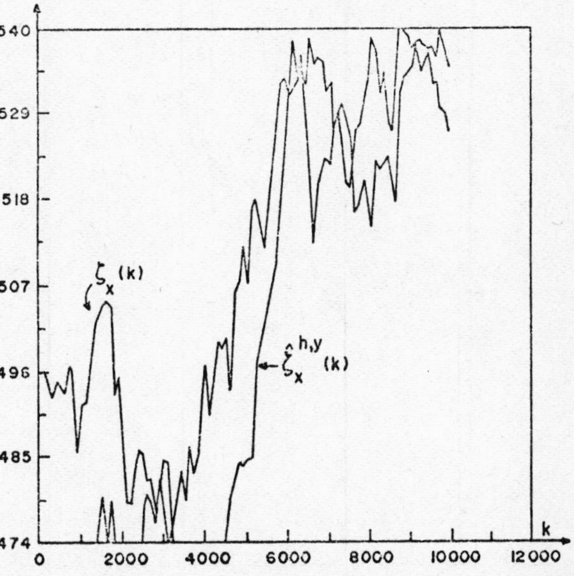

Fig.3.2: $\zeta_x(k)$ and $\hat{\zeta}_x^{h,y}(k)$ as functions of k for the system given by equations (3.57).

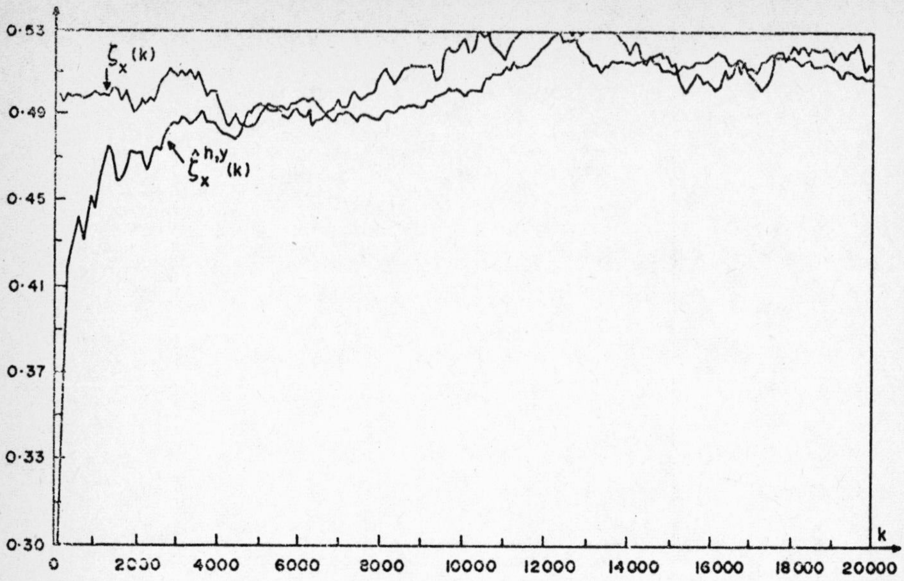

Fig.3.3: $\zeta_x(k)$ and $\hat{\zeta}_x^{h,y}(k)$ as functions of k for the system given by equations (3.58).

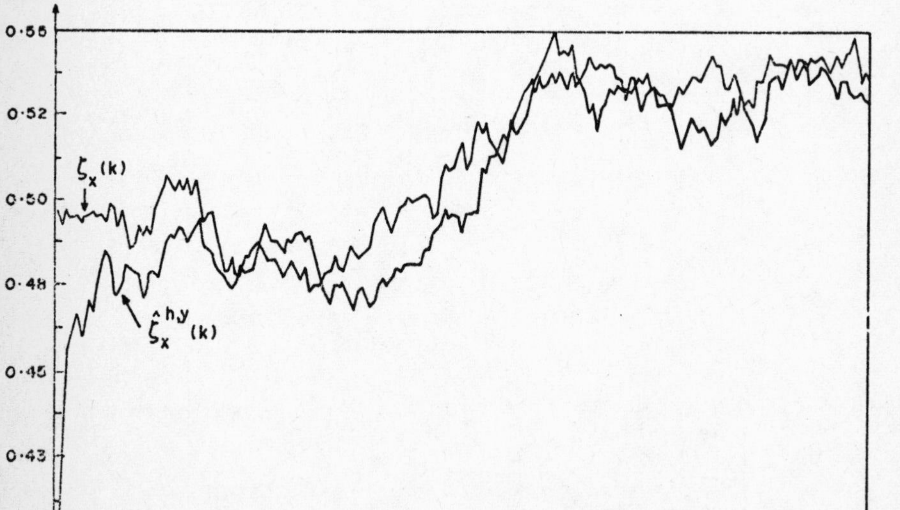

Fig.3.4: $\zeta_x(k)$ and $\hat{\zeta}_x^{h,y}(k)$ as functions of k for the system given by equations (3.59).

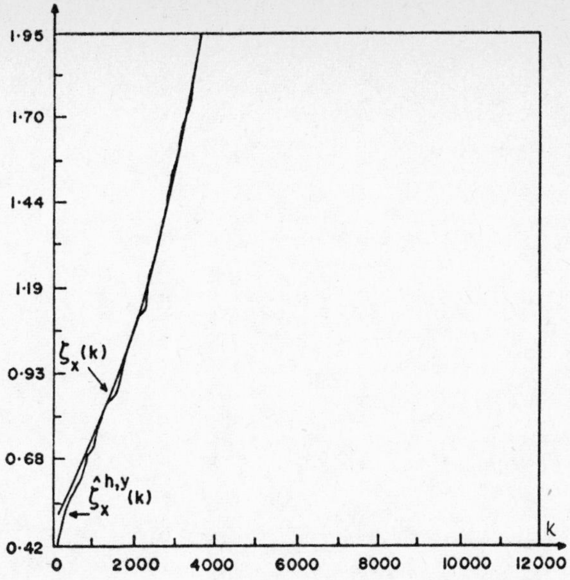

Fig.3.5: $\zeta_x(k)$ and $\hat{\zeta}_x^{h,y}(k)$ as functions of k for the system given by equations (3.60).

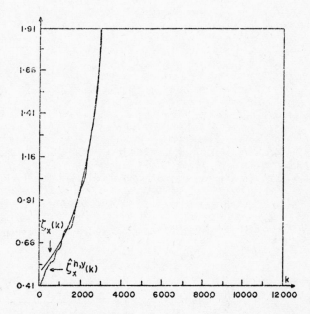

Fig.3.6: $\zeta_x(k)$ and $\hat{\zeta}_x^{h,y}(k)$ as functions of k for the system given by equations (3.61).

Note that in cases (a) and (b) (Tables 3.1-3.2) $\hat{\zeta}_x^{h,y}$ is an approximation to $E[\zeta_x(0)|F^y_{t\wedge\tau_T^-}]$.

The second set of systems consisted of the following cases:

(i) $dx = -0.2(x-0.5)dt + \sigma dW \quad , \quad x \in \mathbb{R} \quad , \quad t > 0$ (3.63)

(j) $dx = -0.2(x-0.5)^3 dt + \sigma dW \quad , \quad x \in \mathbb{R} \quad , \quad t > 0$ (3.64)

(k) $dx = 0.5x^2 dt + \sigma dW \quad , \quad x \in \mathbb{R} \quad , \quad t > 0$ (3.65)

(l) $dx = 0.5\mathrm{sign}(x)dt + \sigma dW \quad , \quad x \in \mathbb{R} \quad , \quad t > 0$ (3.66)

(m) $dx = -0.5x dt/(1 + x^2) + \sigma dW, \quad x \in \mathbb{R} \quad , \quad t > 0$ (3.67)

(n) $dx = 0.2(x^2 - 0.25)dt + \sigma dW, \quad x \in \mathbb{R} \quad , \quad t > 0$ (3.68)

(o) $dx = -0.2(x^2 - 0.25)dt + \sigma dW, \quad x \in \mathbb{R} \quad , \quad t > 0$ (3.69)

and for each of the cases (i)-(o)

$$\begin{cases} dy_1 = \sqrt{h_o^2 + x^2}\, dt + \gamma_1 dv_1 \\ \\ dy_2 = \arctan(x/h_o)dt + \gamma_2 dv_2 \end{cases} \qquad x, y_1, y_2 \in \mathbb{R}, \quad t > 0 \tag{3.70}$$

where $\sigma = 0.01$, $\gamma_1 = 0.02$, $\gamma_2 = 0.01$, $h_o = 2$; $h = 0.01$, $L = 200$ and $\Delta = 10^{-3}$. (Here the function sign(x) is an expression used for tanh(ax), for some $a \gg 1$). Also, for this set of systems the domain D is given by (3.62). Some of the results of the corresponding runs are given in Figs.3.7-3.13.

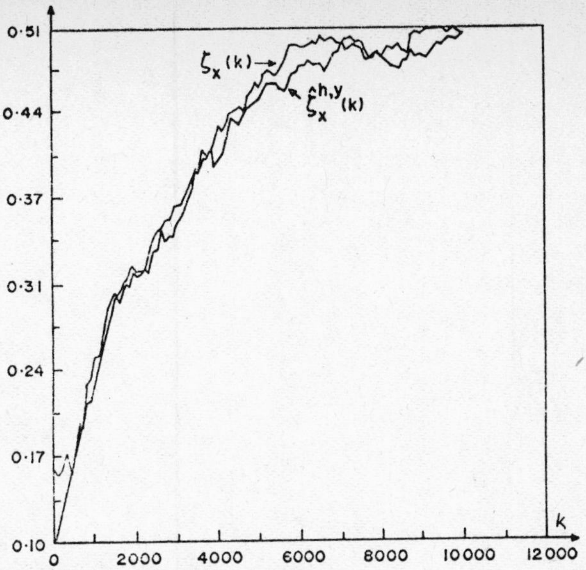

Fig.3.7: $\zeta_x(k)$ and $\hat{\zeta}_x^{h,y}(k)$ as functions of k for the system given by equations (3.63) and (3.70).

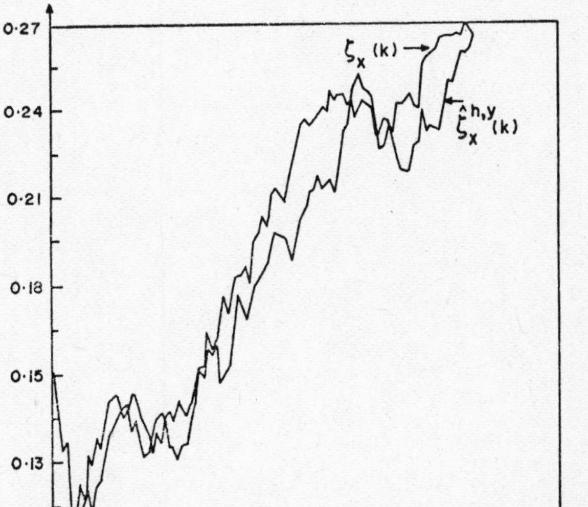

Fig.3.8: $\zeta_x(k)$ and $\hat{\zeta}_x^{h,y}(k)$ as functions of k for the system given by equations (3.64) and (3.70).

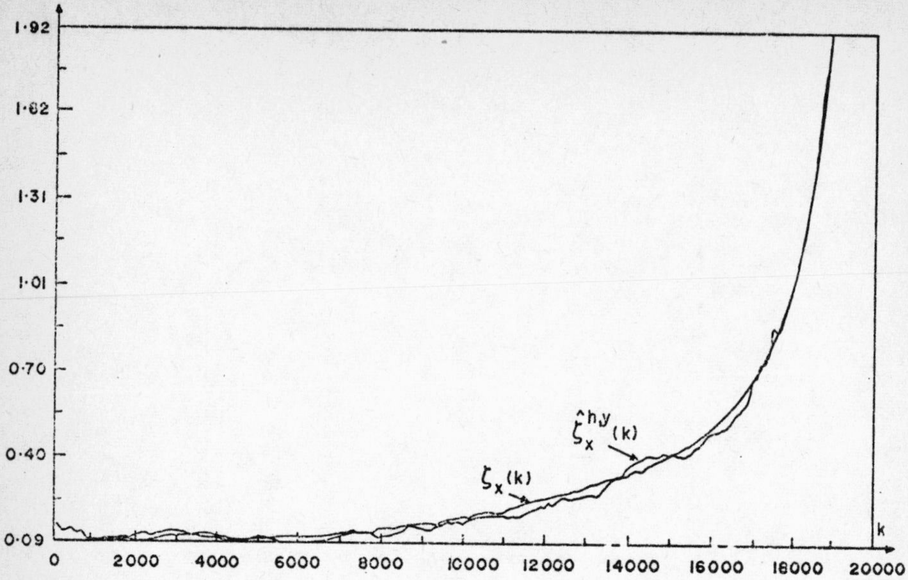

Fig.3.9: $\zeta_x(k)$ and $\hat{\zeta}_x^{h,y}(k)$ as functions of k for the system given by equations (3.65) and (3.70).

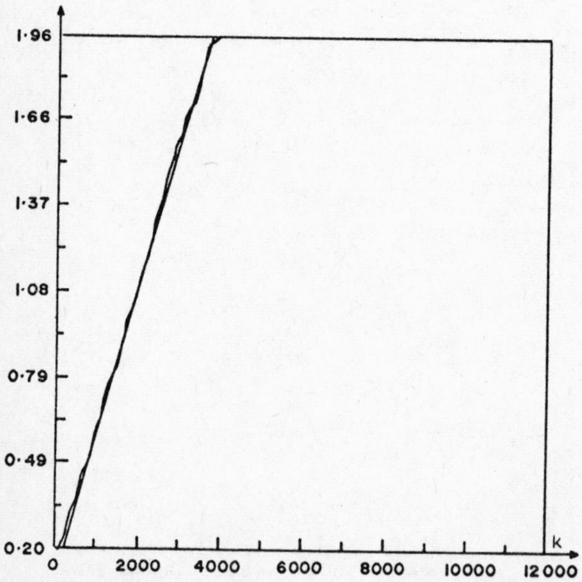

Fig.3.10: $\zeta_x(k)$ and $\hat{\zeta}_x^{h,y}(k)$ as functions of k for the system given by equations (3.66) and (3.70).

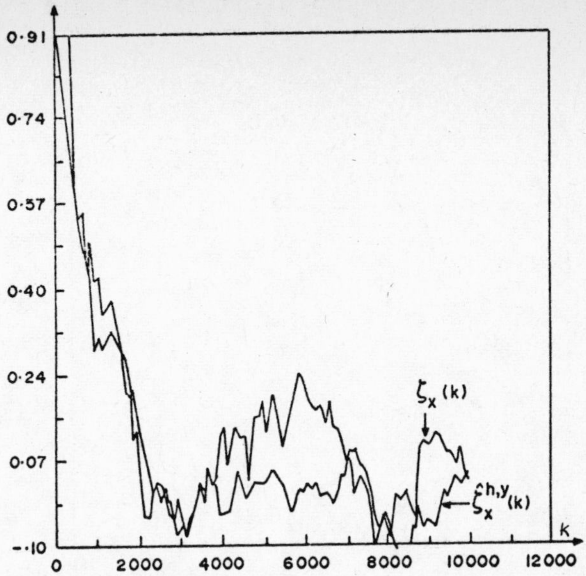

Fig.3.11: $\zeta_x(k)$ and $\hat{\zeta}_x^{h,y}(k)$ as functions of k for the system given by equations (3.67) and (3.70).

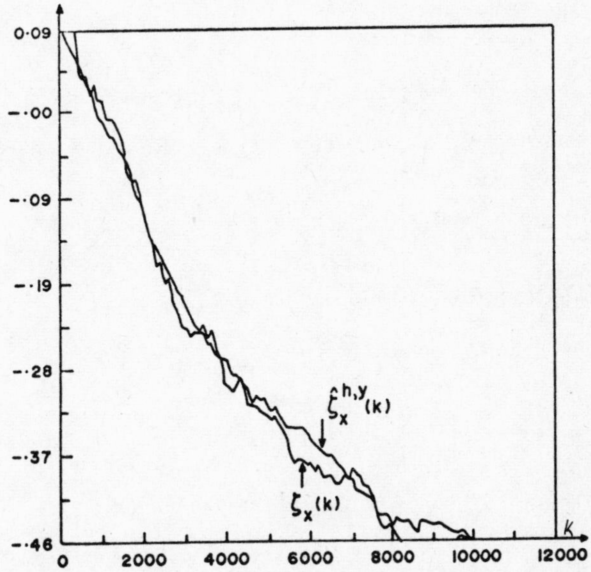

Fig.3.12: $\zeta_x(k)$ and $\hat{\zeta}_x^{h,y}(k)$ as functions of k for the system given by equations (3.68) and (3.70).

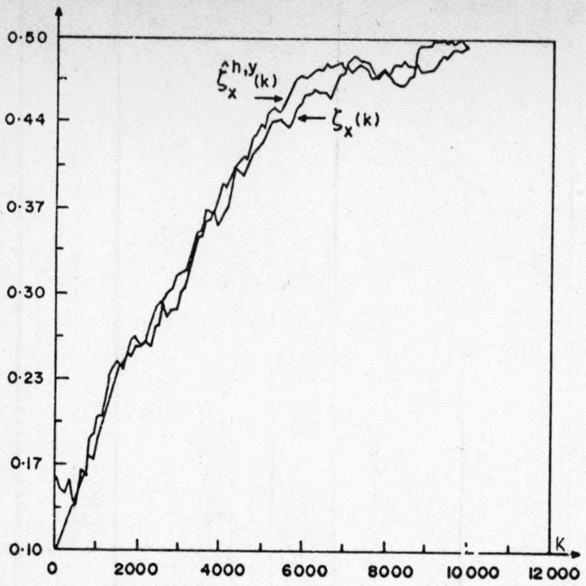

Fig.3.13: $\zeta_x(k)$ and $\hat{\zeta}_x^{h,y}(k)$ as functions of k for the system given by equations (3.69)-(3.70).

3.6 EXAMPLES : THE CASE m=2

In this section cases where $\zeta_x = \{\zeta_x(t),\ t \geq 0\}$ (equations (3.1)) is an $\mathbb{R}^2$-valued Markov process, are considered. In these cases

$$D_h = \{(ih,jh):\ \ i=0,\pm1,\ldots,\pm L_1\ ,\ \ j=0,\pm1,\ldots,\pm L_2\} \tag{3.71}$$

and equations (3.41) of Algorithm 3.4 reduce to:

For $i=-L_1,\ldots,L_1$ and $j=-L_2,\ldots,L_2$ calculate

$$\begin{aligned} P^y_{i,j}(k+1) := P^y_{i,j}(k) &+ [\lambda(i-1,j;\ i,j)P^y_{i-1,j}(k) \\ &+ \lambda(i+1,j;i,j)P^y_{i+1,j}(k) + \lambda(i,j;i,j)P^y_{i,j}(k) \\ &+ \lambda(i,j+1;i,j)P^y_{i,j+1}(k) + \lambda(i,j-1;i,j)P^y_{i,j-1}(k)]\Delta \\ &+ P^y_{i,j}(k) \sum_{\ell=1}^{p} \gamma_\ell^{-2}(g_\ell(ih,jh) - \hat{g}_\ell(k))(y_\ell(k+1) - y_\ell(k) - \hat{g}_\ell(k)\Delta) \end{aligned} \tag{3.72}$$

where

$$\hat{g}_\ell(k) = \sum_{i=-L_1}^{L_1} \sum_{j=-L_2}^{L_2} g_\ell(ih,jh)P^y_{i,j}(k) \tag{3.73}$$

and

$$\lambda(i-1,j;i,j) = [\sigma_1^2/2 + hf_1^+(ih-h,jh)]/h^2, -L_1+1 \le i \le L_1 \quad ; -L_2 \le j \le L_2 \tag{3.74}$$

$$\lambda(i+1,j;i,j) = [\sigma_1^2/2 + hf_1^-(ih+h,jh)]/h^2, \ -L_1 \le i \le L_1-1; -L_2 \le j \le L_2 \tag{3.75}$$

$$\lambda(i,j;i,j) = -[\sigma_1^2 + \sigma_2^2 + h(|f_1(ih,jh)| + |f_2(ih,jh)|)]/h^2, \ -L_1 \le i \le L_1, \ -L_2 \le j \le L_2 \tag{3.76}$$

$$\lambda(i,j+1;i,j) = [\sigma_2^2/2 + hf_2^- (ih,jh+h)]/h^2, \ -L_1 \le i \le L_1; \ -L_2 \le j \le L_2-1 \tag{3.77}$$

$$\lambda(i,j-1;i,j) = [\sigma_2^2/2 + hf_2^+(ih,jh-h)]/h^2, \ -L_1 \le i \le L_1; \ -L_2+1 \le j \le L_2 \tag{3.78}$$

$$\lambda(-L_1-1,j;-L_1,j) = \lambda(L_1+1,j;L_1,j) = 0 \ \forall j \tag{3.79}$$

$$\lambda(i,L_2+1;i,L_2) = \lambda(i,-L_2-1;i,-L_2) = 0 \ \forall i. \tag{3.80}$$

The following set of systems was considered:

(a)

$$\begin{cases} dx_1 = 0.5x_1^3dt + 0.01\,dW_1 \\ \qquad\qquad\qquad\qquad t > 0 \\ dx_2 = 0 \quad , \ \zeta_{x2}(0) = 0.2 \end{cases} \tag{3.81}$$

$$dy_i = x_i^3dt + 0.001dv_i \quad , \quad i=1,2 \quad , \quad t > 0 \tag{3.82}$$

(b)

$$\begin{cases} dx_1 = 0.1x_1dt + 0.01dW_1 \\ \qquad\qquad\qquad\qquad t > 0 \\ dx_2 = 0.2x_2dt + 0.01dW_2 \end{cases} \tag{3.83}$$

$$dy_i = x_i^3dt + 0.001dv_i \quad , \quad i=1,2 \quad , \quad t > 0 \tag{3.84}$$

(c)
$$\begin{cases} dx_1 = [-50x_2 + 50x_1(0.36 - x_1^2 - x_2^2)]dt + 0.001dW_1 \\ \qquad\qquad\qquad\qquad\qquad\qquad\qquad\qquad t > 0 \\ dx_2 = [50x_1 + 50x_2(0.36 - x_1^2 - x_2^2)]dt + 0.001dW_2 \end{cases} \qquad (3.85)$$

$$dy_i = x_i dt + 0.0001dv_i \quad , \quad i=1,2 \quad , \quad t > 0 \qquad (3.86)$$

(See Section 2.4 for more details on the system given by eq (3.85)).

(d) The system given by equations (3.85) but, where

$$dy_i = x_i^3 dt + 0.0001dv_i \quad , \quad i=1,2 \quad , \quad t > 0 \qquad (3.87)$$

(e)
$$\begin{cases} dx_1 = [70\text{sign}(x_2) + 50x_1(0.6 - |x_1| - |x_2|)]dt + 0.001dW_1 \\ \qquad\qquad\qquad\qquad\qquad\qquad\qquad\qquad t > 0 \\ dx_2 = [-70\text{sign}(x_1) + 50x_2(0.6 - |x_1| - |x_2|)]dt + 0.001dW_2 \end{cases} \qquad (3.88)$$

$$dy_i = x_i dt + 0.005dv_i \quad , \quad i=1,2, \quad t > 0 \qquad (3.89)$$

(see Section 2.5 for more details on the system given by eq. (3.88)).

(f) The system given by equations (3.88) but where

$$dy_i = \text{sign}(x_i)dt + 0.005dv_i \quad , \quad i=1,2 \quad , \quad t > 0. \qquad (3.90)$$

(g)
$$\begin{cases} dx_1 = ([\text{sign}(x_2-x_1) + \text{sign}(x_2+x_1)] + [0.3 - (|x_1+x_2|+|x_2-x_1|)]x_1)dt \\ \qquad + \sigma_1 dW_1 \qquad\qquad\qquad\qquad\qquad\qquad\qquad (3.91) \\ dx_2 = ([\text{sign}(x_2-x_1) - \text{sign}(x_2+x_1)]+[0.3 -(|x_1+x_2|+|x_2-x_1|)]x_2)dt \\ \qquad + \sigma_2 dW_2 \qquad\qquad \sigma_1 = \sigma_2 = 0.001 \end{cases}$$

$$dy_i = \text{sign}(x_i)dt + 0.005dv_i \quad , \quad i=1,2 \quad , \quad t > 0 \qquad (3.92)$$

Equations (3.91), where $\sigma_1 = \sigma_2 = 0$, are treated in Kaplan [3.18]. It is shown there that equations (3.91) (where $\sigma_1 = \sigma_2 = 0$) represent a gene=

rator of trapezoidal waves. For more details see [3.18]. (It is tacitly assumed here that $|x_i|$ and $\mathrm{sign}(x_i)$ are expressions used for $x_i \tanh(ax_i)$ and $\tanh(ax_i)$ respectively, for some $a >> 1$).

(h)
$$\begin{cases} dx_1 = x_2 x_1 dt + 0.01 dW_1 \\ \qquad\qquad\qquad\qquad , \quad t > 0 \\ dx_2 = 0 \quad , \zeta_{x2}(0) = 0.5 \end{cases} \tag{3.93}$$

$$dy_i = x_i^3 dt + 0.01 dv_i \quad , \quad i=1,2 \quad , \quad t > 0 \tag{3.94}$$

Note that in cases (a) and (h) $\hat{\zeta}^{h,y}_{x2}$ serves as an approximation to $E[\zeta_{x2}(0)|F^y_{t\wedge\tau_T-}]$.

(i)
$$dx_1 = x_1 x_2 dt + 0.01 dW_1 \tag{3.95a}$$
$$, \quad t > 0$$
$$dx_2 = 0 \quad , \quad \zeta_{x2}(0) = 0.2 \tag{3.95b}$$

$$dy_i = x_i dt + 0.001 dv_i \quad , \quad i=1,2, \quad t > 0 \tag{3.96}$$

(j)
$$dx_1 = x_2 \sin(x_1/2) dt + 0.01 dW_1 \tag{3.97a}$$
$$t > 0$$
$$dx_2 = 0 \quad , \quad \zeta_{x2}(0) = 0.2 \tag{3.97b}$$

$$dy_i = x_i dt + 0.001 dv_i \quad , \quad i=1,2, \quad t > 0 \tag{3.98}$$

Note that equations (3.95a) and (3.97a) can be written as

$$dx_1 = ax_1 dt + 0.01 dW_1 \quad , \quad t > 0 \tag{3.95a'}$$

$$dx_1 = a\sin(x_1/2) dt + 0.01 dW_1 \ , \quad t > 0 \tag{3.97a'}$$

respectively, where in both cases a=0.2.

Hence, in cases (i) and (j) $\hat{\zeta}^{h,y}_{x2}$ serves as an approximation to $E[a|F^y_{t\wedge\tau_T-}]$.

Numerical experimentation was carried out for the following set of para= meters: $L_1 = L_2 = 12$, $h = 0.1$ and $\Delta = 10^{-3}$. Thus, the set D in this section is given by

$$D = \{x \in \mathbb{R}^2 : |x_i| < 1.2 + \delta, \quad i=1,2\} \ , \ \delta < 0.1 \tag{3.99}$$

and the number of the states of ζ_x^h in D_h is 625.

Some of the results from the numerical experimentation are given in Tables 3.3-3.6 and Figs. 3.14-3.19. The graphs in these figures were plotted using the set of points $\{t_k' = 100k\Delta : k=0,1,\dots,100\}$.

The results obtained for equations (3.88)-(3.89) and for equations (3.88) and (3.90) indicate that $\hat{\zeta}_x^{h,y}$ is, in these cases, a very good estimator of ζ_x. Actually, when $\zeta_{xi}(k)$ and $\hat{\zeta}_{xi}^{h,y}(k)$ are plotted as functions of k, for i=1,2, one cannot distinguish between ζ_x and $\hat{\zeta}_x^{h,y}$, and the plots are omitted therefore.

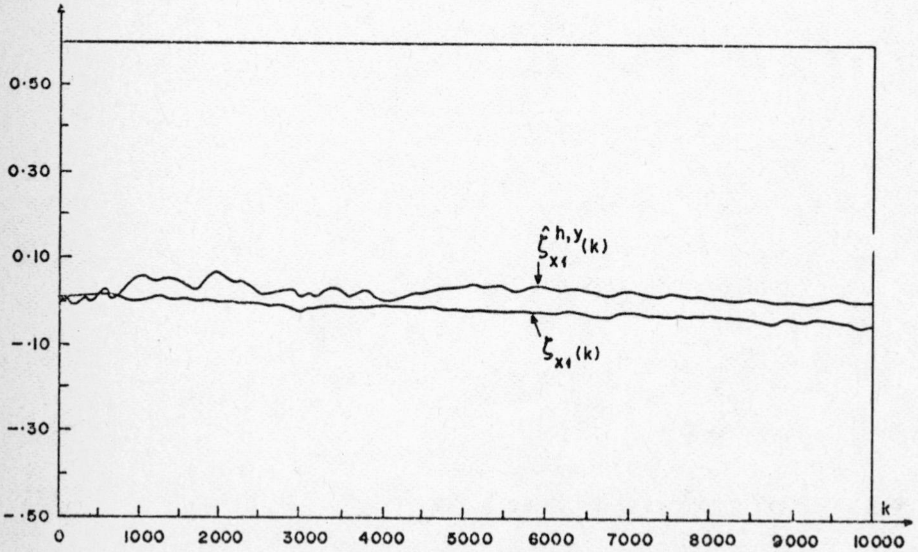

Fig.3.14: $\zeta_{x1}(k)$ and $\hat{\zeta}_{x1}^{h,y}(k)$ as functions of k for the system given by equations (3.81)-(3.82).

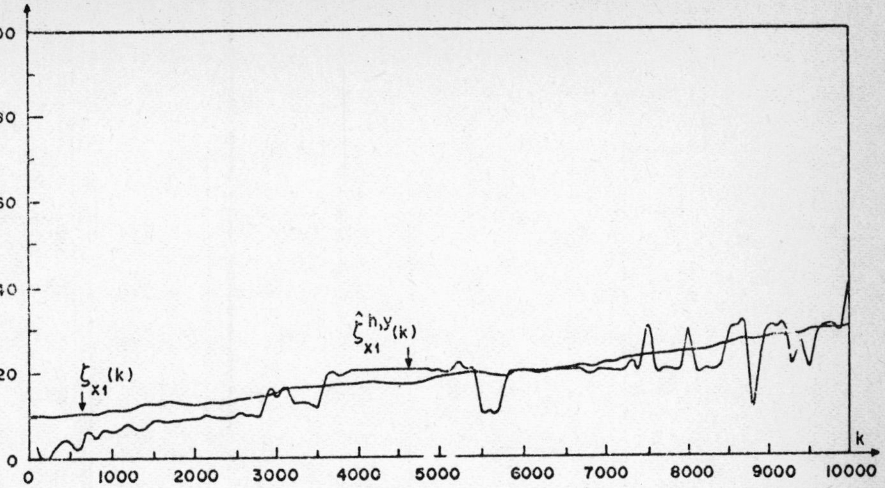

Fig.3.15a: $\zeta_{x1}(k)$ and $\hat{\zeta}^{h,y}_{x1}(k)$ as functions of k for the system given by equations (3.83)-(3.84).

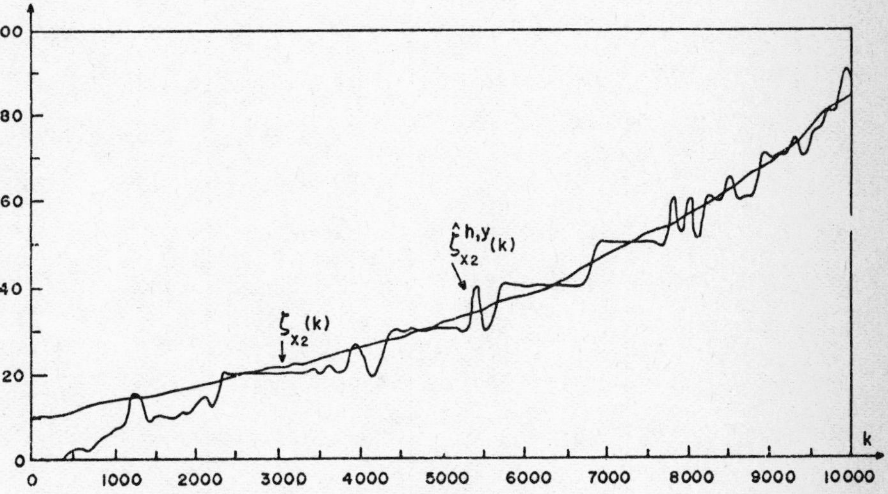

Fig.3.15b: $\zeta_{x2}(k)$ and $\hat{\zeta}^{h,y}_{x2}(k)$ as functions of k for the system given by equations (3.83)-(3.84).

TABLE 3.3: $\zeta_x(k)$ and $\hat{\zeta}_x^{h,y}(k)$ as functions of k, for the system given by equations (3.85)-(3.86)

k	$\zeta_{x1}(k)$	$\hat{\zeta}_{x1}^{h,y}(k)$	$\zeta_{x2}(k)$	$\hat{\zeta}_{x2}^{h,y}(k)$
100	.5206	.5802	.1547	.1759
500	.0735	.0640	.5949	.6011
1000	.1458	.1385	.5816	.5608
1500	.2146	.2246	.5599	.5753
2000	.2814	.3107	.5296	.5665
2500	.3432	.2989	.4917	.4062
3000	.4016	.4052	.4453	.4457
3500	.4527	.4887	.3931	.3894
4000	.4973	.5000	.3349	.3538
4500	.5346	.5282	.2719	.2964
5000	.5636	.5002	.2045	.1041
5500	.5846	.5074	.1334	.1124
6000	.5965	.5997	.0612	.0080
6500	.5997	.6393	-.0126	.0583
7000	.5934	.6001	-.0856	-.1018
7500	.5786	.5998	-.1579	-.1784
8000	.5553	.5801	-.2265	-.3576
8500	.5232	.5004	-.2933	-.2993
9000	.4840	.4680	-.3539	-.3578
9500	.4376	.3116	-.4099	-.4038
10000	.3845	.3290	-.4602	-.4285

and $|\zeta_x(k)| \in [.5980,.5998]$ for all $k \geq 200$

TABLE 3.4: $\zeta_x(k)$ and $\hat{\zeta}_x^{h,y}(k)$ as functions of k, for the system given by equations (3.85) and (3.87)

k	$\zeta_{x1}(k)$	$\hat{\zeta}_{x1}^{h,y}(k)$	$\zeta_{x2}(k)$	$\hat{\zeta}_{x2}^{h,y}(k)$
100	.5208	.4240	.1547	.2274
500	.0737	.0725	.5949	.6001
1000	.1445	.0986	.5822	.5014
1500	.2141	.1998	.5601	.5956
2000	.2806	.3890	.5299	.4072
2500	.3431	.3883	.4918	.5149
3000	.4008	.3942	.4463	.3859
3500	.4517	.3929	.3944	.3511
4000	.4967	.5234	.3357	.2798
4500	.5342	.6222	.2722	.2969
5000	.5639	.5104	.2041	.2345
5500	.5842	.6000	.1352	.0697
6000	.5961	.5919	.0636	-.1351
6500	.5996	.5985	-.0105	-.0420
7000	.5939	.6950	-.0833	-.1192
7500	.5797	.6000	-.1542	-.2002
8000	.5565	.5001	-.2230	-.2222
8500	.5251	.5001	-.2899	-.1041
9000	.4855	.6256	-.3521	-.2522
9500	.4383	.5441	-.4092	-.4436
10000	.3854	.5076	-.4593	-.3225

and $|\zeta_x(k)| \in [.5993,.6000]$ for all $k \geq 2000$

TABLE 3.5: $\zeta_{x2}(k)$ and $\hat{\zeta}_{x2}^{h,y}(k)$ as functions of k, for the system given by equations (3.81)-(3.82)

k	$\zeta_{x2}(k)$	$\hat{\zeta}_{x2}^{h,y}(k)$
100	.2000	.0170
200	.2000	.1440
300	.2000	.1840
400	.2000	.1994
500	.2000	.1999
600	.2000	.2000
700	.2000	.2000

for $k \geq 600$ $\zeta_{x2}(k) = \hat{\zeta}_{x2}^{h,y}(k) = .2000$.

TABLE 3.6: $\zeta_{x2}(k)$ and $\hat{\zeta}_{x2}^{h,y}(k)$ as functions of k, for the system given by equations (3.93)-(3.94)

k	$\zeta_{x2}(k)$	$\hat{\zeta}_{x2}^{h,y}(k)$
100	.5000	.4809
200	.5000	.4998
300	.5000	.4999
400	.5000	.5000
500	.5000	.5000
600	.5000	.4999
700	.5000	.4997

for $k \geq 800$ $\zeta_{x2}(k) = \hat{\zeta}_{x2}^{h,y}(k) = .5000$.

In all the runs carried out on equations (3.95)-(3.96) and on equations (3.97)-(3.98) it turned out that $\hat{\zeta}_{x2}^{h,y}(k) = a$ for all $k \geq 200$.

Hence, in these cases, $\hat{\zeta}_{x2}^{h,y}(k) = E[a|F^{y}_{k\Delta\wedge\tau^{h}_{T}-}]$, $k \geq 200$.

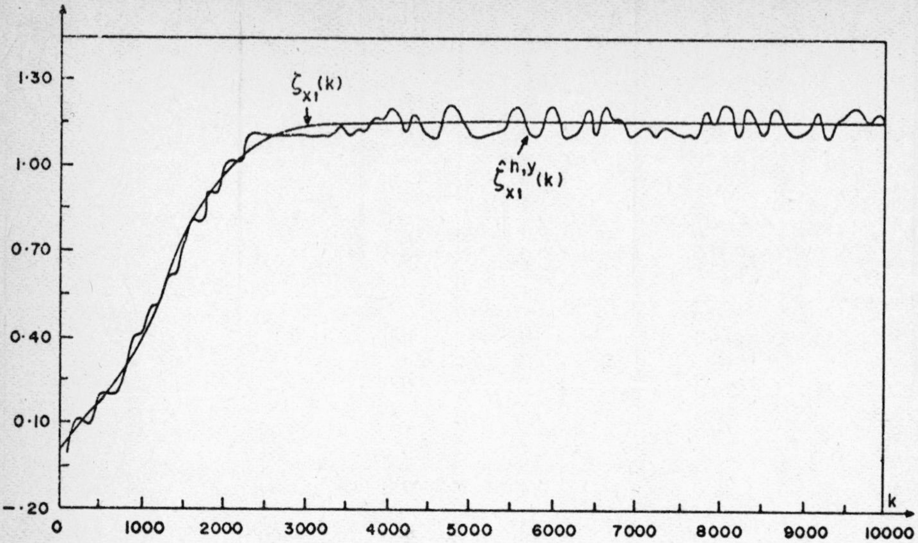

Fig.3.16a: $\zeta_{x1}(k)$ and $\hat{\zeta}^{h,y}_{x1}(k)$ as functions of k for the system given by equations (3.91)-(3.92).

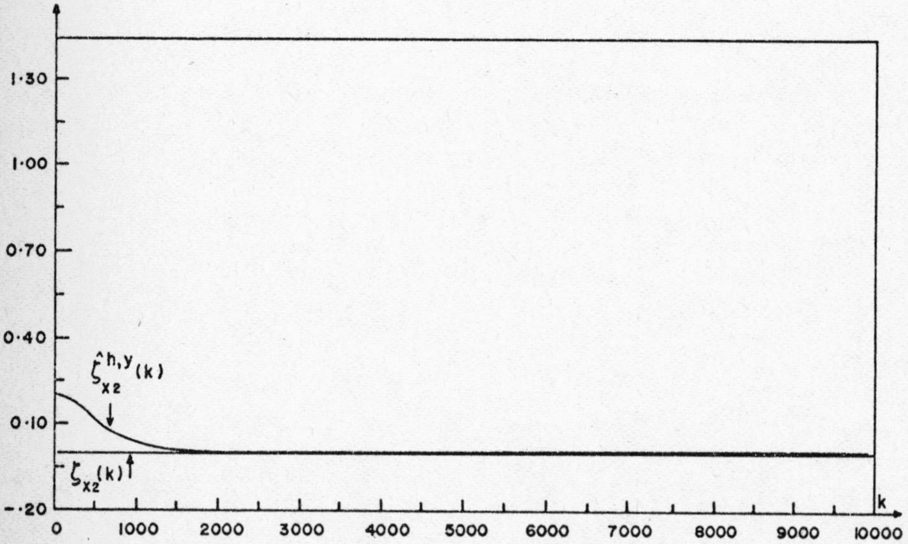

Fig.3.16b: $\zeta_{x2}(k)$ and $\hat{\zeta}^{h,y}_{x2}(k)$ as functions of k for the system given by equations (3.91)-(3.92).

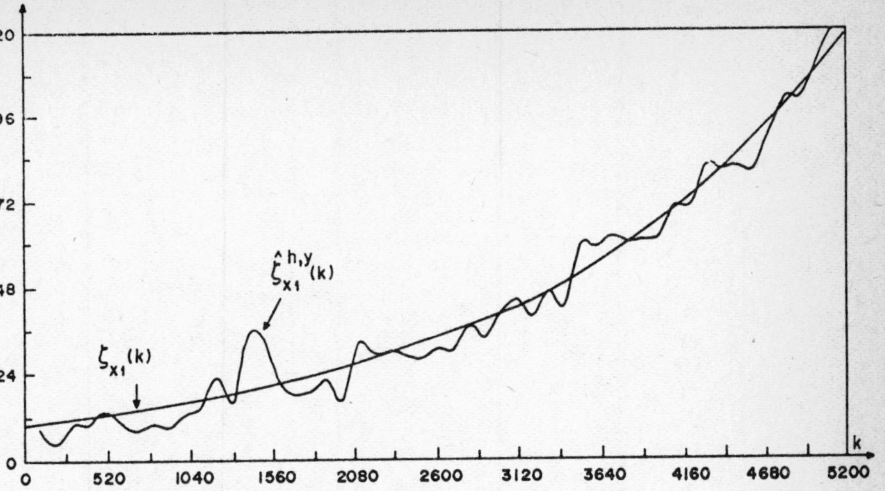

Fig.3.17: $\zeta_{x1}(k)$ and $\hat{\zeta}_{x1}^{h,y}(k)$ as functions of k for the system given by equations (3.93)-(3.94).

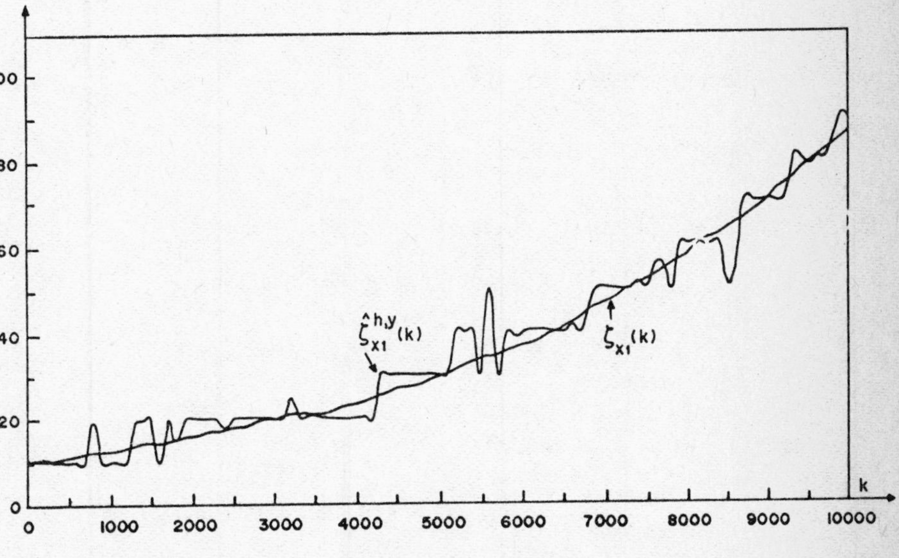

Fig.3.18: $\zeta_{x1}(k)$ and $\hat{\zeta}_{x1}^{h,y}(k)$ as functions of k for the system given by equations (3.95)-(3.96).

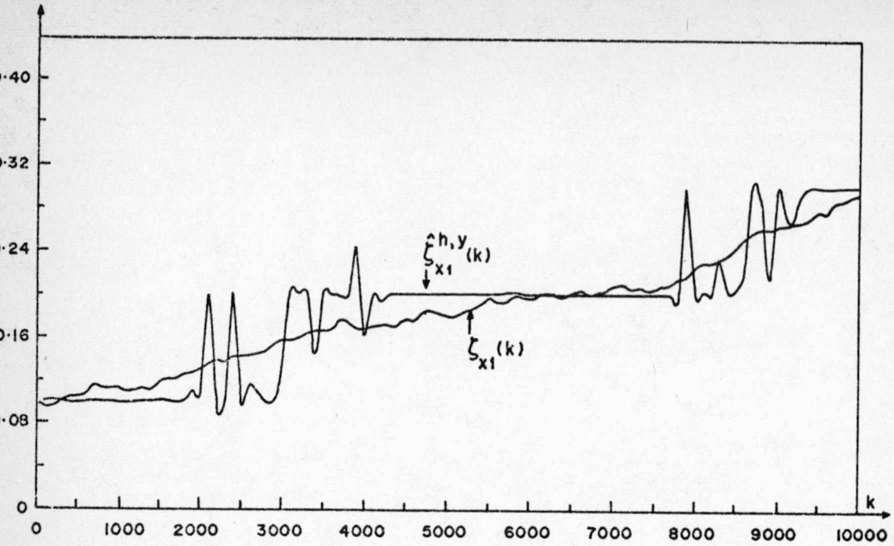

Fig.3.19: $\zeta_{x1}(k)$ and $\hat{\zeta}_{x1}^{h,y}(k)$ as functions of k for the system given by equations (3.97)-(3.98).

3.7 PARTIALLY OBSERVABLE SYSTEMS

3.7.1 Introduction

Consider an $\mathbb{R}^m$-valued Markov process $\zeta_x = (\eta_\alpha, \chi_\beta) = \{(\eta_{\alpha,1}(t), \ldots, \eta_{\alpha,\ell}(t), \chi_{\beta,1}(t), \ldots, \chi_{\beta,m-\ell}(t)),\ t \geq 0\}$ satisfying the equations

$$\eta_{\alpha,i}(t) = \alpha_i + \int_0^t f_i(\eta_\alpha(s))ds + \sigma_i W_i(t),\ t \geq 0,\ i=1,\ldots,\ell \tag{3.100}$$

$$\chi_{\beta,i}(t) = \beta_i + \int_0^t g_i(\eta_\alpha(s), \chi_\beta(s))ds + \gamma_i\ v_i(t),\ t \geq 0,\ i=1,\ldots,m-\ell, \tag{3.101}$$

$x = (\alpha,\beta) = (\alpha_1,\ldots,\alpha_\ell,\beta_1,\ldots,\beta_{m-\ell})$, where $f : \mathbb{R}^\ell \to \mathbb{R}^\ell$ and $g : \mathbb{R}^m \to \mathbb{R}^{m-\ell}$ are given functions satisfying conditions similar to that stated in Section 3.1 on f and g respectively; σ_i, $i=1,\ldots,\ell$, and γ_i, $i=1,\ldots,m-\ell$, are given positive numbers, $\mathbf{W} \triangleq \{W(t) = (W_1(t),\ldots,W_\ell(t)),\ t \geq 0\}$ and $\mathbf{V} \triangleq \{v(t) = (v_1(t),\ldots,v_{m-\ell}(t)),\ t \geq 0\}$ are $\mathbb{R}^\ell$-valued and $\mathbb{R}^{m-\ell}$-valued standard Wiener processes respectively. It is assumed that $\mathbf{W}$ and $\mathbf{V}$ are mutually independent.

Suppose that only the component $\chi_\beta = \{\chi_\beta(t),\ t \geq 0\}$ of ζ_x is observable while $\eta_\alpha = \{\eta_\alpha(t),\ t \geq 0\}$ is not available for observation. The problem dealt with in this section is to find an approximation $\hat{\eta}_\alpha^{h,y}(k)$ to

$$\hat{\eta}_\alpha(t) = E[\eta_\alpha(t \wedge \tau_T -) | \chi_\beta(s) \quad , \quad 0 \leq s \leq t \wedge \tau_T -], \quad t \in [0,T] \tag{3.102}$$

at the instants $t_k = k\Delta$, $k\Delta \in [0,T]$; where $\tau_T = \tau_T(\alpha)$ is the first exit time of $\eta_\alpha(t)$ from an open and bounded domain $D \subset \mathbb{R}^\ell$. By using the notation

$$y_i(t) = \chi_{\beta,i}(t),\ t \geq 0,\ i=1,\ldots,m-\ell;\ y_i \triangleq \beta_i,\ i=1,\ldots,m-\ell, \tag{3.103}$$

equations (3.100)-(3.101) and (3.102) can be written as

$$\eta_{\alpha,i}(t) = \alpha_i + \int_0^t f_i(\eta_\alpha(s))ds + \sigma_i W_i(t),\ t \geq 0,\ i=1,\ldots,\ell \tag{3.104}$$

$$y_i(t) = y_i + \int_0^t g_i(\eta_\alpha(s), y(s))ds + \gamma_i\ v_i(t),\ t \geq 0,\ i=1,\ldots,m-\ell, \tag{3.105}$$

and

$$\hat{\eta}_\alpha(t) = E[\eta_\alpha(t \wedge \tau_T -) | y(s)\ ;\ 0 \leq s \leq t \wedge \tau_T -],\ t \in [0,T]. \tag{3.106}$$

Hence, the problem posed here can be treated by the same methods deve= loped in Sections 3.2-3.4. Define

$$D \triangleq \{x \in \mathbb{R}^\ell : |x_i| < a_i + \delta,\quad i=1,\ldots,\ell\}\ ,\ \delta < h \tag{3.107}$$

and let $\{\eta_\alpha^h(t),\ t \in [0,T]\}$ and $\{\eta_\alpha^h(t \wedge \tau_T^h),\ t \in [0,T]\}$ be continuous-time Markov chains defined on $\mathbb{R}_h^\ell$ by using the same procedures as described in Section 3.2, and let τ_T and τ_T^h be defined in the same manner (with respect to D, (3.106), and $D_h \triangleq \mathbb{R}_h^\ell \cap D$) as in equations (3.20) and (3.21) respectively. Define

$$P_a(t) \triangleq P(\eta_\alpha^h(t \wedge \tau_T^h -) = a | F^{y,h}_{t \wedge \tau_T^h -}),\ \alpha,\ a \in D_h \quad ,\ t \in [0,T] \tag{3.108}$$

then, it can be easily shown that $\{P_a\}$ satisfy the following set of equations

$$dP_a(t) = \sum_{c \in D_h} \lambda(c,a)P_c(t)dt + P_a(t) \sum_{i=1}^{m-\ell} \gamma_i^{-2}(g_i(a,y^h(t)) - \hat{g}_i(t,y^h(t))(dy_i^h(t) - \hat{g}_i(t,y^h(t))dt) \tag{3.109}$$

$$a \in D_h \quad , \quad t \in (0,T)$$

$$\hat{g}_i(t,y^h(t)) = \sum_{a \in D_h} g_i(a,y^h(t))P_a(t), \; t \in [0,T], \quad i=1,\dots,m-\ell, \tag{3.110}$$

$$\hat{\eta}_\alpha^h(t) = \sum_{a \in D_h} a \, P_a(t) = E[\eta_\alpha^h(t \wedge \tau_T^h -) \mid F_{t\wedge\tau_T^h -}^{y,h}] \tag{3.111}$$

where

$$y^h(t) = y + \int_0^t g(\eta_\alpha^h(s),y^h(s))ds + \Gamma \, v(t) \quad , \quad t \geq 0; \tag{3.112}$$

$$\Gamma_{ij} = \gamma_i \, \delta_{ij}, \; i,j=1,\dots,m-\ell; \quad F_t^{y,h} \triangleq \sigma(y^h(s) \; ; \; 0 \leq s \leq t);$$

and $\{\lambda(c,a),\, a,\, c \in D_h\}$ are defined in a similar manner to equations (3.6)-(3.10) and (3.31) (but with respect to $\{\eta_\alpha^h(t),\, t \in [0,T]\}$).

Also, in the same manner as in Section 3.3, a process $\{\hat{\eta}_\alpha^{h,y}(t),\, t \in [0,T]\}$ is defined by

$$\hat{\eta}_\alpha^{h,y}(t) \triangleq \sum_{a \in D_h} a \, P_a^y(t) \quad , \quad t \in [0,T] \tag{3.113}$$

where $\{P_a^y(t),\, a \in D_h,\, t \in [0,T]\}$ denote the solution to equations (3.109)-(3.110) but where in these equations, the increment dy replaces dy^h. Also, a procedure for computing $\hat{\eta}_\alpha^{h,y}$, analogous to Algorithm 3.4, can be constructed.

3.7.2 Examples

(i) Consider the following equations describing a motion of a point in the (x_1,x_2)-plane.

$$\begin{cases} dv_1 = -k_o v_1^2 dt + \sigma_1 dW_1 \\ dv_2 = -k_o v_2^2 dt + \sigma_2 dW_2 \\ dx_1 = v_1 dt + \gamma_1 dW_3 \\ dx_2 = v_2 dt + \gamma_2 dW_4 \end{cases} \tag{3.114}$$

where $\mathbf{W} = \{W(t) = (W_1(t),W_2(t),W_3(t),W_4(t)),\ t \geq 0\}$ is a $\mathbb{R}^4$-valued standard Wiener process. Denote $\eta_\alpha(t) = (v_1(t),v_2(t))$, $t \geq 0$, and $\chi_\beta(t) = (x_1(t),x_2(t))$, $t \geq 0$. The problem dealt with in this example is to find an approximation $\hat{\eta}_\alpha^{h,y}(k)$ to $\hat{\eta}_\alpha(t)$ (equation (3.102)) where here D is given by

$$D \triangleq \{v = (v_1,v_2) : |v_i| < 360+\delta\}, \delta < h \tag{3.115}$$

Numerical experimentation has been carried out for the following set of parameters: $\Delta = 10^{-3}$, $\sigma_1 = \sigma_2 = 7$, $\gamma_1 = \gamma_2 = 3$, $k_o = 5\cdot10^{-5}$, $h = 30$ $L = 12$, and $x_1(0) = x_2(0) = 0$. Some of the numerical results are illu= strated in Figs. 3.20-3.21.

(ii) Consider the frequency perturbed sine waves oscillator given by (see Section 4.4 for more details)

$$\begin{cases} dx_1 = [-a_o(1 + x_3)x_2 + bx_1(\rho^2 - x_1^2 - x_2^2)]dt + \gamma_1 dv_1 \\ dx_2 = [a_o(1 + x_3)x_1 + bx_2(\rho^2 - x_1^2 - x_2^2)]dt + \gamma_2 dv_2 \quad t > 0 \\ dx_3 = -ax_3 dt + \sigma dW \end{cases} \tag{3.116}$$

where $\{(v_1(t),\ v_2(t), W(t)),\ t \geq 0\}$ is an $\mathbb{R}^3$-valued standard Wiener process. Denote $\eta_\alpha(t) = x_3(t)$, $t \geq 0$, and $\chi_\beta(t) = (x_1(t),x_2(t))$, $t \geq 0$. The problem dealt with in this example is to find an approximation $\hat{\eta}_\alpha^{h,y}(k)$ to

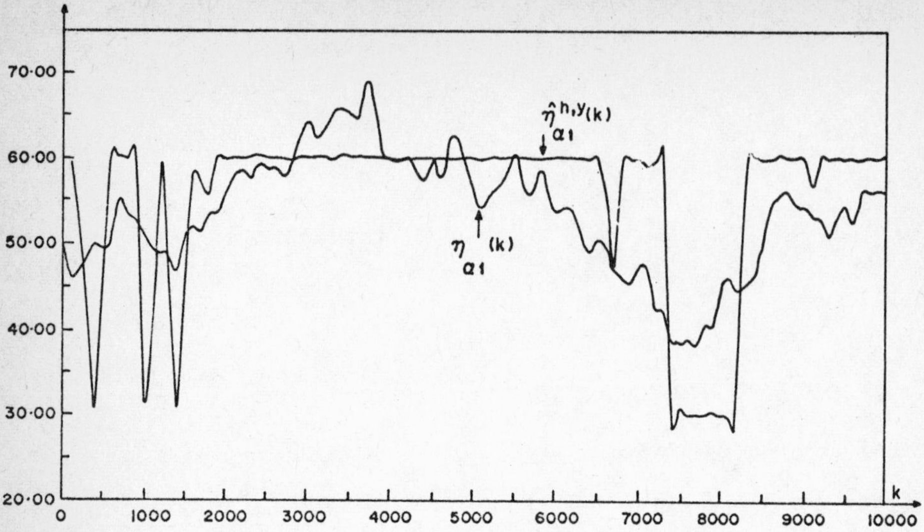

Fig.3.20a: $\eta_{\alpha 1}(k)$ and $\hat{\eta}^{h,y}_{\alpha 1}(k)$ as functions of k for the system given by equations (3.114) and where $v_1(0) = v_2(0) = 50$.

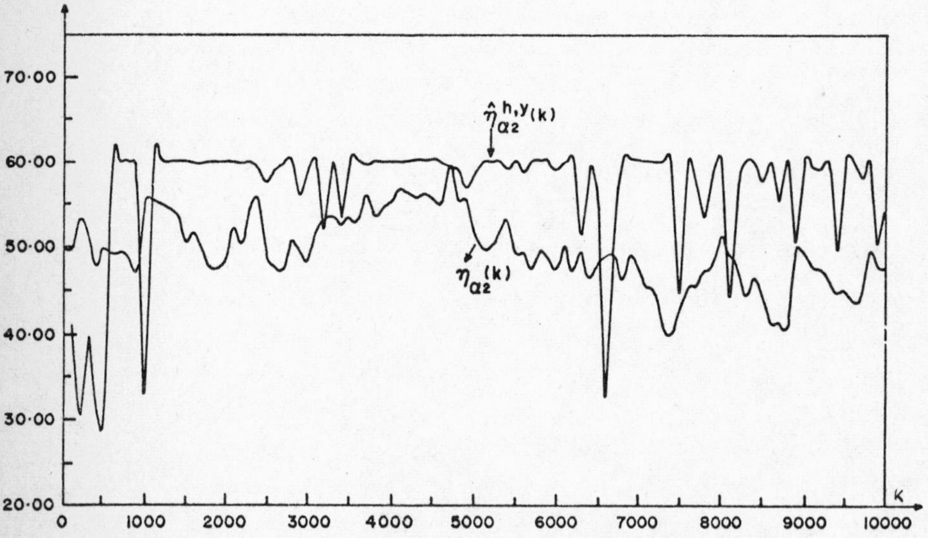

Fig.3.20b: $\eta_{\alpha 2}(k)$ and $\hat{\eta}^{h,y}_{\alpha 2}(k)$ as functions of k for the system given by equations (3.114) and where $v_1(0) = v_2(0) = 50$.

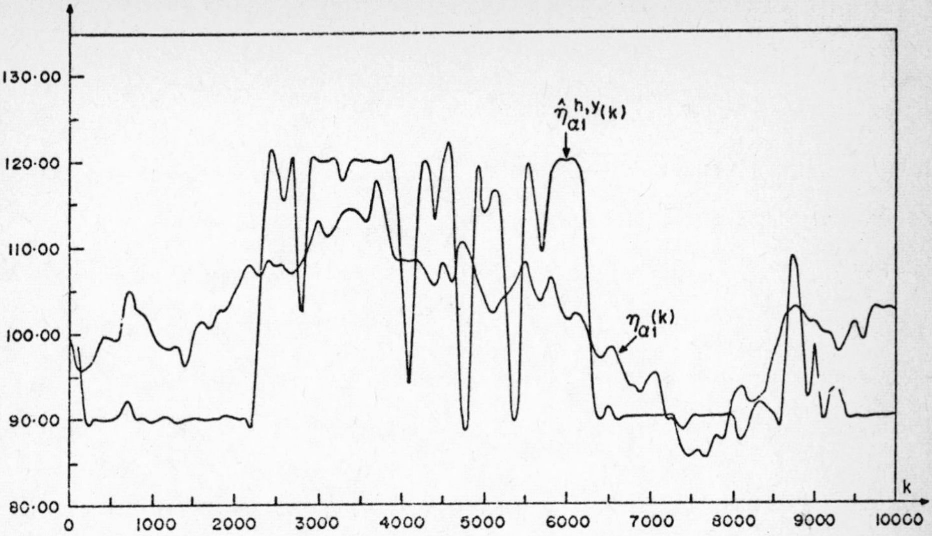

<u>Fig.3.21a</u>: $\eta_{\alpha 1}(k)$ and $\hat{\eta}_{\alpha 1}^{h,y}(k)$ as functions of k for the system given by equations (3.114) and where $v_1(0) = v_2(0) = 100$.

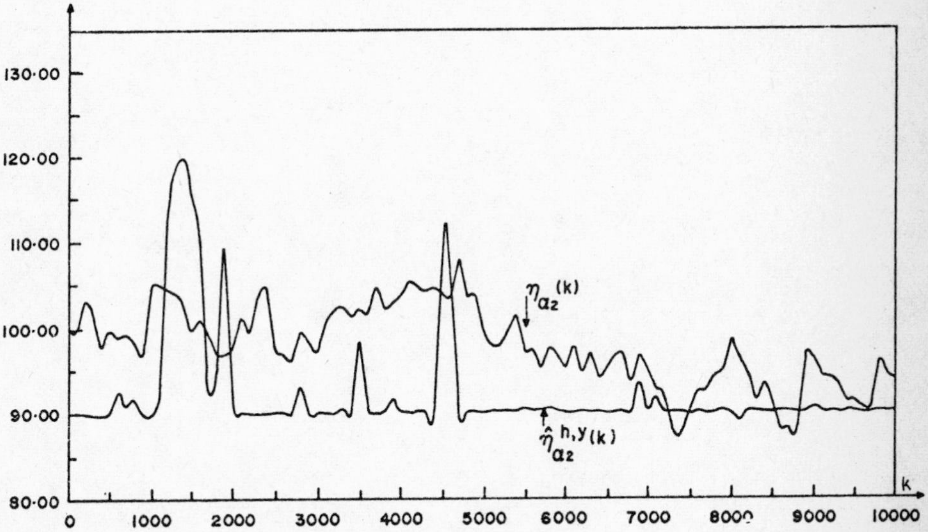

<u>Fig.3.21b</u>: $\eta_{\alpha 2}(k)$ and $\hat{\eta}_{\alpha 2}^{h,y}(k)$ as functions of k for the system given by equations (3.114) and where $v_1(0) = v_2(0) = 100$.

$\hat{\eta}_\alpha(t)$ (equation (3.102)) where here D is given by

$$D \overset{\Delta}{=} \{x_3 : |x_3| < 2 + \delta\} \ , \ \delta < \ 0.01. \tag{3.117}$$

Numerical experimentation has been carried out for the following set of parameters: $a_0 = 50,100$; $\rho = 0.5$; $b = 50$; $a = 0.2$; $\gamma_1 = \gamma_2 = 0.005$; $\sigma = 0.04$; $\Delta = 10^{-3}$; $h = 0.01$ and $L = 200$. Some of the numerical results are illustrated in Tables 3.7-3.8. In these Tables, for the sake of comparison, the same sets of random elements $\{v(k)\}$ and $\{W(k)\}$, were used during the simulation of equations (3.116).

All the graphs in this section were plotted using the set of points $\{t'_k = 100k\Delta : k=0,1,\dots,100\}$.

TABLE 3.7: $\eta_\alpha(k)$ and $\hat{\eta}_\alpha^{h,y}(k)$ as functions of k, for the system given by equations (3.116) and where a_o = 50

TABLE 3.8: $\eta_\alpha(k)$ and $\hat{\eta}_\alpha^{h,y}(k)$ as functions of k, for the system given by equations (3.116) and where a_o = 100

k	$\eta_\alpha(k)$	$\hat{\eta}_\alpha^{h,y}(k)$	$\eta_\alpha(k)$	$\hat{\eta}_\alpha^{h,y}(k)$
99	.08042	.1604	.08042	.2358
899	.06540	.06052	.06540	.06989
1 699	.06028	.05799	.06028	.04827
2499	.05630	.05963	.05630	.06014
3299	.07871	.06997	.07871	.08048
4099	-.00318	-.00512	-.00318	-.000871
4899	-.00089	-.00527	-.00089	-.00161
5699	-.03043	-.02968	-.03043	-.02913
6499	-.03582	-.03028	-.03582	-.03264
7299	.00998	.00945	.00998	.00995
8099	.04890	.04931	.04890	.04977
8899	.05049	.05776	.05049	.04850
9699	.07487	.06991	.07487	.07289
10 499	.1024	.1040	.1024	.1004
11 299	.1044	.1006	.1044	.1001
12 099	.1217	.1201	.1217	.1201
12 899	.1098	.1070	,1098	.1167
13 699	.1033	.09849	.1033	.1051
14 499	.06252	.05883	.06252	.06979
15 299	.00599	.00993	.00590	.00554
16 099	-.00235	-.00334	-.00235	.00005
16 899	.03022	.02632	.03022	.02988
17 699	.06561	.07004	.06561	.07021
18 499	.07418	.07260	.07418	.07827
19 299	.05575	.06106	.05575	.05270
19 999	.03535	.03924	.03535	.03993

3.8 REMARKS

The technique of replacing the 'signal process' X by an approximating Markov chain X^h and then computing the corresponding filtering equations has been considered in Refs. [3.14-3.15] and [3.19-3.21]. Refs.[3.14-3.15] deal only with systems given by equations (3.1) - (3.2), while Refs. [3.19-3.21] deal with systems given by equation (3.1) with a measurement process given by

$$y(t) = \int_0^t C(X_s)ds + v(t) + N(t), \quad t \geq 0, \tag{3.118}$$

where $\{X_t, t \in [0,T]\}$ is a diffusion process (the treatment there holds for some other types of the 'signal process' X), v is a standard Wiener process and N is a doubly stochastic Poisson process. In these references it is shown that the approximate filters, constructed via Markov chain approxima= tion and using the measurement signal y (given by (3.2) or (3.118)) as an input, satisfy

$$E[\psi(\bar{X}^h(t\wedge\tau_h-)) \mid F^y_{t\wedge\tau_h-}] \underset{h\downarrow 0}{\rightarrow} E[\psi(\bar{X}(t\wedge\tau-)) \mid F^y_{t\wedge\tau-}], \tag{3.119}$$

where $\psi : \mathbb{R}^m \rightarrow \mathbb{R}$ is a measurable function, $\bar{X}$ is a version of X ($\bar{X}$ is inde= pendent of y) and $\bar{X}^h$ is the approximate Markov chain for $\bar{X}$. The random times τ and τ_h are defined in the same manner as τ_T and τ_T^h here (equations (3.20) and (3.21) respectively). However, in any digital implementation of the approximate filters, one always uses only a quantized version of y, and not the process y itself, as an input to the filter. Thus, an inte= resting problem is to establish conditions for the convergence of the ap= proximate filters, which use a quantized version of y in the input, to the appropriate limit. This convergence problem is still open.

In this work, using the results of [1.1] (or Themrem 1.1 here), the filter equations derived for the approximate Markov chain use a 'quantified' ver= sion y^h of y (eq. (3.22)) as input. Given h, the parameter of quantization of ζ_x, the filter equations (eqns. (3.36) - (3.38)) obtained, constitute the true filter (the minimum variance) for computing $E[\zeta_x^h(t\wedge\tau_T^h-) \mid F^{y,h}_{t\wedge\tau_T^h-}]$ ($F_t^{y,h} = \sigma(y^h(s), 0 \le s \le t)$). In the implementation, y^h is replaced by a quantized version of y obtained from digital realization of y (obtained by using finite length representation of real numbers). These are the characteristics of the approach used here in Chapters 3, 5, 7, 8 and 9.

Note that the procedure applied here, for constructing approximate filters, can be applied to the problem dealt with in Refs. [3.19-3.21].

The examples appearing in this chapter are divided into two classes:

(a) ζ_x is a one-dimensional Markov process: In this class of examples, different kinds of nonlinear first-order stochastic differential equations satisfied by ζ_x were treated numerically in order to expe= riment with the filtering algorithm. Most of these systems include a basic nonlinearity, e.g., x^2, x^3, sign (x), arctan (x) etc., which is common in engineering.

(b) ζ_x is a two-dimensional Markov process: Most of the computational work was done on this kind of system. The examples include problems of state and parameter estimation (i.e. the cases given by (3.81)-(3.82), (3.93)-(3.94), (3.95)-(3.96), (3.97)-(3.98) where different forms of nonlinearities are present either in the stochastic differential equations satisfied by ζ_x or in the measurement process; estimation of the state of nonlinear oscillators using measurement processes with nonlinearities (i.e., the cases given by (3.85)-(3.86), (3.85) and (3.87), (3.88)-(3.89), (3.88) and (3.90), (3.91)-(3.92) and (3.116)); and estimation of the velocity of a body moving in a re= sisting medium using measurements of the body's displacement (eqns. (3.114)).

Numerical experimentation with these and other examples suggests that $\{\hat{\zeta}_x^{h,y}(t),\ t \in [0,T]\}$ (eq. (3.39)) is a good state estimator.

3.9 REFERENCES

3.1 R. Stratonovich, On the theory of optimal nonlinear filtration of random functions, *Theory of Probability and its Applications*, 4, pp 223-225, 1959.

3.2 H.J. Kushner, On the dynamical equations of conditional probability density functions with applications to optimal stochastic con= trol theory, *J. Math. Anal. App.*, 8, pp 332-344, 1964.

3.3 H.J.Kushner, On the differential equations satisfied by conditional probability densities of Markov processes, with applications, *J.SIAM Control*, 2, pp 106-119, 1964.

3.4 W.M. Wonham, Some applications of stochastic differential equations to optimal nonlinear filtering, *J. SIAM Control*, 2, pp 347-369, 1965.

3.5 R.S. Bucy, Nonlinear filtering theory, IEEE Trans. on Automatic Control, 10, p 198, 1965.

3.6 R.S.Bucy and P.D.Joseph, *Filtering for Stochastic Processes with Applications to Guidance*, Interscience, New York, 1968.

3.7 M. Zakai, On the optimal filtering of diffusion processes, *Z. Wahr. Verw. Geb.*, 11, pp 230-243, 1969.

3.8 A.H. Jazwinski, *Stochastic Processes and Filtering Theory*, Academic Press, New York, 1970.

3.9 P.A. Frost and T.K. Kailath, An innovations approach to least-square estimation - Part III : Nonlinear estimation in white Gaussian noise, *IEEE Trans. Automat. Contr.*, 16, pp 217-226, 1971.

3.10 T.P.McGarty, *Stochastic Systems and State Estimation*, John Wiley & Sons, New York, 1974.

3.11 M. Fujisaki, G. Kallianpur and H. Kunita, Stochastic differential equations for the non linear filtering problem, *Osaka J. Math.*, 9, pp 19-40, 1972.

3.12 G. Kallianpur, *Stochastic Filtering Theory*, Springer-Verlag, New York, 1980.

3.13 J.H. Van Schuppen, Stochastic filtering theory: A discussion of concepts, methods and results, in *Stochastic Control Theory and Stochastic Differential Systems*, M. Kohlmann and W. Vogel, eds., pp 209-226, Springer-Verlag, New York, 1979.

3.14 H.J. Kushner, *Probability Methods for Approximations in Stochastic Control and for Elliptic Equations*, Academic Press, New York, 1977.

3.15 G.B. Di Masi and W.J. Runggaldier, Continuous-time approximations for the nonlinear filtering problem, *App. Math. Optim.*, 7, pp 233-245, 1981.

3.16 G. Kallianpur and C. Striebel, Estimation of stochastic systems: arbitrary system process with additive white noise observation errors, *Ann.Math.Stat.*, 39, pp 785-801, 1968.

3.17 R.S.Liptser and A.N. Shiryayev, Statistics of Random Processes, Springer-Verlag, New York, Vol.I, 1977; Vol.II, 1978.

3.18 B.Z.Kaplan, Rotation of a waveform generator, *Electronics Letters* *15*,pp 158-159,1979.

3.19 G.B. Di Masi and W.J. Runggaldier, An approximation to optimal non=linear filtering with discontinuous observation, in M. Hazewinkel and J.C. Willems (eds.) *Stochastic Systems: The Mathematics of Filtering and Identification with Applications*, pp 583-590, D. Reidel Publishing Company, Dordracht, 1981.

3.20 G.B. Di Masi and W.J. Runggaldier, On robust approximations in non=linear filtering, in M. Kohlmann and N. Christopeit (eds.) *Stochastic Differential Systems*, pp 179-186, Lecture Notes in Control and Information Sciences, 43, Berlin, 1982.

3.21 G.B. Di Masi and W.J. Runggaldier, Non-linear filtering with dis=continuous observations and applications to life sciences, *Bulletin of Mathematical Biology*, 45, pp 571-577, 1983.

CHAPTER 4

A KALMAN FILTER FOR A CLASS OF NONLINEAR STOCHASTIC SYSTEMS

4.1 INTRODUCTION

Let a nonlinear stochastic system be given by

$$dx = A(x)xdt + BdW \quad , \quad t > 0 \quad , \quad x \in \mathbb{R}^m \quad , \tag{4.1}$$

and suppose that the measurements of the state X(t) are given by

$$y(t_k) = H(t_k)X(t_k) + v(t_k), \quad t_o < t_1 < t_2 < \dots \; ; \; y(t_k) \in \mathbb{R}^p, \text{ for all } k, \tag{4.2}$$

where A(x) and B are given m×m matrices and $\{H(t_k)\}$ are given p×m ma= trices. $\mathbf{W} = \{W(t) = (W_1(t),\dots,W_m(t)),\ t \geq 0\}$ is an $\mathbb{R}^m$-valued standard Wiener process and $\mathbf{V} = \{v(t_k),\ t_o < t_1 < t_2 < \dots\}$, where $\{t_k\}$ is a given sequence, is an $\mathbb{R}^p$-valued white Gaussian sequence with

$$Ev(t_k) = 0 \text{ and } E[v(t_k)v'(t_\ell)] = \delta_{k\ell}R(t_k);\ k,\ell=0,1,2,\dots \tag{4.3}$$

where $\{R(t_k)\}$ are given p×p symmetric and positive definite matrices. It is assumed here that **W** and **V** are mutually independent, and that $\sup_k (t_{k+1}-t_k) \ll 1$.
Let $A(x) = \{a_{ij}(x)\}_{i,j=1}^m$. It is assumed here that a_{ij}, i,j=1,...,m are continuously differentiable on $\mathbb{R}^m$, and that $|A(x)x|^2 \leq M(1 + |x|^2)$ for some $0 < M < \infty$ (here $|x|^2 = \sum_{i=1}^m x_i^2$). Given $x \in \mathbb{R}^m$, then by using the results of Gihman and Skorohod [4.1] it can be shown that equation (4.1) has an unique solution denoted here by $\zeta_x = \{\zeta_x(t),\ t \geq 0\}$ with continuous sample paths and such that $\zeta_x(0) = x$. (It is tacitly assumed here that $\zeta_x(0)$ is independent of W and V and that $E|\zeta_x(0)|^2 < \infty$). Let

$$Y^k = \{y(t_o),y(t_1),\ldots,y(t_k)\} \tag{4.4}$$

and let $\sigma(Y^k)$ denote the smallest σ-field containing Y^k. The problem dealt with in this chapter is to find an approximation $\hat{\zeta}(k|k)$ to

$$\hat{\zeta}_x(t_k|t_k) = E[\zeta_x(t_k)|\sigma(Y^k)],\ k=1,2,\ldots\ . \tag{4.5}$$

The problem of computing $\hat{\zeta}_x(t|t_k)$, $t_k \le t < t_{k+1}$, in the case where (4.1) represents a linear system (i.e. $A = A(t)$ in (4.1)) is treated in Jazwinski [4.2]. In this chapter an entirely different approach towards the filtering problem is followed. First a procedure for the discretization (in time) of the system given by (4.1) is suggested. Second, a version of the discrete Kalman filter is proposed for the computation of $\hat{\zeta}(k|k)$. The discretization procedure together with the filter are described also in Yavin [4.3].

4.2 THE DISCRETE-TIME MODEL

Consider the following set of m×m+m stochastic differential equations

$$\begin{cases} dx = A(x)xdt + BdW\ , & x \in \mathbb{R}^m \\ & \qquad t_k < t < t_{k+1} \\ d\Phi(t) = A(x)\Phi(t)dt\ , & \Phi(t) \in \mathbb{R}^{m\times m} \end{cases} \tag{4.6}$$

with the initial conditions $x(t_k) = \zeta_x(t_k)$ and $\Phi(t_k) = I_m$. Under the assumptions on $A(x)$ (given in Section 4.1) it follows that equations (4.6) have an unique solution $(\zeta_x,\Phi^{(k)}(\cdot,\cdot;\zeta_x)) = \{(\zeta_x(t),\Phi(t,t_k;\zeta_x)),$ $t \in [t_k,t_{k+1}]\}$, $k=0,1,2,\ldots$.

Furthermore, by using (4.1) and (4.6) it can be shown that ζ_x can be written, for each $k=0,1,2,\ldots$, in the following form

$$\zeta_x(t) = \Phi(t,t_k;\zeta_x)\zeta_x(t_k) + \int_{t_k}^{t} \Phi(t,s;\zeta_x)BdW(s),\ t_k \le t \le t_{k+1}. \tag{4.7}$$

Note that (4.7) is not an explicit expression for the solution of (4.1) but merely a representation for ζ_x. Hence

$$\zeta_x(t_{k+1}) = \Phi(t_{k+1},t_k;\zeta_x)\zeta_x(t_k) + \int_{t_k}^{t_{k+1}} \Phi(t_{k+1},s;\zeta_x)BdW(s), \tag{4.8}$$

$k=0,1,\ldots,$

where, without loss of generality, $t_o = 0$.

Equations (4.8) constitute a discrete-time representation of the system given by (4.1). An approximation $X = \{x(t_k), k=0,1,\ldots\}$ to ζ_x, at the points $\{t_k\}$, can be computed from (4.8) by applying the following pro= cedure:

1. Let $x(t_o) = \bar{x}(t_o) = \zeta_x(t_o)$

2. Given $x(t_k)$, construct the process $\bar{X}$ by

$$\bar{x}(t) = x(t_k) + A(x(t_k))x(t_k)(t-t_k) + B[W(t)-W(t_k)], \quad t_k \le t \le t_{k+1} \tag{4.9}$$

3. Construct the process $\Phi_X^{(k)} = \{\Phi(t,t_k;X),\ t \in [t_k,t_{k+1}]\}$ by solving the following problem:

$$\begin{cases} d\Phi(t,t_k) = A(\bar{x}(t))\Phi(t,t_k)dt \quad , t_k < t < t_{k+1} \\ \\ \Phi(t_k,t_k) = I_m \end{cases} \tag{4.10}$$

The solution to (4.10) is denoted by $\Phi(t,t_k;X)$.

4. Compute $x(t_{k+1})$ by

$$x(t_{k+1}) = \Phi(t_{k+1},t_k;X)x(t_k) + \int_{t_k}^{t_{k+1}} \Phi(t_{k+1},s;X)BdW(s) \tag{4.11}$$

$k:=k+1$ and go to 2.

Throughout this chapter the points $\{t_k\}$ have been taken to be of the form

$t_k = k\Delta$, $k = 0,1,2,\ldots$, where Δ is a given positive number.

For computational purposes, with Δ chosen sufficiently small, the following approximations have been applied:

(i) Upon integrating (4.10) and using the trapezoidal rule it follows that

$$\Phi(t_{k+1},t_k;X) - I_m = \int_{t_k}^{t_{k+1}} A(\bar{x}(t))\Phi(t,t_k;X)dt \tag{4.12}$$

$$\cong (\Delta/2)[A(x(t_k)) + A(\bar{x}(t_{k+1}))\Phi(t_{k+1},t_k;X)]$$

which implies

$$\Phi(t_{k+1},t_k;X) \cong [I_m - (\Delta/2)A(\bar{x}(t_{k+1}))]^{-1}[I_m + (\Delta/2)A(x(t_k))] \tag{4.13}$$

provided that $[I_m - (\Delta/2)A(\bar{x}(t_{k+1}))]^{-1}$ exists.

(ii) $$\int_{t_k}^{t_{k+1}} \Phi(t_{k+1},s;X)BdW(s) \cong \Phi(t_{k+1},t_k;X)B[W(t_{k+1})-W(t_k)]. \tag{4.14}$$

Thus the following sequences $\{X(k)\}$ and $\{\bar{X}(k)\}$ are computed by applying the procedure described below

(a) $X(0) = \bar{X}(0) = \zeta_x(0)$ (4.15)

(b) $\bar{X}(k+1) = X(k) + A(X(k))X(k)\Delta + \sqrt{\Delta}\, BW(k)$ (4.16)

(c) $\Phi(k+1,k) = [I_m - (\Delta/2)A(\bar{X}(k+1))]^{-1}[I_m + (\Delta/2)A(X(k))]$ (4.17)

(d) $X(k+1) = \Phi(k+1,k)X(k) + \sqrt{\Delta}\,\Phi(k+1,k)BW(k)$ (4.18)

(e) $k:=k+1$ and go to (b), (4.19)

where $\{W(k)\}_{k=0}^{\infty}$ is a sequence of independent $\mathbb{R}^m$-valued Gaussian random elements with

$$EW(k) = 0 \text{ and } E[W(k)W'(\ell)] = \delta_{k\ell}\, I_m\,, \quad k,\ell=0,1,\ldots\ .$$

4.3 THE DISCRETE-TIME FILTER

In this chapter, using the notations $H(k) \triangleq H(t_k)$ and $R(k) \triangleq R(t_k)$, $k=0,1,\ldots$, the following algorithm is suggested for the computation of $\hat{\zeta}(k|k)$ (the approximation to $\hat{\zeta}_x(t_k|t_k)$), given the sequence $\{y(t_k)\}_{k=0}^{N}$, (4.2).

Algorithm 4.3

For $k=0,1,\ldots$

1. Given $\hat{\zeta}(k|k)$ compute $\hat{\zeta}(k+1|k)$ by

$$\hat{\zeta}(k+1|k) = \hat{\zeta}(k|k) + A(\hat{\zeta}(k|k))\hat{\zeta}(k|k)\Delta \tag{4.20}$$

2. Compute $\psi(k+1,k)$ by

$$\psi(k+1,k) = [I_m - (\Delta/2)A(\hat{\zeta}(k+1|k))]^{-1}[I_m + (\Delta/2)A(\hat{\zeta}(k|k))] \tag{4.21}$$

3. Compute $P(k+1|k)$ by

$$\Gamma(k) = \sqrt{\Delta}\,\psi(k+1,k)B \tag{4.22}$$

$$P(k+1|k) = \psi(k+1,k)P(k|k)\psi'(k+1,k) + \Gamma(k)\Gamma'(k) \tag{4.23}$$

4. Compute

$$K(k+1) = P(k+1|k)H'(k+1)[H(k+1)P(k+1|k)H'(k+1) + R(k+1)]^{-1} \tag{4.24}$$

5. Compute $\hat{\zeta}(k+1|k+1)$ by

$$\hat{\zeta}(k+1|k+1) = \hat{\zeta}(k+1|k) + K(k+1)[y(t_{k+1}) - H(k+1)\hat{\zeta}(k+1|k)] \tag{4.25}$$

6. Compute $P(k+1|k+1)$ by

$$P(k+1|k+1) = P(k+1|k) - K(k+1)\,H(k+1)P(k+1|k) \tag{4.26}$$

7. $k:=k+1$ and go to 1.

Note that $P(k+1|k)$, $K(k+1)$, $\hat{\zeta}(k+1|k+1)$ and $P(k+1|k+1)$ are computed in the same manner as in the discrete-time Kalman filter, [4.2]. This has been made possible by the form of equations (4.1)-(4.2) and was suggested by the procedure for computing $\{X(k)\}$ (equations (4.15)- (4.18)).

In the sequel a numerical study of Algorithm 4.3 will be carried via numerical experimentation. Each experiment is called a *run*. Here a run always consists of two stages. In the first stage equation (4.1) is simulated by using the procedure given by (4.15)-(4.19), and the sequence $\{y(k\Delta)\}_{k=1}^{N}$ is computed and stored. In the second stage Algorithm 4.3 is applied, where the sequence $\{y(k\Delta)\}_{k=1}^{N}$ acts as the input to the filter, and $\{\hat{\zeta}(k|k)\}_{k=1}^{N}$ constitutes the output of the filter.

4.4 EXAMPLE 4.1 : FREQUENCY PERTURBED SINE WAVES OSCILLATOR

Consider the noise-driven nonlinear stochastic system given by

$$\begin{cases} dx_1 = [-a_o(1 + x_3)x_2 + bx_1(\rho^2 - x_1^2 - x_2^2)]dt + \sigma_1 dW_1 \\ dx_2 = [a_o(1 + x_3)x_1 + bx_2(\rho^2 - x_1^2 - x_2^2)]dt + \sigma_2 dW_2 \\ dx_3 = -ax_3 dt + \sigma_3 dW_3 \end{cases} \tag{4.27}$$

with the observations

$$\begin{cases} y_i(t_k) = x_i(t_k) + v_i(t_k) \quad , \quad i=1,2, \\ y_3(t_k) = v_3(t_k) \end{cases} \qquad t_k = k\Delta \ , \ k=0,1,2,\dots \tag{4.28}$$

where a_o, a, b, ρ, σ_1, σ_2 and σ_3 are given positive numbers; $\mathbf{W} \triangleq \{W(t) = (W_1(t), W_2(t), W_3(t)), t \geq 0\}$ is an $\mathbb{R}^3$-valued standard Wiener process; $\mathbf{V} \triangleq \{v(t_k), k=0,1,\dots\}$ is a sequence of independent $\mathbb{R}^3$-valued Gaussian random elements with

$$Ev(t_k) = 0 \text{ and } E[v(t_k)v'(t_\ell)] = \delta_{k\ell}R \quad , \quad k,\ell=0,1,\dots \tag{4.29}$$

where

$$R = \begin{pmatrix} \gamma_1^2 & 0 & 0 \\ 0 & \gamma_1^2 & 0 \\ 0 & 0 & \gamma_2^2 \end{pmatrix} \tag{4.30}$$

and γ_i, i=1,2, are given positive numbers. It is assumed that **W** and **V** are mutually independent.

In order to indicate the physical meaning of equations (4.27) consider the case where $\sigma_1 = \sigma_2 = 0$. Also, substitute $x_1 = r\cos\theta$ and $x_2 = r\sin\theta$. Then the first two equations of (4.27) reduce to

$$dr/dt = br(\rho^2 - r^2) \quad , \quad d\theta/dt = a_o(1 + x_3) \quad . \tag{4.31}$$

The solution to equations (4.31) leads to

$$r(t) = \rho\, \exp(\rho^2 bt)/[c + \exp(2\rho^2 bt)]^{\frac{1}{2}},\ t > 0,\ c = (\rho/r(0))^2-1, \tag{4.32}$$

$$\theta(t) = \theta_o + a_o \int_o^t (1 + x_3(s))ds \tag{4.33}$$

By using equation (4.32) it follows that

$$\lim_{t \to \infty} r(t) = \rho\ . \tag{4.34}$$

Thus, equations (4.27) (for the case $\sigma_i \neq 0$, i=1,2,3) are interpreted as representing a noise-driven nonlinear stochastic frequency perturbed sine waves oscillator (For frequency perturbed oscillators see for example Bird and Folchi [4.4]).

Equations (4.27)-(4.28) can be written as

$$\begin{pmatrix} dx_1 \\ dx_2 \\ dx_3 \end{pmatrix} = \begin{pmatrix} b(\rho^2 - x_1^2 - x_2^2) & -a_o(1 + x_3) & 0 \\ a_o(1 + x_3) & b(\rho^2 - x_1^2 - x_2^2) & 0 \\ 0 & 0 & -a \end{pmatrix} \begin{pmatrix} x_1 \\ x_2 \\ x_3 \end{pmatrix} dt$$

$$+ \begin{pmatrix} \sigma_1 & 0 & 0 \\ 0 & \sigma_2 & 0 \\ 0 & 0 & \sigma_3 \end{pmatrix} \begin{pmatrix} dW_1 \\ dW_2 \\ dW_3 \end{pmatrix}, \quad t > 0 \tag{4.35}$$

and

$$\begin{pmatrix} y_1(t_k) \\ y_2(t_k) \\ y_3(t_k) \end{pmatrix} = \begin{pmatrix} 1 & 0 & 0 \\ 0 & 1 & 0 \\ 0 & 0 & 0 \end{pmatrix} \begin{pmatrix} x_1(t_k) \\ x_2(t_k) \\ x_3(t_k) \end{pmatrix} + \begin{pmatrix} v_1(t_k) \\ v_2(t_k) \\ v_3(t_k) \end{pmatrix}, t_k = k\Delta, k = 0,1,\ldots \tag{4.36}$$

which is of the form of equations (4.1)-(4.2).

Numerical experimentation was carried out for the following set of para= meters: $t_k = k\Delta$, $k = 0,1,2,\ldots,5000$; $\Delta = 10^{-3}$; $\sigma_1 = \sigma_2 = \sigma_3 = 0.01$; $\gamma_1 = \gamma_2 = 5\cdot 10^{-3}$; $a_o = 0.2, 1, 50$; $b = 50$, $\rho = 1$; $a = 0.1, 1.0$; $X(0) = (1,1,0)'$, $\hat{\zeta}(0|0) = (0,0,0)'$ and

$$P(0|0) = \begin{pmatrix} 1 & 1 & 0 \\ 1 & 1 & 0 \\ 0 & 0 & 0 \end{pmatrix} \tag{4.37}$$

Typical extracts from the numerical results are presented in Tables 4.1-4.2. Here, the following notations

$$|X(k)| = [\sum_{i=1}^{3} X_i^2(k)]^{\frac{1}{2}} \text{ and } |\hat{\zeta}(k|k)| = [\sum_{i=1}^{3} \hat{\zeta}_i^2(k|k)]^{\frac{1}{2}}. \tag{4.38}$$

are used.

TABLE 4.1: $X(k)$, $\hat{\zeta}(k|k)$, $|X(k)|$, $|\hat{\zeta}(k|k)|$ as functions of k for a = 0.1

k	$X_1(k)$	$\hat{\zeta}_1(k\|k)$	$X_2(k)$	$\hat{\zeta}_2(k\|k)$	$X_3(k)$	$\|X(k)\|$	$\|\hat{\zeta}(k\|k)\|$
50	-.9907	-.9907	-.1452	-.1980	.0014	1.0013	1.0103
250	.7423	.7164	.6680	.7086	.0041	.9986	1.0076
500	.7495	.7239	.6629	.7010	.0079	1.0006	1.0077
750	.7134	.6882	.6975	.7349	.0066	.9977	1.0068
1000	.7394	.7124	.6718	.7125	.0035	.9990	1.0075
1250	.7522	.7263	.6568	.6978	.0037	.9986	1.0072
1500	.7359	.7097	.6765	.7142	.0087	.9996	1.0069
1750	.6822	.6601	.7304	.7617	.0151	.9995	1.0079
2000	.5809	.5566	.8121	.8393	.0173	.9986	1.0071
2250	.5072	.4738	.8606	.8907	.0094	.9990	1.0089
2500	.4684	.4353	.8837	.9087	.0137	.9997	1.0076
2750	.3823	.3495	.9230	.9441	.0162	.9992	1.0067
3000	.2791	.2421	.9609	.9785	.0138	1.0007	1.0080
3250	.1749	.1384	.9863	.9979	.0176	1.0019	1.0075
3500	.0246	-.0095	.9996	1.0074	.0220	1.0002	1.0075
3750	-.1900	-.2225	.9796	.9830	.0211	.9981	1.0078
4000	-.4105	-.4432	.9109	.9053	.0223	.9994	1.0079
4250	-.5626	-.5986	.8265	.8108	.0191	1.0000	1.0078
4500	-.7022	-.7322	.7123	.6929	.0207	1.0004	1.0081
4750	-.8370	-.8622	.5463	.5211	.0214	.9998	1.0074
5000	-.9138	-.9154	.4043	.4195	.0186	.9993	1.0079

and $\hat{\zeta}_3(k|k) = 0.0$ for all k.

In all of the runs it turned out that

$$|X(k)| \cong 1 \text{ and } |\hat{\zeta}(k|k)| \cong 1 \text{ for } k \geq 50. \tag{4.39}$$

TABLE 4.2: $X(k)$, $\hat{\zeta}(k|k)$, $|X(k)|$, $|\hat{\zeta}(k|k)|$ as functions of k for a=1

k	$X_1(k)$	$\hat{\zeta}_1(k\|k)$	$X_2(k)$	$\hat{\zeta}_2(k\|k)$	$X_3(k)$	$\|X(k)\|$	$\|\hat{\zeta}(k\|k)\|$
50	-.9907	-.9907	-.1453	-.1981	.0014	1.0013	1.0103
250	.7432	.7171	.6670	.7079	.0037	.9986	1.0076
500	.7575	.7316	.6537	.6929	.0065	1.0006	1.0077
750	.7406	.7153	.6685	.7086	.0036	.9977	1.0068
1000	.7891	.7638	.6127	.6570	.0006	.9990	1.0075
1250	.8225	.7999	.5664	.6122	.0006	.9987	1.0073
1500	.8340	.8125	.5511	.5946	.0047	.9996	1.0069
1750	.8245	.8072	.5648	.6037	.0096	.9994	1.0080
2000	.7943	.7755	.6051	.6429	.0098	.9986	1.0074
2250	.8025	.7784	.5946	.6408	.0010	.9988	1.0083
2500	.8379	.8177	.5455	.5884	.0053	.9999	1.0073
2750	.8459	.8280	.5306	.5719	.0071	.9986	1.0063
3000	.8520	.8326	.5242	.5680	.0039	1.0004	1.0079
3250	.8607	.8430	.5118	.5519	.0070	1.0014	1.0076
3500	.8511	.8349	.5256	.5656	.0102	1.0004	1.0082
3750	.8185	.8005	.5707	.6125	.0074	.9978	1.0080
4000	.7910	.7702	.6106	.6508	.0068	.9993	1.0083
4250	.8054	.7821	.5940	.6365	.0032	1.0008	1.0084
4500	.8106	.7896	.5854	.6264	.0044	.9999	1.0078
4750	.8140	.7928	.5799	.6212	.0043	.9995	1.0072
5000	.8447	.8510	.5341	.5405	.0015	.9994	1.0079

and $\hat{\zeta}_3(k|k) = 0.0$ for all k.
Throughout all the runs it turned out that the relative errors

$$\varepsilon_{ri}(k) \triangleq |X_i(k) - \hat{\zeta}_i(k|k)|/|X_i(k)| \ , \ i=1,2,3, \ \ k=0,1,2,.. \qquad (4.40)$$

obtained their minimum values at instants k where $|X_i(k)|$ obtained their maximum values.

Since for $a_o = 50$, $\{X(k)\}$ and $\{\hat{\zeta}(k|k)\}$ are signals with very large variations, it was preferable to represent the results in tables rather than in graphs.

4.5 EXAMPLE 4.2: A THREE PHASE SINE WAVES GENERATOR

Consider the noise-driven nonlinear stochastic system given by

$$\begin{cases} dx_1 = [a_o(x_2-x_3) + bF(x)x_1]dt + \sigma_1 dW_1 \\ dx_2 = [a_o(x_3-x_1) + bF(x)x_2]dt + \sigma_2 dW_2 \qquad t > 0 \\ dx_3 = [a_o(x_1-x_2) + bF(x)x_3]dt + \sigma_3 dW_3 \end{cases} \tag{4.41}$$

$$F(x) = 1 - \mu[(x_1-x_2)^2 + (x_2-x_3)^2 + (x_3-x_1)^2] \tag{4.42}$$

with the observation process given by

$$y_i(t_k) = x_i(t_k) + v_i(t_k) \ , \ i=1,2,3 \quad , \quad t_k=k\Delta, \ k=0,1,2,... \tag{4.43}$$

where a_o, b, μ, σ_1, σ_2 and σ_3 are given positive numbers; $\mathbf{W} \triangleq \{W(t) = (W_1(t), W_2(t), W_3(t)), t \geq 0\}$ is an $\mathbb{R}^3$-valued standard Wiener process; $\mathbf{V} \triangleq \{v(t_k), k=0,1,2,...\}$ is a sequence of independent $\mathbb{R}^3$-valued Gaussian random elements with

$$Ev(t_k) = 0 \text{ and } E[v(t_k)v'(t_\ell)] = \delta_{k\ell}\, \gamma_o^2 I_3 \tag{4.44}$$

where γ_o is a given positive number. It is assumed that $\mathbf{W}$ and $\mathbf{V}$ are mutually independent.

Equations (4.41), where $\sigma_1 = \sigma_2 = \sigma_3 = 0$, have been dealt with in Kaplan and Bachar [4.5] and Kaplan [4.6]. It is shown there that $X_t = (x_1(t), x_2(t), x_3(t))$, the solution to (4.41) (with $\sigma_1 = \sigma_2 = \sigma_3 = 0$) has a stable steady state solution (except for the case $x_1(0) = x_2(0) = x_3(0)$)

$$\begin{cases} x_1(t) = B + A\sin(a_o\sqrt{3}\,t + \phi) \\ x_2(t) = B + A\sin(a_o\sqrt{3}\,t - 2\pi/3 + \phi) \\ x_3(t) = B + A\sin(a_o\sqrt{3}\,t - 4\pi/3 + \phi) \end{cases} \tag{4.45}$$

Hence, equations (4.41), in the case where $\sigma_i \neq 0$, i=1,2,3, represent a noise-driven three phase sine waves generator.

Equations (4.41) and (4.43) can be written in the following form

$$\begin{pmatrix} dx_1 \\ dx_2 \\ dx_3 \end{pmatrix} = \begin{pmatrix} bF(x) & a_o & -a_o \\ -a_o & bF(x) & a_o \\ a_o & -a_o & bF(x) \end{pmatrix} \begin{pmatrix} x_1 \\ x_2 \\ x_3 \end{pmatrix} dt + \begin{pmatrix} \sigma_1 & 0 & 0 \\ 0 & \sigma_2 & 0 \\ 0 & 0 & \sigma_3 \end{pmatrix} \begin{pmatrix} dW_1 \\ dW_2 \\ dW_3 \end{pmatrix}$$

$t > 0$ (4.46)

and

$$\begin{pmatrix} y_1(t_k) \\ y_2(t_k) \\ y_3(t_k) \end{pmatrix} = \begin{pmatrix} 1 & 0 & 0 \\ 0 & 1 & 0 \\ 0 & 0 & 1 \end{pmatrix} \begin{pmatrix} x_1(t_k) \\ x_2(t_k) \\ x_3(t_k) \end{pmatrix} + \begin{pmatrix} v_1(t_k) \\ v_2(t_k) \\ v_3(t_k) \end{pmatrix},$$

$$t_k = k\Delta, \quad k=0,1,2,\ldots, \tag{4.47}$$

which is of the form of equations (4.1)-(4.2).

Numerical experimentation was carried out for the following set of para= meters : $t_k = k\Delta$, $k = 0,1,2,\ldots,5000$; $\Delta = 10^{-3}$; $\sigma_1 = \sigma_2 = \sigma_3 = 0.01$; $\gamma_o = 5.10^{-3}$; $\mu = 1$; $a_o = 0.2, 1, 50$; $b = 10$; $x(0) = (1,1,0)'$, $\hat{\zeta}(0|0) = (0,0,0)'$ and $P(0|0)$ is given by (4.37).

Typical extracts from the numerical results are presented in Fig.4.2 and Table 4.3. As in the previous example, since for $a_o = 50$, $\{X(k)\}$ and $\{\hat{\zeta}(k|k)\}$ are signals with very large variations, it was preferable to re= present the results in tables rather than in graphs.

TABLE 4.3 : X(k), $\hat{\zeta}(k|k)$, |X(k)|, $|\hat{\zeta}(k|k)|$ and $\varepsilon(k)$ as functions of k for a_o = 50

k	$X_1(k)$	$\hat{\zeta}_1(k\|k)$	$X_2(k)$	$\hat{\zeta}_2(k\|k)$	$X_3(k)$	$\hat{\zeta}_3(k\|k)$	$\|X(k)\|$	$\|\hat{\zeta}(k\|k)\|$	$\varepsilon(k)$
50	.0041	-.0213	.8395	.8815	.7174	.6676	1.1043	1.1060	.0700
250	.3893	.3390	.1142	.1171	.9162	.9438	1.0020	1.0096	.0575
500	.3926	.4228	.9097	.9070	.1062	.0555	.9965	1.0023	.0590
750	.7083	.6748	.0039	-.0257	.7066	.7456	1.0005	1.0060	.0594
1000	.1049	.1064	.9066	.9349	.3877	.3348	.9916	.9987	.0600
1250	.9111	.9110	.1099	.0617	.3784	.4097	.9927	1.0008	.0574
1500	.0062	-.0218	.7048	.7441	.7182	.6897	1.0063	1.0148	.0560
1750	.9199	.9481	.4102	.3573	.1140	.1173	1.0137	1.0199	.0599
2000	.1273	.0762	.3874	.4177	.9283	.9286	1.0139	1.0211	.0594
2250	.7085	.7515	.7328	.7024	.0165	-.0134	1.0194	1.0287	.0606
2500	.4288	.3788	.1205	.1228	.9306	.9588	1.0317	1.0382	.0574
2750	.4078	.4363	.9471	.9475	.1533	.1017	1.0425	1.0481	.0589
3000	.7523	.7195	.0336	.0020	.7261	.7641	1.0461	1.0495	.0593
3250	.1265	.1281	.9345	.9636	.4353	.3837	1.0387	1.0451	.0593
3500	.9482	.9496	.1548	.1052	.4008	.4278	1.0408	1.0468	.0568
3750	.0316	-.0007	.7154	.7536	.7566	.7279	1.0418	1.0478	.0577
4000	.9351	.9635	.4421	.3898	.1255	.1248	1.0419	1.0468	.0595
4250	.1513	.1007	.3859	.4154	.9449	.9464	1.0318	1.0384	.0586
4500	.7025	.7431	.7513	.7213	.0218	-.0099	1.0288	1.0356	.0591
4750	.4444	.3925	.1105	.1104	.9214	.9510	1.0289	1.0348	.0598
4950	.9543	.9637	.2020	.1483	.3118	.3366	1.0240	1.0315	.0599

TABLE 4.4: $X(k)$, $\hat{\zeta}(k|k)$, $|X(k)|$, $|\hat{\zeta}(k|k)|$ and $\varepsilon(k)$ as functions of k for $a_o = 0.2$

k	$X_1(k)$	$\hat{\zeta}_1(k\|k)$	$X_2(k)$	$\hat{\zeta}_2(k\|k)$	$X_3(k)$	$\hat{\zeta}_3(k\|k)$	$\|X(k)\|$	$\|\hat{\zeta}(k\|k)\|$	$\varepsilon(k)$
50	.7911	.7870	.7731	.7724	.0003	-.0007	1.1061	1.1027	.0043
250	.7486	.7497	.6685	.6684	.0059	.0058	1.0036	1.0044	.0011
500	.7859	.7848	.6294	.6285	.0134	.0128	1.0070	1.0056	.0015
750	.8130	.8121	.5901	.5896	.0216	.0203	1.0048	1.0038	.0017
1000	.8409	.8395	.5531	.5531	.0327	.0332	1.0070	1.0060	.0014
1250	.8572	.8579	.5048	.5063	.0414	.0430	.9957	.9971	.0023
1500	.8787	.8804	.4663	.4658	.0612	.0635	.9966	.9980	.0029
1750	.8985	.8984	.4272	.4268	.0900	.0896	.9990	.9987	.0005
2000	.9191	.9196	.3958	.3954	.1169	.1181	1.0075	1.0079	.0013
2250	.9305	.9332	.3635	.3654	.1413	.1407	1.0089	1.0120	.0034
2500	.9496	.9498	.3280	.3292	.1777	.1780	1.0202	1.0208	.0012
2750	.9578	.9573	.3010	.3013	.2069	.2074	1.0250	1.0248	.0008
3000	.9554	.9540	.2605	.2594	.2354	.2337	1.0178	1.0159	.0024
3250	.9499	.9505	.2189	.2192	.2723	.2730	1.0122	1.0129	.0009
3500	.9496	.9499	.1864	.1879	.3163	.3160	1.0181	1.0186	.0016
3750	.9402	.9393	.1555	.1529	.3522	.3553	1.0160	1.0158	.0041
4000	.9259	.9255	.1241	.1231	.3907	.3891	1.0126	1.0115	.0020
4250	.9010	.9016	.0888	.0891	.4209	.4218	.9984	.9994	.0011
4500	.8837	.8843	.0662	.0668	.4592	.4602	.9981	.9991	.0013
4750	.8608	.8600	.0472	.0465	.4993	.5002	.9962	.9960	.0014
4950	.8376	.8388	.0285	.0295	.5315	.5324	.9924	.9939	.0018

4.6 ESTIMATION WITH UNCERTAIN OBSERVATIONS

4.6.1 Introduction

Consider an $\mathbb{R}^m$-valued Markov process $\zeta_x = \{\zeta_x(t),\ t \geq 0\}$ satisfying the equation

$$\zeta_x(t) = x + \int_0^t A(\zeta_x(s))\zeta_x(s)ds + BW(t),\ t \geq 0,\ x \in \mathbb{R}^m \qquad (4.48)$$

where $A : \mathbb{R}^m \to \mathbb{R}^{m\times m}$ and $B \in \mathbb{R}^{m\times m}$ are given matrices satisfying the con= ditions stated in Section 4.1, and suppose that the observation of ζ_x is given by

$$y(t_k) = \gamma(t_k)H(t_k)\zeta_x(t_k) + v(t_k),\ t_k = k\Delta,\ k=0,1,2,\ldots,y(t_k) \in \mathbb{R}^p, \qquad (4.49)$$

where, on a probability space (Ω,F,P):(i) $\mathbf{W} = \{W(t) = (W_1(t),\ldots,W_m(t)),\ t \geq 0\}$ is an $\mathbb{R}^m$-valued standard Wiener process; (ii) $\mathbf{V} = \{v(t_k) : t_k = k\Delta,\ k=0,1,2,\ldots\}$ is an $\mathbb{R}^p$-valued white Gaussian sequence with

$$Ev(t_k) = 0 \text{ and } E[v(t_k)v'(t_\ell)] = \delta_{k\ell}R(t_k);\ k,\ell=0,1,2,\ldots \qquad (4.50)$$

where $\{R(t_k)\}$ are given m×m symmetric and positive definite matrices; (iii) $\{\gamma(t_k);\ t_k = k\Delta,\ k=0,1,2,\ldots\}$ is an $\{0,1\}$-valued sequence of inde= pendent random variables satisfying

$$p(k) \triangleq P(\gamma(t_k) = 1)\quad,\quad k=0,1,2,\ldots \qquad (4.51)$$

$$q(k) = 1 - p(k) \triangleq P(\gamma(t_k) = 0)\quad,\quad k=0,1,2,\ldots\ . \qquad (4.52)$$

It is assumed that **W**, **V** and $\{\gamma(t_k),\ k=0,1,2,\ldots\}$ are mutually independent, and that the sequence $\{p(k)\}_{k=0}^{\infty}$ is known. Let

$$Y^k = \{y(t_0),\ y(t_1),\ldots,y(t_k)\}\ . \qquad (4.53)$$

The problem dealt with in this section is to find an approximation $\hat{\zeta}(k|k)$ to

$$\hat{\zeta}_x(t_k|t_k) = E[\zeta_x(t_k)|\sigma(Y^k)], \quad k=1,2,\dots \quad . \tag{4.54}$$

The problem of computing approximations to $\hat{\zeta}_x(t_k|t_k)$ has been treated mainly in the case where the state $\{\zeta_x(t_k),\ k=1,2,\dots\}$ is determined by a discrete-time linear system. In Nahi, [4.7], the best linear mean-square estimator of $\zeta_x(t_k)$ given Y^k is derived, and its asymptotic beha= viour is discussed in Tugnait, [4.8]. In Sawaragi et al. [4.9], an approximate minimum variance estimator of $\zeta_x(t_k)$ given Y^k is derived and the case where $p(k)=p$ for all k, but p is unknown, is also treated. Other approaches to the problem can be found for example in Askar et al. [4.10], and in Haddad and Tugnait, [4.11].

In this section, using the results of [4.7] and Section 4.2, an algorithm for computing an approximation $\hat{\zeta}(k|k)$ to $\hat{\zeta}_x(t_k|t_k)$ (for the system and measurements given by (4.48)-(4.52)), is suggested. It is assumed here that $E\zeta_x(0) = 0$.

4.6.2 An Algorithm for Computing $\hat{\zeta}(k|k)$

Note that $H(k) \triangleq H(t_k)$ and $R(k) \triangleq R(t_k)$, $k=0,1,\dots$.

Algorithm 4.6

$$\hat{\zeta}(0|0) = E\zeta_x(0) \quad , \quad P(0|0) = P(0) = S(0) = E[\zeta_x(0)\zeta_x'(0)]$$

1. $k=0$

   Compute the following elements

2. $$\hat{\zeta}(k+1|k) = \hat{\zeta}(k|k) + A(\hat{\zeta}(k|k))\hat{\zeta}(k|k)\Delta \tag{4.55}$$

3. $$\psi(k+1,k) = [I_m - (\Delta/2)A(\hat{\zeta}(k+1|k))]^{-1}[I_m + (\Delta/2)A(\hat{\zeta}(k|k))] \tag{4.56}$$

4. $$\Gamma(k) = \sqrt{\Delta}\,\psi(k+1,k)B \tag{4.57}$$

5. $$P(k+1|k) = \psi(k+1,k)P(k|k)\psi'(k+1,k) + \Gamma(k)\Gamma'(k) \tag{4.58}$$

6. $S(k+1) = \psi(k+1,k)S(k)\psi'(k+1,k) + \Gamma(k)\Gamma'(k)$ (4.59)

7. $K(k+1) = p(k+1)P(k+1|k)H'(k+1)[R(k+1) + p^2(k+1)H(k+1)P(k+1|k)H'(k+1)$ (4.60)

$+ p(k+1)(1 - p(k+1))H(k+1)S(k+1)H'(k+1)]^{-1}$

8. $\hat{\zeta}(k+1|k+1) = \hat{\zeta}(k+1|k) + K(k+1)[y(t_{k+1}) - p(k)H(k)\hat{\zeta}(k+1|k)]$ (4.61)

9. $P(k+1|k+1) = P(k+1|k) - p(k+1)K(k+1)H(k+1)P(k+1|k)$ (4.62)

10. If k=N stop. Otherwise k:=k+1 and go to 2.

The numerical experimentation was carried out in this section in the same manner as that described at the end of Section 4.3. However, in order to compute the sequence $\{y(t_k)\}_{k=0}^{N}$ we need (in addition to $\{X(k)\}_{k=0}^{N}$) to generate the sequence $\{\gamma(t_k)\}_{k=0}^{N}$. This is done by applying the following procedure:

Let $\{u(k)\}_{k=0}^{\infty}$ be a sequence of independent Gaussian random variables with

$$Eu(k) = 0 \text{ and } E[u(k)u'(\ell)] = \delta_{k\ell}\ , \quad k,\ell=0,1\ldots \tag{4.63}$$

and assume that $\{u(k)\}_{k=0}^{\infty}$, $\{W(k)\}_{k=0}^{\infty}$ and $\{v(t_k)\}_{k=0}^{\infty}$ are mutually inde= pendent. Then $\{\gamma(t_k), k=0,1,\ldots\}$ are determined by

$$\gamma(t_k) = \begin{cases} 1 & \text{if } u(k) \le z(k) \\ \\ 0 & \text{if } u(k) > z(k) \end{cases} \qquad k=0,1,\ldots \tag{4.64}$$

where, for each $k \ge 0$

$$p(k) = (1/\sqrt{2\pi}) \int_{-\infty}^{z(k)} \exp(-u^2/2)du. \tag{4.65}$$

4.6.3 Examples

(i) Consider the system given by equations (4.41)-(4.42) with the ob= servation process given by

$$y_i(t_k) = \gamma(t_k)x_i(t_k) + v_i(t_k), \quad t_k = k\Delta, \; k=0,1,2,\ldots,i=1,2,3 \tag{4.66}$$

where $(R(t_k))_{ij} = \delta_{ij}\, 25\cdot 10^{-6}$, i,j=1,2,3.

Numerical experimentation was carried out for the following set of para= meters: $\sigma_1 = \sigma_2 = \sigma_3 = 0.01$, $a_o = 0.2$; $b = 10$, $\mu = 1$, $\Delta = 10^{-3}$; $p(k)=p$ for all $k \geq 0$, where p=0.6,0.8,1. A typical extract from the numerical results is illustrated in Figs. 4.1a and 4.1b, where the plots of $\varepsilon(k) = [\sum_{i=1}^{3} (\hat{\zeta}_i(k|k) - X_i(k))^2]^{\frac{1}{2}}$, as functions of k, are given. The graphs in these figures have been plotted on the set of points $\{t'_k = 50k\Delta : k=0,1,\ldots,100\}$.

(ii) Consider the dynamical system (see also (3.114))

$$dx_i = -\alpha x_i^2 dt + \sigma_i dW_i, \; t > 0, \; i=1,2 \tag{4.67}$$

with the uncertain and noisy observations given by

$$y_i(t_k) = \gamma(t_k)x_i(t_k) + v_i(t_k), \; t_k = k\Delta, \; k=0,1,\ldots, \quad i=1,2. \tag{4.68}$$

Numerical experimentation was carried out for the following set of para= meters: $\zeta_{x1}(0) = \zeta_{x2}(0) = 100$; $\sigma_1 = \sigma_2 = 7$; $(R(t_k))_{ij} = 9\delta_{ij}$, i,j=1,2; $\alpha = 5\cdot 10^{-5}$; $\Delta = 10^{-3}$; $p(k) = p$ for all $k \geq 0$ where p = 0.6, 0.8,1. Typi= cal extracts from the numerical results are given in Figs. 4.2-4.4. In these figures $\varepsilon(k) \triangleq [\sum_{i=1}^{3} (\hat{\zeta}_i(k|k) - X_i(k))^2]^{\frac{1}{2}}$. The graphs in these figures have been plotted on the set of points $\{t'_k = 20k\Delta : k=0,1,\ldots,250\}$.

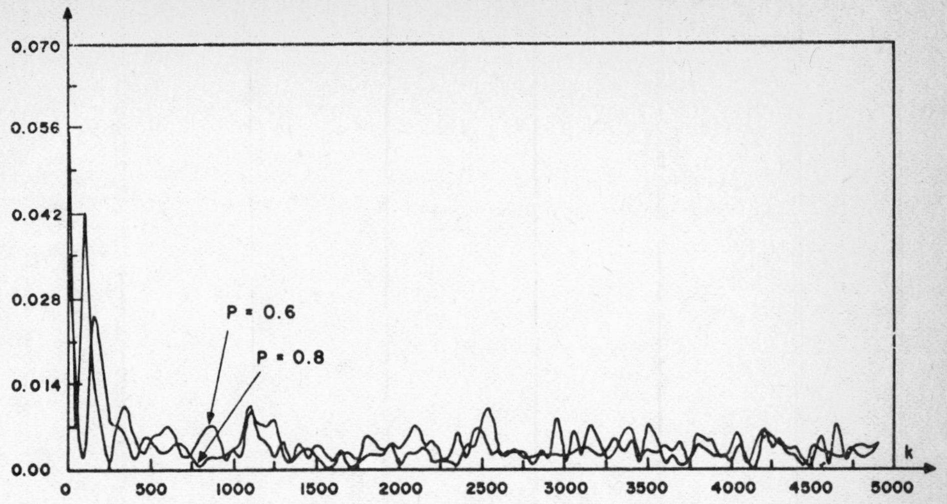

Fig.4.1a: $\varepsilon(k)$ as functions of k where p=0.6,0.8; for the system given by equations (4.41)-(4.42) and (4.66).

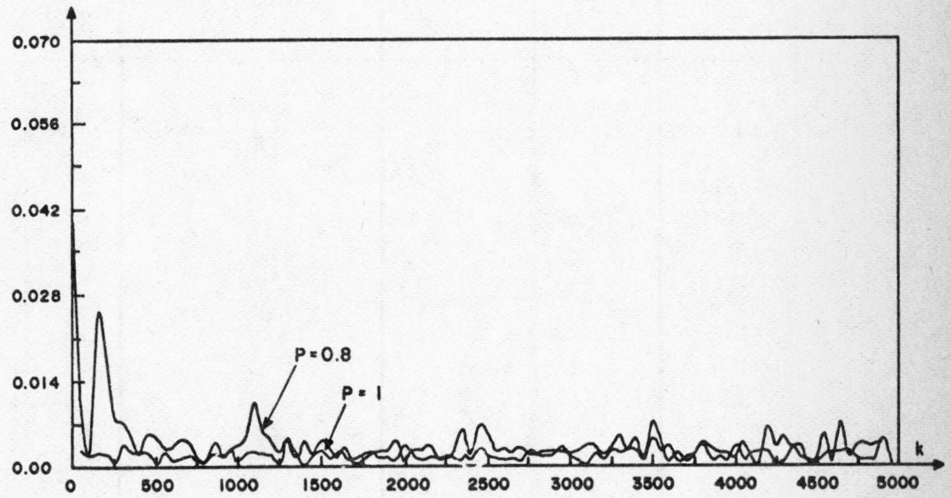

Fig.4.1b: $\varepsilon(k)$ as functions of k where p=0.8,1; for the system given by equations (4.41)-(4.42) and (4.66).

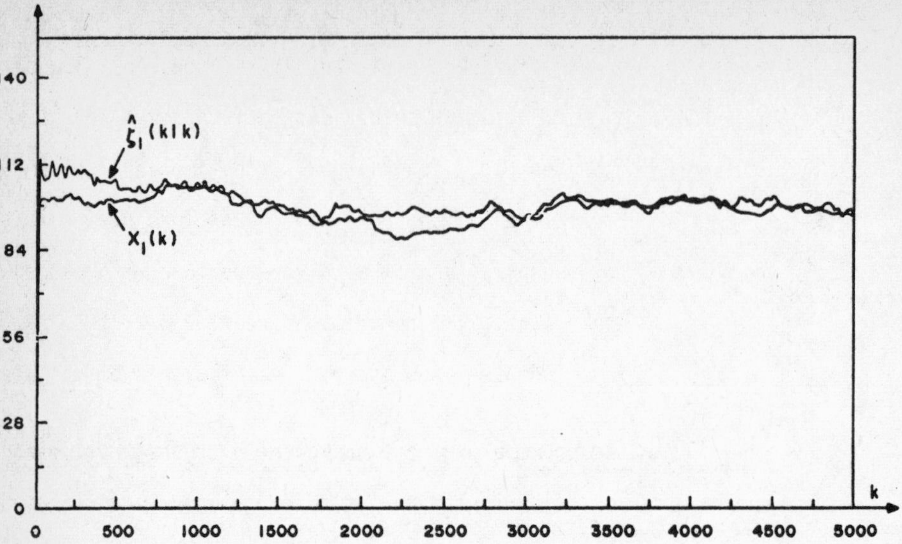

Fig.4.2a: $X_1(k)$ and $\hat{\zeta}_1(k|k)$ as functions of k for the system given by equations (4.67)-(4.68) and where p=0.6.

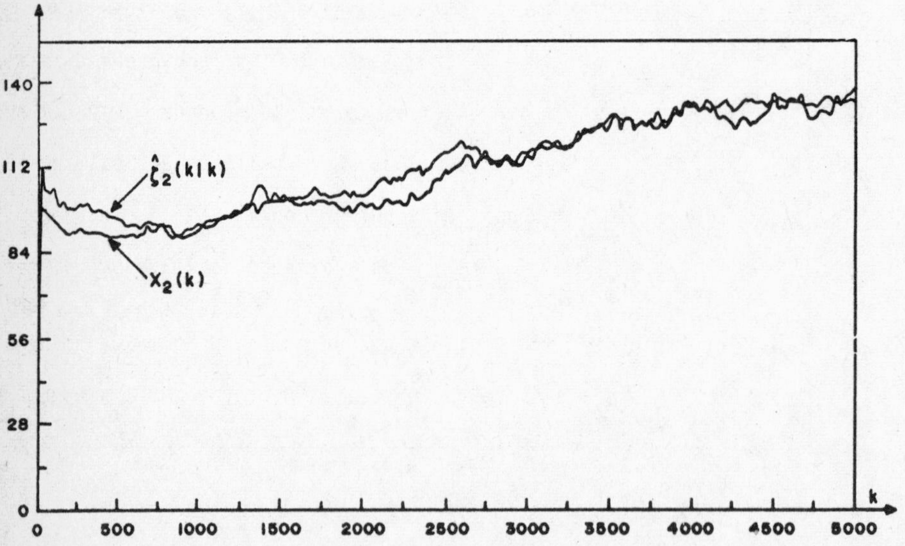

Fig.4.2b: $X_2(k)$ and $\hat{\zeta}_2(k|k)$ as functions of k for the system given by equations (4.67)-(4.68) and where p=0.6.

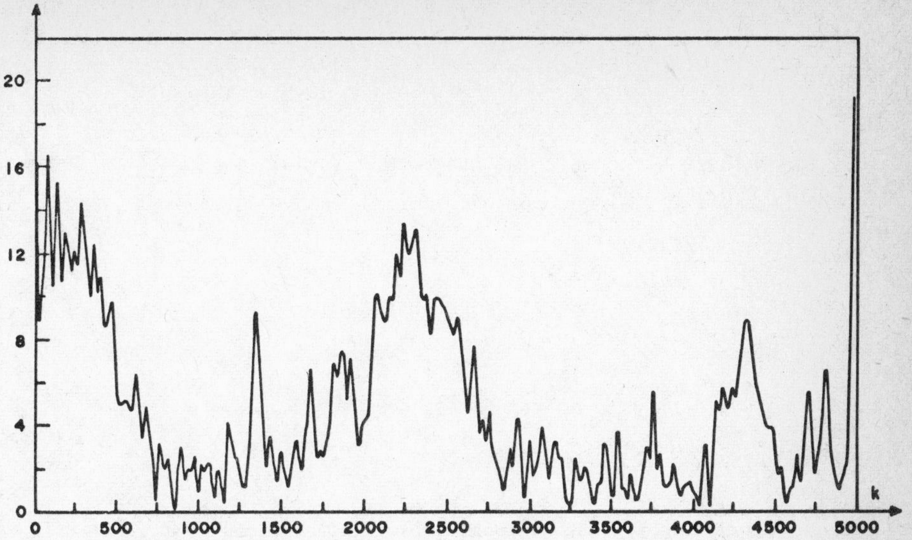

Fig.4.2c: $\varepsilon(k)$ as a function of k for the system given by equations (4.67)-(4.68) and where p=0.6.

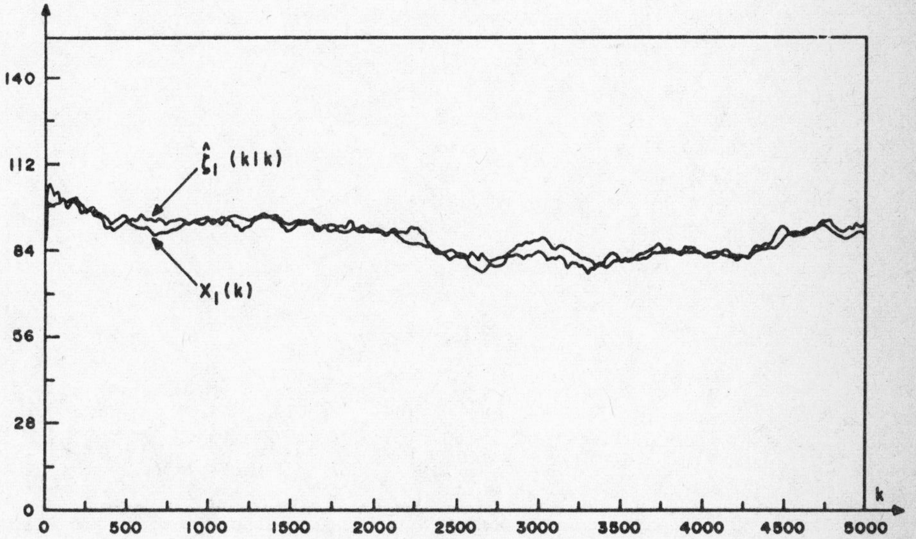

Fig.4.3a: $X_1(k)$ and $\hat{\zeta}_1(k|k)$ as functions of k for the system given by equations (4.67)-(4.68) and where p=0.8.

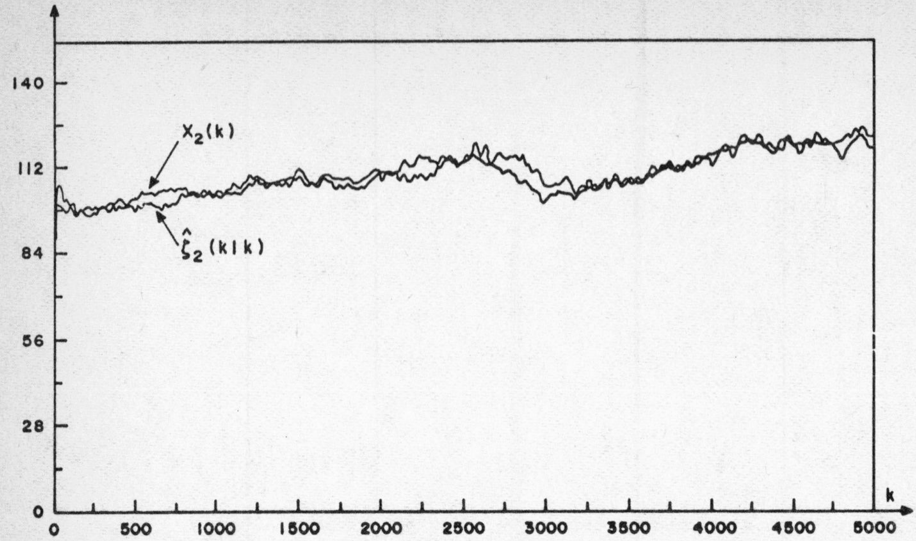

Fig.4.3b: $X_2(k)$ and $\hat{\zeta}_2(k|k)$ as functions of k for the system given by equations (4.67)-(4.68) and where p=0.8.

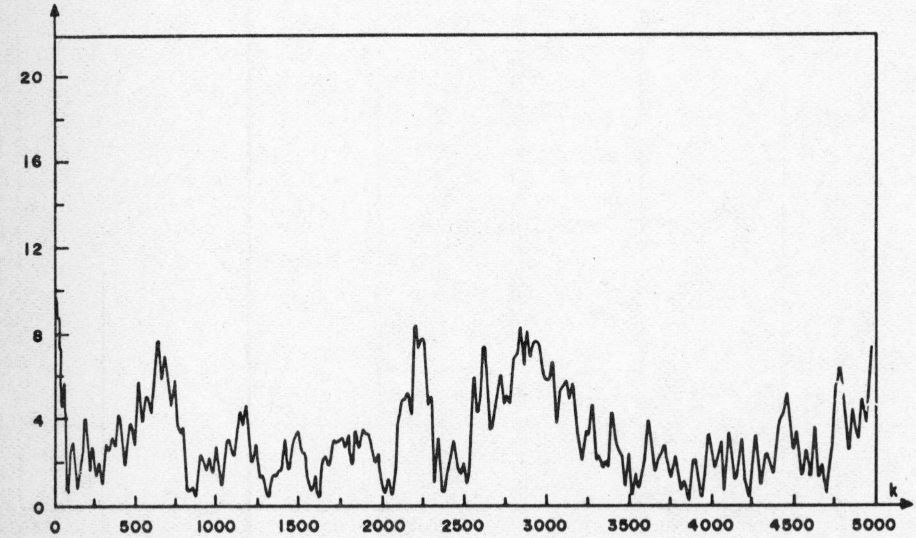

Fig.4.3c: $\varepsilon(k)$ as a function of k for the system given by equations (4.67)-(4.68) and where p=0.8.

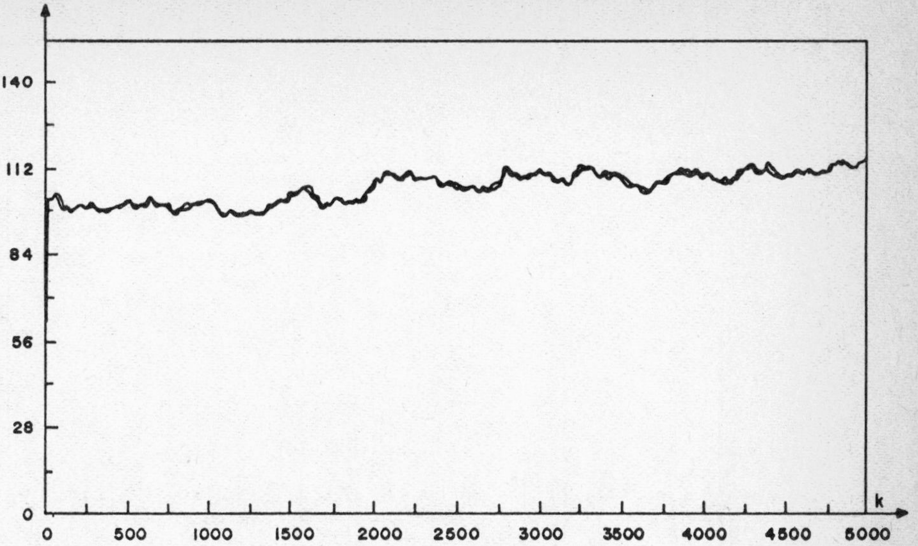

Fig.4.4a: $X_1(k)$ and $\hat{\zeta}_1(k|k)$ as functions of k for the system given by equations (4.67)-(4.68) and where p=1.

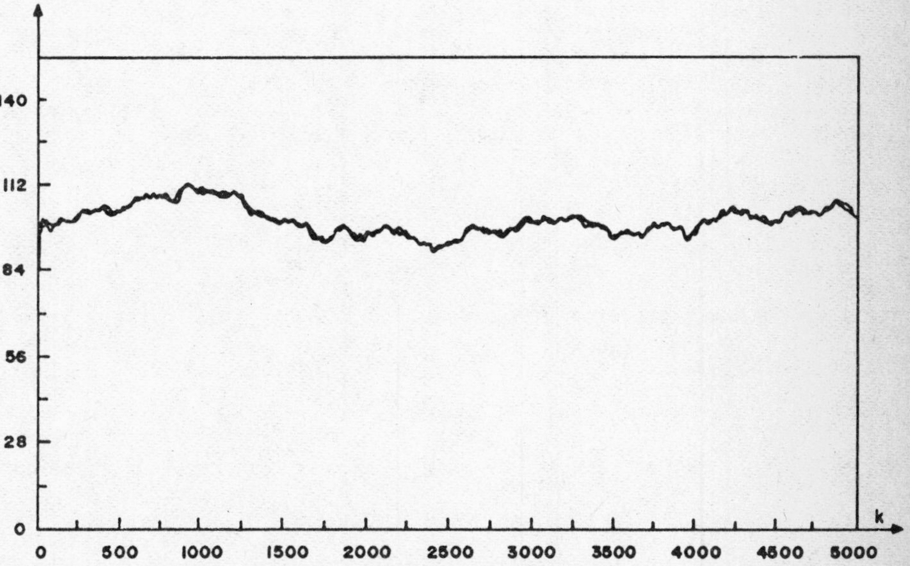

Fig.4.4b: $X_2(k)$ and $\hat{\zeta}_2(k|k)$ as functions of k for the system given by equations (4.67)-(4.68) and where p=1.

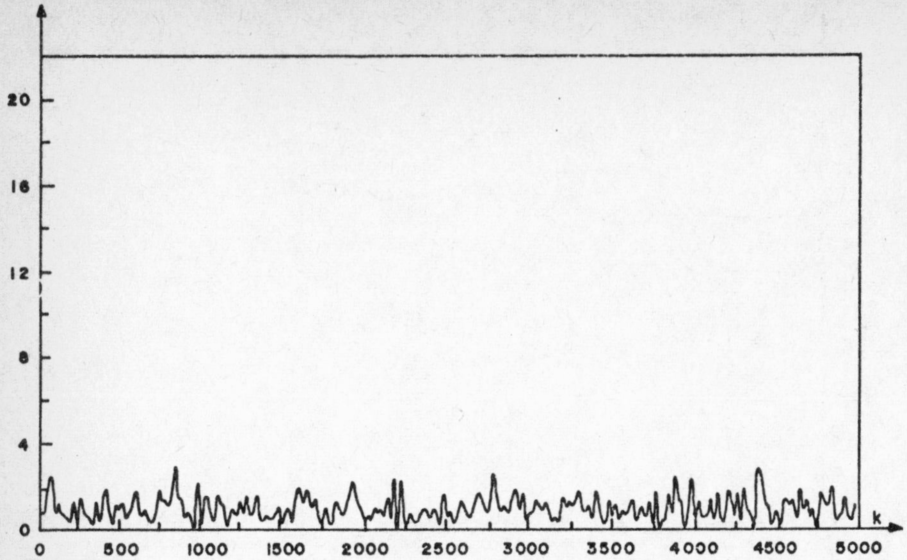

Fig.4.4c: $\varepsilon(k)$ as function of k for the system given by equations (4.67)-(4.68) and where p=1.

4.7 REMARKS

The results obtained throughout the numerical experimentation suggest that $\{\hat{\zeta}(k \mid k)\}_{k=1}^{N}$ is a good state estimator for systems given by (4.1)-(4.2) for which the equation

$$dx/dt = A(x)x, \quad t > 0 \tag{4.69}$$

has a stable limit cycle. This property is demonstrated in Example 4.1 (equations (4.31)-(4.34)). The three-phase sine-wave generator introduced in Section 4.5 also possesses this property, as is shown below.

Denote $u_1 \triangleq x_2 - x_3$, $u_2 \triangleq x_3 - x_1$ and $u_3 \triangleq x_1 - x_2$ and consider equations (4.41) for $\sigma_1 = \sigma_2 = \sigma_3 = 0$, i.e.

$$\begin{cases} dx_1/dt = a_o u_1 + bF(u)x_1 \\ dx_2/dt = a_o u_2 + bF(u)x_2 \qquad t > 0 \\ dx_3/dt = a_o u_3 + bF(u)x_3, \end{cases} \tag{4.70}$$

where $F(u) = 1 - \mu u^2$, $u^2 = \sum_{i=1}^{3} u_i^2$; or

$$\begin{cases} du_1/dt = a_o(u_2 - u_3) + bF(u)u_1 \\ du_2/dt = a_o(u_3 - u_1) + bF(u)u_2 \quad , \quad t > 0 \\ du_3/dt = a_o(u_1 - u_2) + bF(u)u_3. \end{cases} \tag{4.71}$$

By using (4.71), it follows that

$$dF(x)/dt = dF(u)/dt = -2\mu bF(u)u^2 \quad , \quad t > 0, \tag{4.72}$$

from which the following equation is obtained

$$du/dt = bu(1 - \mu u^2), \quad t > 0 \ . \tag{4.72}$$

The solution to equation (4.72) leads to

$$u(t) = u_o e^{bt}/[\mu u_o^2(e^{2bt} - 1) + 1]^{\frac{1}{2}} \ , t \geq 0 \quad , \tag{4.73}$$

where $u_o = u(0)$. Hence, for $u_o \neq 0$

$$\lim_{t \to \infty} u(t) = 1/\sqrt{\mu} \ , \tag{4.74}$$

and, consequently,

$$\lim_{t \to \infty} A(X(t)) = \begin{pmatrix} 0 & a_o & -a_o \\ -a_o & 0 & a_o \\ a_o & -a_o & 0 \end{pmatrix} \tag{4.75}$$

Three-phase sine-wave generators are used as references for modern ac power convertors [4.12-4.13]. In addition, their model (eqns. (4.41)-(4.42) where $\sigma_1 = \sigma_2 = \sigma_3 = 0$) seems important for simulation work, especially for the simulation of power systems and multimachine systems. The model enables a simplified simulation of such systems, which nevertheless em=

braces a completely three-phase dynamics. This model may also be of help to designers of three-phase electronic generators [4.14]. These generators are found applicable for the development, investigation and measurements of control and power electronic systems.

4.8 REFERENCES

4.1 I.I.Gihman and A.V. Skorohod, *Stochastic Differential Equations*, Springer-Verlag, Berlin, 1972.

4.2 A.H. Jazwinski, *Stochastic Processes and Filtering Theory*, Academic Press, New York, 1970.

4.3 Y.Yavin, A discrete Kalman filter for a class of nonlinear stochastic systems, *Int.J.Systems Sci.*, 13, pp 1233-1246, 1982.

4.4 S.C. Bird and J.A. Folchi, Time base requirements for a waveform recorder, *Hewlett-Packard J.*, 33, No.11, pp 29-34, November 1982.

4.5 B.Z.Kaplan and S.T.Bachar, A new stabilized genrator model for three phase sinewaves, *Mathematics and Computers in Simulation XXI*, pp 207-208, 1979.

4.6 B.Z.Kaplan, Stabilized generators of precise triangular waves, square waves and trapezoidal waves in three phases, *J. of the Franklin Institute*, 309, pp 379-387, 1980.

4.7 N.E.Nahi, Optimal recursive estimation with uncertain observation, *IEEE Trans. on Information Theory*, IT-15, pp 457-462, 1969.

4.8 J.K.Tugnait, Asymptotic stability of the MMSE linear filter for systems with uncertain observations, *IEEE Trans. on Information Theory*, IT-27, pp 247-250, 1981.

4.9 Y.Sawaragi, T. Katayama and S.Fujishige, Sequential state estimation with interrupted observation, *Information and Control*, 21, pp 56-71, 1972.

4.10 M. Askar, H. Derin and H.O.Yurtseven, On joint detection and esti= mation of Gauss-Markov processes, *Int.J. Control*, 30, pp 1031-1042, 1979.

4.11 A.H. Haddad and J.K. Tugnait, On state estimation using detection-estimation schemes for uncertain systems, *Proceedsings JACC*, Denver, Colorado, pp 514-519, 1979.

4.12 D.J. Clark and P.C. Sen, A versatile three phase oscillator, *IEEE Trans. Contr. Instrum.*, Vol IECI-24, pp 57-60, 1977.

4.13 S.K. Datta, A novel three phase oscillator for the speed control of AC motors, *IEEE Trans. Ind. Gen. Appl.*, Vol. IGA-7, pp 61-68, 1971.

4.14 Catalogue of waveform generators, Prosser Scientific Instruments Ltd., Industrial Estate, Hadleigh, Ipswich, Suff. England.

CHAPTER 5

APPROXIMATING FILTERS FOR CONTINUOUS-TIME SYSTEMS WITH INTER=RUPTED OBSERVATIONS

5.1 INTRODUCTION

Let (Ω,F,P) be a probability space. Consider the $\mathbb{R}^m$-valued Markov pro=cess $\zeta_x = \{\zeta_x(t),\ t \geq 0\}$ satisfying the equation

$$\zeta_x(t) = x + \int_0^t f(\zeta_x(s))ds + BW(t), \quad t \geq 0, \quad x \in \mathbb{R}^m \tag{5.1}$$

and let the observation process $Y = \{y(t),\ t \geq 0\}$ be given by

$$y(t) = \int_0^t \theta(s)g(\zeta_x(s))ds + \Gamma v(t), \quad t \geq 0, \quad y(t) \in \mathbb{R}^p \tag{5.2}$$

where $f : \mathbb{R}^m \to \mathbb{R}^m$ and $g : \mathbb{R}^m \to \mathbb{R}^p$ are given functions satisfying the conditions stated in Section 3.1. $B \in \mathbb{R}^{m\times m}$ and $\Gamma \in \mathbb{R}^{p\times p}$ are matrices satisfying $B_{ij} = \sigma_i\ \delta_{ij}$, $i,j=1,\dots,m$ and $\Gamma_{ij} = \gamma_i\ \delta_{ij}$, $i,j,\dots,p$, where σ_i, $i=1,\dots,m$ and $\gamma_i = 1,\dots,p$ are given positive numbers.
$\mathbf{W} = \{W(t) = (W_1(t),\dots,W_m(t)),\ t \geq 0\}$ and $\mathbf{V} = \{v(t) = (v_1(t),\dots,v_p(t)),\ t \geq 0\}$ are $\mathbb{R}^m$-valued and $\mathbb{R}^p$-valued standard Wiener processes respectively on (Ω,F,P). $\Theta = \{\theta(t),\ t \geq 0\}$ is a homogeneous jump Markov process on (Ω,F,P) with state space $S_\theta = \{0,1\}$ and transition probabilities

$$P(\theta(t+\Delta)=j\,|\,\theta(t)=i) = \begin{cases} q\Delta + 0(\Delta^2) & \text{if } j\neq i \\ 1 - q\Delta + 0(\Delta^2) & \text{if } j=i \end{cases} \tag{5.3}$$

$i,j=0,1$

where q is a given positive number. It is assumed that the processes $\mathbf{W}$, $\mathbf{V}$ and Θ are mutually independent.

Denote by F_t^y the smallest σ-field generated by the family of random ele= ments $Y^t = \{y(s);\ 0 \le s \le t\}$. The problem dealt with in this chapter is to find approximations $\hat{\zeta}_x^{h,y}(k)$ and $\hat{\theta}^{h,y}(k)$ to $\hat{\zeta}_x(t)$ and $\hat{\theta}(t)$ respec= tively, at the instants $t_k = k\Delta$, $k\Delta \in [0,T]$, where

$$\hat{\zeta}_x(t) \triangleq E[\zeta_x(t\wedge\tau_T-)|F^y_{t\wedge\tau_T-}], \qquad t \in [0,T] \quad , \tag{5.4}$$

$$\hat{\theta}(t) \triangleq E[\theta(t\wedge\tau_T-)|F^y_{t\wedge\tau_T-}] , \qquad t \in [0,T] \quad , \tag{5.5}$$

and $\tau_T = \tau_T(x)$ is the first exit time of $\zeta_x(t)$ from an open and bounded domain $D \subset \mathbb{R}^m$.

It is well known that $E[\zeta_x(t)|F_t^y]$ is the minimum variance estimate of $\zeta_x(t)$ given Y^t. The problem of finding $E[\zeta_x(t)|F_t^y]$, in the case where Y is determined by equation (5.2), is called *state estimation for systems with interrupted observation*. In Sawaragi et al. [5.1], the case where equations (5.1)-(5.2) are linear (in $\zeta_x(t)$) stochastic differential equa= tions is considered and an infinite-dimensional filter for the computa= tion of $E[\zeta_x(t)|F_t^y]$, is derived. Also, presented are feasible approxi= mate estimator algorithms for the practical implementation.

In this chapter, the problem of state estimation from interrupted obser= vations is treated by using methods different from those used in [5.1]. Here, methods similar to those used in Chapter 3, are applied. Given an open and and bounded domain D in $\mathbb{R}^m$. Let $\tau_T = \tau_T(x)$ be the first exit time of $\zeta_x(t)$ from D, during the time interval [0,T]. First, the process $\{(\zeta_x(t\wedge\tau_T),\theta(t\wedge\tau_T)),\ t \in [0,T]\}$ is approximated by a continuous-time Markov chain $\{(\zeta_x^h(t\wedge\tau_T^h),\tilde{\theta}(t\wedge\tau_T^h)),\ t \in [0,T]\}$ with a finite state space $S = D_h \times \{0,1\}$, where $D_h = \mathbb{R}_h^m \cap D$ ($\mathbb{R}_h^m$ is given by equation (3.4)). Second, an optimal least-squares filter is derived for the on-line com= putation of $(\hat{\zeta}_x^h(t),\hat{\theta}^h(t)) \triangleq (E[\zeta_x^h(t\wedge\tau_T^h-)|F^{y,h}_{t\wedge\tau_T^h-}],\ E[\tilde{\theta}(t\wedge\tau_T^h-)|F^{y,h}_{t\wedge\tau_T^h-}])$,

(τ_T^h and $F_t^{y,h}$ are defined in a similar manner as in Section 3.2). Third, an estimator $\{(\hat{\zeta}_x^{h,y}(k),\hat{\theta}^{h,y}(k)),\ k\Delta \in [0,T]\}$ is constructed as an approxi= mation to $\{(\hat{\zeta}_x(k\Delta),\hat{\theta}(k\Delta)),\ k\Delta \in [0,T]\}$ (equations (5.4)-(5.5)) and this estimator is simulated for a variety of examples.

5.2 CONSTRUCTION OF THE MARKOV CHAIN

Let $\mathbb{R}_h^m$ be a grid on $\mathbb{R}^m$ with a constant mesh size h along all axes as defined by equation (3.4), and denote by e^i the unit vector along the i-th axis, i=1,...,m.

Define the following function $\lambda : (\mathbb{R}_h^m \times \{0,1\}) \times (\mathbb{R}_h^m \times \{0,1\}) \to \mathbb{R}$ by

$$\lambda(x,0;x,0) = \lambda(x,1;x,1) \triangleq -\left(\sum_{i=1}^{m} \sigma_i^2 + h \sum_{i=1}^{m} |f_i(x)| + qh^2\right)/h^2 \qquad x \in \mathbb{R}_h^m \tag{5.6}$$

$$\begin{cases} \lambda(x,0;x + e^i h,0) = \lambda(x,1;x + e^i h,1) \triangleq (\sigma_i^2/2 + h\, f_i^+(x))/h^2 \\ i=1,\dots,m \quad , \quad x \in \mathbb{R}_h^m \end{cases} \tag{5.7}$$

$$\begin{cases} \lambda(x,0;x - e^i h,0) = \lambda(x,1;x - e^i h,1) \triangleq (\sigma_i^2/2 + h\, f_i^-(x))/h^2 \\ i=1,\dots,m \quad , \quad x \in \mathbb{R}_h^m \end{cases} \tag{5.8}$$

$$\lambda(x,0;x,1) = \lambda(x,1;x,0) \triangleq q \quad , \quad x \in \mathbb{R}_h^m \tag{5.9}$$

$$\lambda(x,i;y,i) = 0 \quad , \quad x \in \mathbb{R}_h^m \quad , \quad y \in U_x \text{ and } i=0,1 \tag{5.10}$$

$$\lambda(x,i;y,j) = 0, \quad x,y \in \mathbb{R}_h^m \quad , \ x\neq y \text{ and } i\neq j, \quad i,j=0,1 \tag{5.11}$$

where for any $\alpha \in \mathbb{R}$, $\alpha^+ = \max(0,\alpha)$, $\alpha^- = -\min(0,\alpha)$ and

$$U_x \triangleq \{y \in \mathbb{R}_h^m : y\neq x \text{ and } y\neq x \pm e^i h \ , \ i=1,\dots,m\}. \tag{3.10}$$

Note that $\lambda(x,i;y,i) \geq 0$ for $x,y \in \mathbb{R}_h^m$, $x \neq y$, and $i=0,1$; $\lambda(x,i;x,j) > 0$, $x \in \mathbb{R}_h^m$, $i,j=0,1$, $i \neq j$, and

$$\begin{cases} \lambda(x,0;x,0) + \sum\limits_{\substack{y \\ y \neq x}} \lambda(x,0;y,0) + \lambda(x,0;x,1) = 0 \\ \\ \lambda(x,1;x,1) + \sum\limits_{\substack{y \\ y \neq x}} \lambda(x,1;y,1) + \lambda(x,1;x,0) = 0. \end{cases} \tag{5.12}$$

Hence, given $(x,i) \in \mathbb{R}_h^m \times \{0,1\}$, we can construct a continuous-time Markov chain $\{(\zeta_x^h(t),\tilde{\theta}(t)),\ t \in [0,T]\}$ with state space $\tilde{S} = \mathbb{R}_h^m \times \{0,1\}$, by defining the following set of transition probabilities

$$\begin{cases} P((\zeta_x^h(t+\Delta),\tilde{\theta}(t+\Delta))=(z,i) \mid (\zeta_x^h(t),\tilde{\theta}(t))=(z,i)) \triangleq 1 + \lambda(z,i;z,i)\Delta + O(\Delta^2) \\ \\ (z,i) \in \mathbb{R}_h^m \times \{0,1\} \end{cases} \tag{5.13}$$

$$\begin{cases} P((\zeta_x^h(t+\Delta),\tilde{\theta}(t+\Delta))=(z \pm e^i h,j) \mid (\zeta_x^h(t),\tilde{\theta}(t))=(z,j)) \triangleq \lambda(z,j;z \pm e^i h,j)\Delta + O(\Delta^2) \\ \\ (z,j) \in \mathbb{R}_h^m \times \{0,1\}, \quad i=1,\dots,m \end{cases} \tag{5.14}$$

$$\begin{cases} P((\zeta_x^h(t+\Delta),\tilde{\theta}(t+\Delta))=(z,j) \mid (\zeta_x^h(t),\tilde{\theta}(t))=(z,i)) \triangleq \lambda(z,i;z,j)\Delta + O(\Delta^2) \\ \\ z \in \mathbb{R}_h^m \ ; \quad i,j \in \{0,1\}, \quad i \neq j. \end{cases} \tag{5.15}$$

$$\begin{cases} \sum\limits_{y \in U_z} P((\zeta_x^h(t+\Delta),\tilde{\theta}(t+\Delta))=(y,i) \mid (\zeta_x^h(t),\tilde{\theta}(t))=(z,i)) \triangleq O(\Delta^2) \\ \\ (z,i) \in \mathbb{R}_h^m \times \{0,1\} \end{cases} \tag{5.16}$$

$$\begin{cases} \sum\limits_{y \in \mathbb{R}_h^m - \{z\}} P((\zeta_x^h(t+\Delta),\tilde{\theta}(t+\Delta))=(y,j) \mid (\zeta_x^h(t),\tilde{\theta}(t))=(z,i)) \triangleq O(\Delta^2) \\ \\ z \in \mathbb{R}_h^m \ ; \quad i,j=0,1, \quad i \neq j. \end{cases} \tag{5.17}$$

$$P(\zeta_x^h(0)=x) = 1, \quad x \in \mathbb{R}_h^m. \tag{5.18}$$

$$P(\tilde{\theta}(0)=i) = \pi_i, \quad i=0,1. \tag{5.19}$$

Thus, using equations (5.13)-(5.17) it follows that

$$\begin{cases} E[\zeta^h_{xi}(t+\Delta) - \zeta^h_{xi}(t) \mid (\zeta^h_x(t),\tilde{\theta}(t)) = (z,\ell)] = f_i(z)\Delta + hO(\Delta^2) \\ i=1,\dots,m \quad \ell=0,1 \quad , \quad x,z \in \mathbb{R}^m_h \end{cases} \tag{5.20}$$

$$E[\tilde{\theta}(t+\Delta) - \tilde{\theta}(t) \mid (\zeta^h_x(t),\tilde{\theta}(t))=(z,i)] = (-1)^i q\Delta + O(\Delta^2),(z,i) \in \mathbb{R}^m_h \times \{0,1\}. \tag{5.21}$$

$$\begin{cases} E[(\zeta^h_{xi}(t+\Delta) - \zeta^h_{xi}(t))(\zeta^h_{xj}(t+\Delta) - \zeta^h_{xj}(t)) \mid (\zeta^h_x(t),\tilde{\theta}(t))=(z,\ell)] \\ = \delta_{ij}(\sigma_i^2 + h|f_i(z)\ |)\Delta + (\delta_{ij} + 1)h^2 O(\Delta^2) \\ i,j=1,\dots,m \quad , \quad \ell=0,1, \quad z \in \mathbb{R}^m_h \end{cases} \tag{5.22}$$

$$\begin{cases} E[(\zeta^h_{xi}(t+\Delta) - \zeta^h_{xi}(t))(\tilde{\theta}(t+\Delta) - \tilde{\theta}(t)) \mid (\zeta^h_x(t),\tilde{\theta}(t)) = (z,\ell)] \\ = hO(\Delta^2) \quad , \quad i=1,\dots,m \quad , \quad (z,\ell) \in \mathbb{R}^m_h \times \{0,1\} \end{cases} \tag{5.23}$$

$$\begin{cases} E[(\tilde{\theta}(t+\Delta) - \tilde{\theta}(t))^2 \mid (\zeta^h_x(t),\tilde{\theta}(t)) = (z,\ell)] = q\Delta + O(\Delta^2) \\ (z,\ell) \in \mathbb{R}^m_h \times \{0,1\}. \end{cases} \tag{5.24}$$

Equations (5.20)-(5.24) illustrate the relations between the Markov chain $\{(\zeta^h_x(t),\tilde{\theta}(t)),\ t \in [0,T]\}$ and the Markov process $\{(\zeta_x(t),\theta(t)),\ t \in [0,T]\}$.

Let the set D and the stopping times τ_T and τ^h_T be defined in the same manner as in equations (3.19),(3.20) and (3.21) respectively, where $D_h \triangleq \mathbb{R}^m_h \cap D$. Define

$$y^h(t) \triangleq \int_0^t \tilde{\theta}(s)g(\zeta^h_x(s))ds + \Gamma v(t) \quad , \quad t \in [0,T] \tag{5.25}$$

and denote by $F^{y,h}_t$ the σ-field generated by $\{y^h(s),\ 0 \le s \le t\}$.

In the next section an optimal minimum variance filter is constructed for the computation of $(E[\zeta_x^h(t\wedge\tau_T^h-)|F^{y,h}_{t\wedge\tau_T^h-}], E[\tilde{\theta}(t\wedge\tau_T^h-)|F^{y,h}_{t\wedge\tau_T^h-}])$.

5.3 THE EQUATIONS OF THE OPTIMAL FILTER

Assume that $\sup_{t\in[0,T]} E|\zeta_x^h(t)|^2 < \infty$, $x \in \mathbb{R}_h^m$, and denote

$$G_t \triangleq \sigma((\zeta_x^h(s),\tilde{\theta}(s)), v(s);\ 0 \le s \le t),\ t \in [0,T] \tag{5.26}$$

$$h_t \triangleq \Gamma^{-1}\tilde{\theta}(t)g(\zeta_x^h(t)) \quad , \quad t \in [0,T] \tag{5.27}$$

$$z^h(t) \triangleq \int_0^t h_s\, ds + v(t) \quad , \quad t \in [0,T] \tag{5.28}$$

$$\nu^h(t) \triangleq z^h(t) - \int_0^t E[h_s|F_s^{y,h}]ds \quad , \quad t \in [0,T] \tag{5.29}$$

$$\begin{cases} \tilde{P}_{\alpha i}(t) \triangleq P((\zeta_x^h(t),\tilde{\theta}(t)) = (\alpha,i)|F_t^{y,h}), \\ t \in [0,T] \quad , (\alpha,i) \in \mathbb{R}_h^m \times \{0,1\} \end{cases} \tag{5.30}$$

$$\begin{cases} P_{\alpha i}(t) \triangleq P((\zeta_x^h(t\wedge\tau_T^h-),\tilde{\theta}(t\wedge\tau_T^h-)) = (\alpha,i)|F^{y,h}_{t\wedge\tau_T^h-}) \\ t \in [0,T] , \quad (\alpha,i) \in D_h \times \{0,1\} \end{cases} \tag{5.31}$$

We further assume that $\int_0^T E|h_t|^2 dt < \infty$.

For each $t \in [0,T]$, the σ-fields G_t and $\sigma(v(s_2)-v(s_1);\ t < s_1 < s_2 \le T)$ are independent and h_t is G_t-measurable. Thus, by following the same development given in Section 2.7 we obtain

$$\begin{cases} d\tilde{P}_{\alpha i}(t) = \sum_{(\gamma,j)\in\mathbb{R}_h^m\times\{0,1\}} \lambda(\gamma,j;\alpha,i)\tilde{P}_{\gamma j}(t)dt \\ + \tilde{P}_{\alpha i}(t) \sum_{\ell=1}^{p} \gamma_\ell^{-2}(ig_\ell(\alpha) - \hat{g}_\ell(t))(dy_\ell^h(t) - \hat{g}_\ell(t)dt) \\ t \in (0,T) \quad , (\alpha,i) \in \mathbb{R}_h^m \times \{0,1\} \end{cases} \tag{5.32}$$

$$\begin{cases} \hat{g}_\ell(t) = \sum_{i=0}^{1} \sum_{\beta \in \mathbb{R}_h^m} i\, g_\ell(\beta)\tilde{P}_{\beta i}(t) = \sum_{\beta \in \mathbb{R}_h^m} g_\ell(\beta)\tilde{P}_{\beta 1}(t) \\ t \in [0,T] \ , \quad \ell=1,\ldots,p \end{cases} \tag{5.33}$$

and

$$\begin{aligned} E[\zeta_x^h(t) \mid F_t^{y,h}] &= \sum_{\alpha \in \mathbb{R}_h^m} \alpha \sum_{i=0}^{1} \tilde{P}_{\alpha i}(t) \quad , \quad t \in [0,T] \ , \\ E[\tilde{\theta}(t) \mid F_t^{y,h}] &= \sum_{i=0}^{1} i \sum_{\alpha \in \mathbb{R}_h^m} \tilde{P}_{\alpha i}(t) = \sum_{\alpha \in \mathbb{R}_h^m} \tilde{P}_{\alpha 1}(t) \quad , \quad t \in [0,T], \end{aligned} \tag{5.34}$$

where $\{\lambda(\gamma,j;\alpha,i)\}$ are defined by equations (5.6)-(5.11).

In order to obtain the filter equations for computing $\{P_{\alpha i}(t),(\alpha,i) \in D_h \times \{0,1\},\ t \in [0,T]\}$ we follow the procedure given in Section 3.3.

Let $\{\lambda(\alpha,i;\beta,j) : (\alpha,i),(\beta,j) \in \mathbb{R}_h^m \times \{0,1\}\}$ be defined by equations (5.6)-(5.11) together with the additional condition

$$\lambda(z,i;y,j) = 0,\ (z,i) \in (\mathbb{R}_h^m - D_h) \times \{0,1\},\ (y,j) \in \mathbb{R}_h^m \times \{0,1\}. \tag{5.35}$$

Then, by following the development given in Section 3.3, a continuous-time Markov chain $\{(Z_x^h(t),\tilde{\tilde{\theta}}(t)),\ t \in [0,T]\}$, $x \in D_h$, with the infinitesi= mal characteristics $\{\lambda(\alpha,i;\beta,j) : (\alpha,i),(\beta,j) \in \mathbb{R}_h^m \times \{0,1\}\}$ given by equations (5.6)-(5.11) and (5.35), is constructed. In the same manner as in Section 3.3 we can choose the sample paths of $\{(Z_x^h(t),\tilde{\tilde{\theta}}(t)),\ t \in [0,T]\}$ to satisfy (with probability 1)

$$(Z_x^h(t),\tilde{\tilde{\theta}}(t)) = \begin{cases} (\zeta_x^h(t),\ \tilde{\theta}(t)) & 0 \le t < \tau_T^h(x) \\ (\zeta_x^h(\tau_T^h),\tilde{\theta}(\tau_T^h)) & \tau_T^h(x) \le t \le T \ . \end{cases} \quad x \in D_h \tag{5.36}$$

Keeping this choice in mind we can write

$$(Z_x^h(t),\tilde{\tilde{\theta}}(t)) = (\zeta_x^h(t\wedge\tau_T^h),\tilde{\theta}(t\wedge\tau_T^h)), \text{ w.p.1 }, \ t \in [0,T]. \tag{5.37}$$

Thus, for $x \in D_h$, the equations for computing $(\hat{\zeta}_x^h(t),\hat{\theta}^h(t)) \triangleq$ $(E[\zeta_x^h(t\wedge\tau_T^h-)|F^{y,h}_{t\wedge\tau_T^h-}], E[\tilde{\theta}(t\wedge\tau_T^h-)|F^{y,h}_{t\wedge\tau_T^h-}])$ (the optimal least-squares estimate of $(\zeta_x^h(t\wedge\tau_T^h-), \tilde{\theta}(t\wedge\tau_T^h-))$ given $(y^h(s)\ ;\ 0 \le s \le t \wedge \tau_T^h-))$ are given by

$$\begin{cases} dP_{\alpha i}(t) = \sum\limits_{(\gamma,j) \in D_h \times \{0,1\}} \lambda(\gamma,j;\alpha,i)P_{\gamma j}(t)dt \\ \qquad + P_{\alpha i}(t) \sum\limits_{\ell=1}^{p} \gamma_\ell^{-2}(ig_\ell(\alpha) - \hat{g}_\ell(t))(dy_\ell^h(t) - \hat{g}_\ell(t)dt) \\ t \in (0,T) \quad , \quad (\alpha,i) \in D_h \times \{0,1\} \end{cases} \tag{5.38}$$

$$\begin{cases} \hat{g}_\ell(t) = \sum\limits_{i=0}^{1} \sum\limits_{\gamma \in D_h} i\ g_\ell(\gamma)P_{\gamma i}(t) = \sum\limits_{\gamma \in D_h} g_\ell(\gamma)P_{\gamma 1}(t) \\ t \in [0,T] \quad , \quad \ell=1,\dots,p \end{cases} \tag{5.39}$$

and

$$\begin{aligned} &E[\zeta_x^h(t\wedge\tau_T^h-)|F^{y,h}_{t\wedge\tau_T^h-}] = \sum_{\alpha \in D_h} \alpha \sum_{i=0}^{1} P_{\alpha i}(t) \quad , \quad t \in [0,T] \\ &E[\tilde{\theta}(t\wedge\tau_T^h-)|F^{y,h}_{t\wedge\tau_T^h-}] = \sum_{i=0}^{1} i \sum_{\alpha \in D_h} P_{\alpha i}(t) = \sum_{\alpha \in D_h} P_{\alpha 1}(t), \ t \in [0,T] \end{aligned} \tag{5.40}$$

where in equation (3.19), $a_i=L_ih$, $i=1,\dots,m$; $\{L_i\}$ are given positive in= tegers and $\{\lambda(\gamma,j;\alpha,i)\}$ are given by equations (5.6)-(5.11) and (5.35).

Given $Y^t = \{y(s);\ 0 \le s \le t\}$, $t \in [0,T]$. Then, in order to compute an approximation to $(\hat{\zeta}_x(t),\hat{\theta}(t))$ (equations (5.4)-(5.5)), equations (5.38)-(5.39) are solved, where in equations (5.38) the increment dy^h is replaced by dy. Let $\{P^y_{\alpha i}(t),\ (\alpha,i) \in D_h \times \{0,1\},\ t \in [0,T]\}$, denote the solution to equations (5.38)-(5.39) (where dy replaces dy^h in (5.38)).

Then, a process $(\hat{\zeta}_x^{h,y},\hat{\theta}^{h,y}) = \{(\hat{\zeta}_x^{h,y}(t),\hat{\theta}^{h,y}(t)),\ t \in [0,T]\}$ is defined by

$$\begin{cases} \hat{\zeta}_x^{h,y}(t) \triangleq \sum\limits_{\alpha \in D_h} \alpha \sum\limits_{i=0}^{1} P_{\alpha i}^{y}(t) \quad , \quad t \in [0,T] \\ \hat{\theta}^{h,y}(t) \triangleq \sum\limits_{\alpha \in D_h} P_{\alpha 1}^{y}(t) \quad , \quad t \in [0,T] . \end{cases} \tag{5.41}$$

$(\hat{\zeta}_x^{h,y},\hat{\theta}^{h,y})$ serves here as an approximation to $(\hat{\zeta}_x,\hat{\theta})$. In the next sec= tion a procedure for computing $\{(\hat{\zeta}_x^{h,y}(k\Delta),\hat{\theta}^{h,y}(k\Delta)),\ k\Delta \in [0,T]\}$ is suggested. We assume that $\pi_{\alpha i} = P_{\alpha i}^{y}(0)$, $(\alpha,i) \in D_h \times \{0,1\}$, are unknown.

5.4 AN ALGORITHM FOR COMPUTING $(\hat{\zeta}_x^{h,y},\hat{\theta}^{h,y})$

In the sequel the following notations are used:
$y(k) \triangleq y(k\Delta)$, $\theta(k) \triangleq \theta(k\Delta)$, $P_{\alpha i}^{y}(k) \triangleq P_{\alpha i}^{y}(k\Delta)$, $\hat{g}_\ell(k) \triangleq \hat{g}_\ell(k\Delta)$, $\hat{\zeta}_x^{h,y}(k) \triangleq \hat{\zeta}_x^{h,y}(k\Delta)$, $\hat{\theta}^{h,y}(k) \triangleq \hat{\theta}^{h,y}(k\Delta)$; $k=0,1,\dots,N$; $(\alpha,i) \in D_h \times \{0,1\}$; $\ell=1,\dots,p$. We choose $P_{\alpha i}^{y}(0) \triangleq (\tfrac{1}{2}) \prod\limits_{i=1}^{m} (2L_i+1)^{-1}$, $(\alpha,i) \in D_h \times \{0,1\}$. Then, $\hat{g}_\ell(0) = P_{\alpha 1}^{y}(0) \sum\limits_{\gamma \in D_h} g_\ell(\gamma)$, $\ell=1,\dots,p$. Let ε be a given positive number.

Algorithm 5.4

1. $k:=0$
2. For $(\alpha,i) \in D_h \times \{0,1\}$ calculate

$$\begin{aligned} P_{\alpha i}^{y}(k+1) &:= P_{\alpha i}^{y}(k) + \sum_{j=0}^{1} \sum_{\gamma \in D_h} \lambda(\gamma,j;\alpha,i) P_{\gamma j}^{y}(k)\Delta \\ &+ P_{\alpha i}^{y}(k) \sum_{\ell=1}^{p} \gamma_\ell^{-2}(i g_\ell(\alpha) - \hat{g}_\ell(k))(y_\ell(k+1) - y_\ell(k) - \hat{g}_\ell(k)\Delta) \end{aligned} \tag{5.42}$$

$$P_{\alpha i}^{y}(k+1) := \max(0, P_{\alpha i}^{y}(k+1)) \tag{5.43}$$

3. $$Z(k+1) := \sum_{i=0}^{1} \sum_{\alpha \in D_h} P_{\alpha i}^{y}(k+1) \tag{5.44}$$
4. If $Z(k+1) \geq \varepsilon$ then : for $(\alpha,i) \in D_h \times \{0,1\}$, $P_{\alpha i}^{y}(k+1) := P_{\alpha i}^{y}(k+1)/Z(k+1)$ (5.45)

   Otherwise: stop.

5. For $\ell=1,\dots,p$ calculate

$$\hat{g}_\ell(k+1) := \sum_{\gamma \in D_h} g_\ell(\gamma) P^y_{\gamma 1}(k+1) \tag{5.46}$$

6. $$\hat{\zeta}^{h,y}_x(k+1) := \sum_{\alpha \in D_h} \alpha \sum_{i=0}^{1} P^y_{\alpha i}(k+1) \tag{5.47}$$

$$\hat{\theta}^{h,y}(k+1) := \sum_{\alpha \in D_h} P^y_{\alpha 1}(k+1) \tag{5.48}$$

7. If $k=N$ or if $\hat{\zeta}^{h,y}_x(k+1) \notin D$ then stop. Otherwise: $k:=k+1$ and go to 2.

The problem of establishing conditions for the weak convergence of $\{(\hat{\zeta}^{h,y}_x(t), \hat{\theta}^{h,y}(t)),\ t \in [0,T]\}$ to $\{(\hat{\zeta}_x(t), \hat{\theta}(t)),\ t \in [0,T]\}$ (equations (5.4)-(5.5)) as $h \downarrow 0$ is out of the scope of this work. Instead, the role of $(\hat{\zeta}^{h,y}_x, \hat{\theta}^{h,y})$, as an approximation to $(\hat{\zeta}_x, \hat{\theta})$, is demonstrated in the sequel throughout numerical experimentation with several examples.

The numerical study of Algorithm 5.4 was carried out via numerical ex= perimentation. Each experiment (run) consisted of the following stages:

(i) Simulate equations (5.1) by applying the same procedure as given by equations (3.48)-(3.49).

(ii) A sample path of $\{\theta(t),\ t \in [0,T]\}$ is constructed by applying the following procedure

1. Read the numbers R_i, $i=1,\dots,M+1$ from a random number genera= tor with a uniform probability density on (0,1).

2. Calculate

$$z_i = (-1/q)\ell n R_i \quad , \quad i=1,\dots,M+1$$

$$T_0 \triangleq 0 \quad , \quad T_i \triangleq \sum_{\ell=1}^{i} z_\ell \quad , \ i=1,\dots,M+1.$$

where M is determined by $T_M \le T < T_M+1$.

3. Define the following function

(a) $\theta(k) \triangleq [1 - (-1)^i]/2 \qquad T_i \le k\Delta < T_{i+1} \quad i=0,1,\ldots,M$

or

(b) $\theta(k) \triangleq [1 + (-1)^i]/2 \qquad T_i \le k\Delta < T_{i+1} \quad i=0,1,\ldots,M$

(iii) Calculate $\{y(k)\}$ in the same manner as in equation (3.50).

(iv) Apply Algorithm 5.4 where $\{y(k+1) - y(k)\}_{k=0}^{N-1}$ serves as the input to the filter and $\{(\hat{\zeta}_x^{h,y}(k), \hat{\theta}^{h,y}(k)\}_{k=1}^{N}$ serves as the filter's output.

5.5 EXAMPLES : THE CASE m=1

In this section cases where $\{\zeta_x(t), t \ge 0\}$ (equations (5.1)) is an $\mathbb{R}$-valued Markov process are considered. In these cases

$$D_h \triangleq \{ih : i=0,\pm 1,\ldots,\pm L\} \tag{3.53}$$

and equations (5.42) of Algorithm 5.4 reduce to:

For i=-L, -L+1,...,L and j=0,1 calculate

$$\begin{aligned} P^y_{i,j}(k+1) :=& P^y_{i,j}(k) + [\lambda(i-1,j;i,j)P^y_{i-1,j}(k) + \lambda(i,j;i,j)P^y_{i,j}(k) \\ &+ \lambda(i+1,j;i,j)P^y_{i+1,j}(k) + \lambda(i,\bar{j};i,j)P^y_{i,\bar{j}}(k)]\Delta \\ &+ P^y_{i,j}(k)\sum_{\ell=1}^{p} \gamma_\ell^{-2}(jg_\ell(ih)-\hat{g}_\ell(k))(y_\ell(k+1)-y_\ell(k)-\hat{g}_\ell(k)\Delta) \end{aligned} \tag{5.49}$$

where $\bar{0} = 1$ and $\bar{1} = 0$;

$$\hat{g}_\ell(k) := \sum_{i=-L}^{L} g_\ell(ih)P^y_{i1}(k) \quad , \quad \ell=1,\ldots,p \tag{5.50}$$

and for i=-L,...,L, j=0,1

$$\begin{cases} \lambda(i\pm 1,j;i,j) \triangleq \lambda(ih\pm h,j;ih,j) \\ \lambda(i,j;i,j) \triangleq \lambda(ih,j;ih,j),\ \lambda(i,\bar{j};i,j) \triangleq \lambda(ih,\bar{j};ih,j). \end{cases} \tag{5.51}$$

The following set of systems were considered:

(a) $dx = -5\cdot 10^{-5}x^2dt + 7dW \quad ; \quad dy = \theta x dt + 3dv \ , \ x,y \in \mathbb{R}, \ t > 0.$ (5.52)

In this case the set D was taken to be

$$D \triangleq \{x \in \mathbb{R} : |x| < 150 + \delta\} \ , \quad \delta < 0.75 \qquad (5.53)$$

and the following set of parameters was used: h=0.75; L=200; $\zeta_x(0)=100$; q=1,5; $\theta(0)=0,1$; $\Delta = 10^{-3}$ and $N=10^4$

(b) $dx = -xdt + 0.01dW \ ; \quad dy = \theta x dt + 0.005dv \ , \ x,y \in \mathbb{R}, \ t > 0.$ (5.54)

In this case the set D was taken to be

$$D \triangleq \{x \in \mathbb{R} : |x| < 1 + \delta\} \ , \quad \delta < 5\cdot 10^{-3} \qquad (5.55)$$

and the following set of parameters was used: h=0.005; L=200; $\zeta_x(0)=1$; q=1,5; $\theta(0)=0,1$; $\Delta = 10^{-3}$ and $N=10^4$.

(c) $dx = 0.5 \ \mathrm{sign}(x)dt + 0.01dW \ ; \ dy = \theta x dt + 0.005dv, \ x,y \in \mathbb{R}, \ t > 0$ (5.56)

In this case the set D was taken to be

$$D \triangleq \{x \in \mathbb{R} : |x| < 1 + \delta\} \ , \ \delta < 0.005 \qquad (5.57)$$

and the following set of parameters was used: h=0.005; L=200; $\zeta_x(0)=0.1$; q=1,5; $\theta(0)=0,1$; $\Delta = 10^{-3}$ and $N=10^4$.

Typical extracts from the results are given in the following figures. All the graphs in this section were plotted using the set of points $\{t'_k = 50k\Delta : k=0,1,\dots,200\}$.

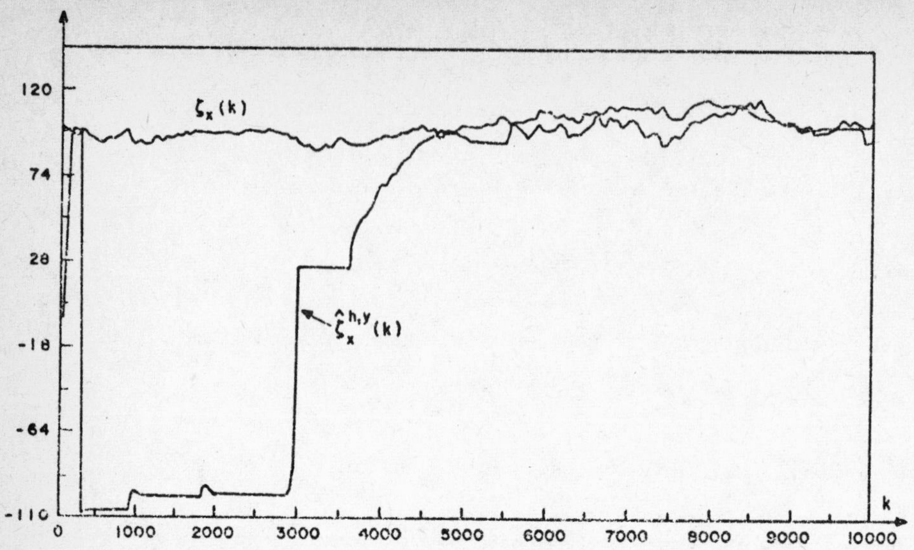

Fig.5.1a: $\zeta_x(k)$ and $\hat{\zeta}_x^{h,y}(k)$ as functions of k for the system given by equations (5.52), where q=1 and $\theta(0)=0$.

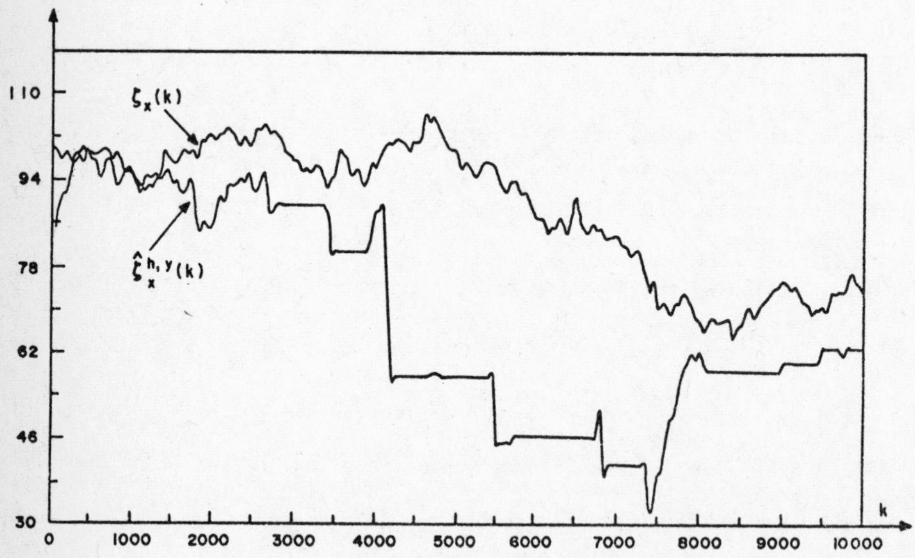

Fig.5.1b: $\zeta_x(k)$ and $\hat{\zeta}_x^{h,y}(k)$ as functions of k for the system given by equations (5.52), where q=1 and $\theta(0)=1$.

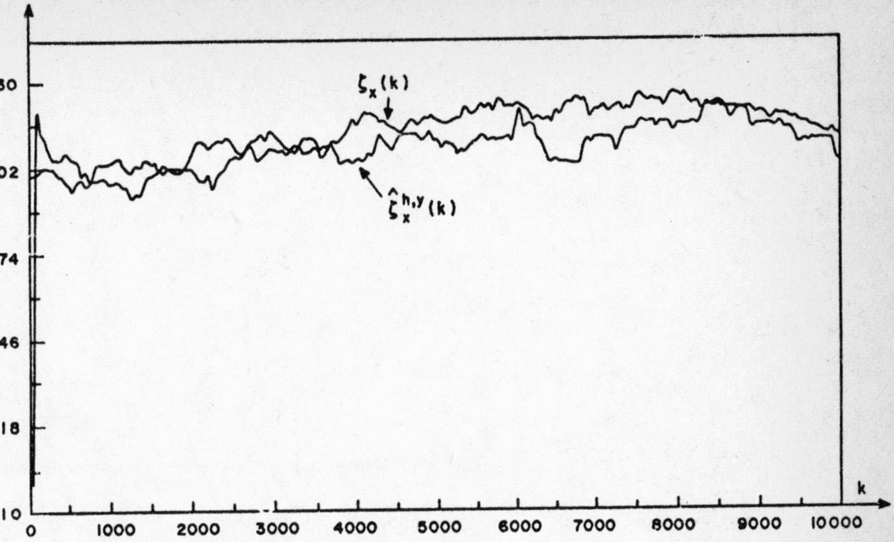

Fig.5.1c: $\zeta_X(k)$ and $\hat{\zeta}_X^{h,y}(k)$ as functions of k for the system given by equations (5.52), where q=5 and $\theta(0)=0$.

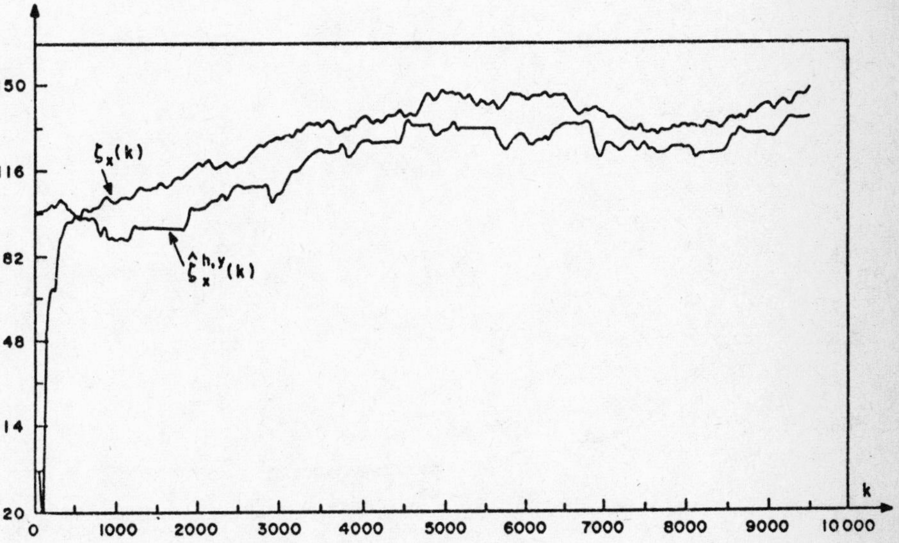

Fig.5.1d: $\zeta_X(k)$ and $\hat{\zeta}_X^{h,y}(k)$ as functions of k for the system given by equations (5.52), where q=5 and $\theta(0)=1$.

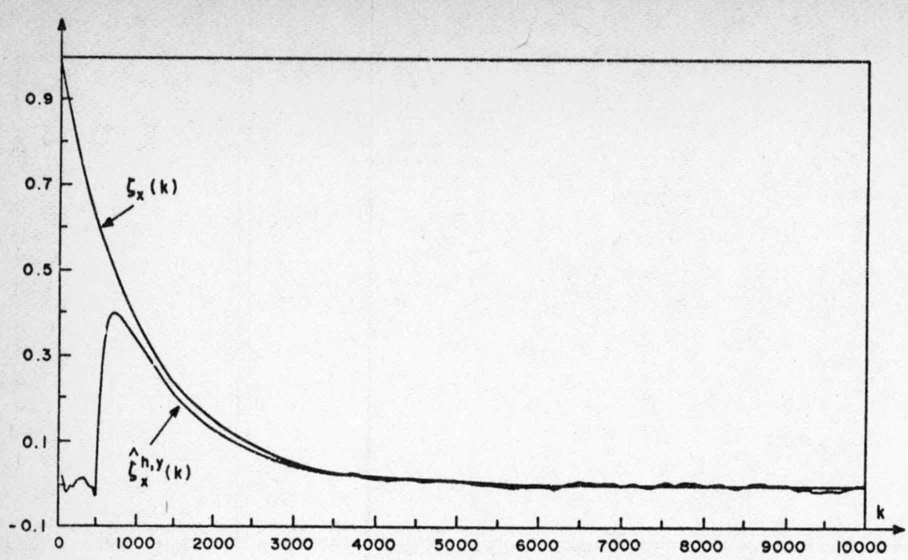

Fig.5.2a: $\zeta_x(k)$ and $\hat{\zeta}_x^{h,y}(k)$ as functions of k for the system given by equations (5.54), where q=1 and $\theta(0)=0$.

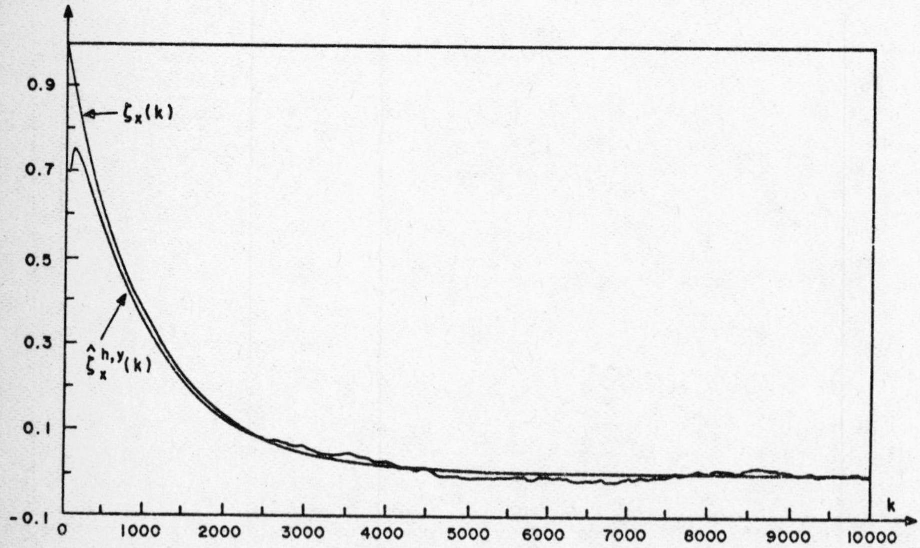

Fig.5.2b: $\zeta_x(k)$ and $\hat{\zeta}_x^{h,y}(k)$ as functions of k for the system given by equations (5.54), where q=1 and $\theta(0)=1$.

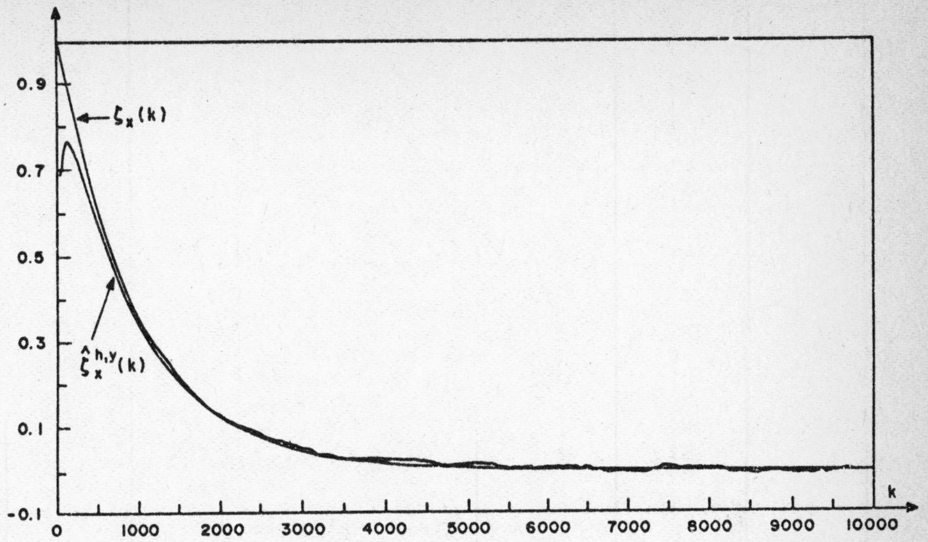

Fig.5.2c: $\zeta_x(k)$ and $\hat{\zeta}_x^{h,y}(k)$ as functions of k for the system given by equations (5.54), where q=5 and $\theta(0)=0$.

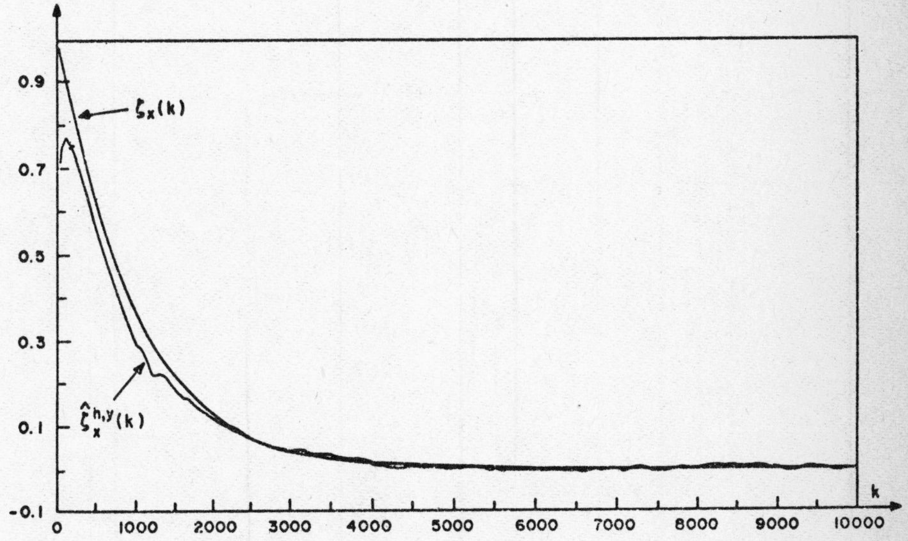

Fig.5.2d: $\zeta_x(k)$ and $\hat{\zeta}_x^{h,y}(k)$ as functions of k for the system given by equations (5.54), where q=5 and $\theta(0)=1$.

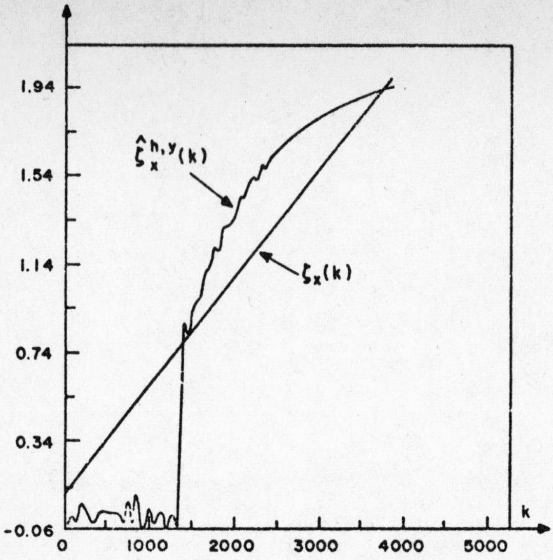

Fig.5.3a: $\zeta_x(k)$ and $\hat{\zeta}_x^{h,y}(k)$ as functions of k for the system given by equations (5.56), where q=1 and $\theta(0)=0$.

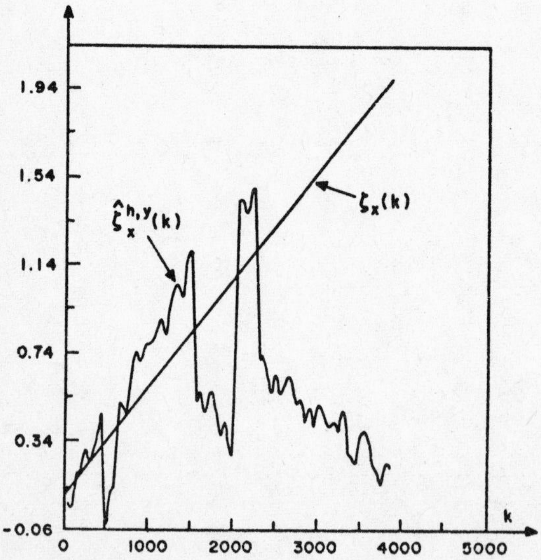

Fig.5.3b: $\zeta_x(k)$ and $\hat{\zeta}_x^{h,y}(k)$ as functions of k for the system given by equations (5.56), where q=1 and $\theta(0)=1$.

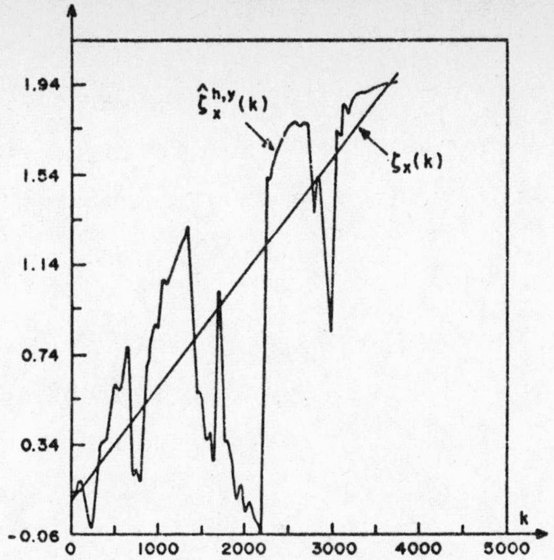

Fig.5.3c: $\zeta_x(k)$ and $\hat{\zeta}_x^{h,y}(k)$ as functions of k for the system given by equations (5.56), where q=5 and $\theta(0)=0$.

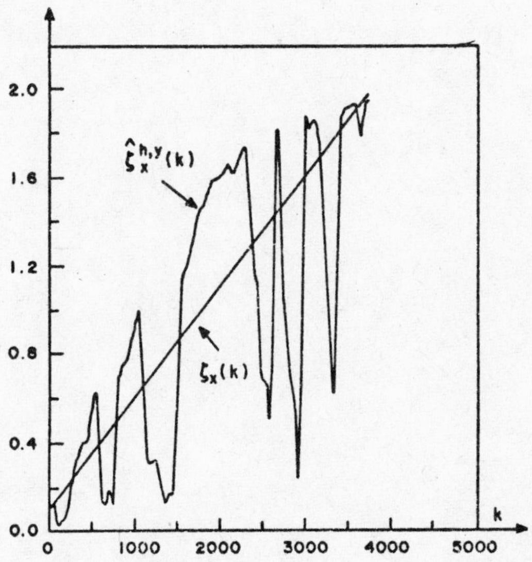

Fig.5.3d: $\zeta_x(k)$ and $\hat{\zeta}_x^{h,y}(k)$ as functions of k for the system given by equations (5.56), where q=5 and $\theta(0)=1$.

5.6 EXAMPLES : THE CASE m=2

In this section cases where $\zeta_x = \{\zeta_x(t),\ t \geq 0\}$ (equations (5.1)) is an $\mathbb{R}^2$-valued Markov process are considered. In these cases the set D_h is given by equation (3.71) and equations (5.42) of Algorithm 5.4 reduce to:

For $((ih,jh),r) \in D_h \times \{0,1\}$ calculate

$$\begin{cases} P^y_{(i,j),r}(k+1) := P^y_{(i,j),r}(k) + \sum_{\nu=0}^{1} \sum_{q=-L_1}^{L_1} \sum_{s=-L_2}^{L_2} \lambda((q,s),\nu;(i,j),r) P^y_{(q,s),\nu}(k)\Delta \\ \quad + P^y_{(i,j),r}(k) \sum_{\ell=1}^{p} \gamma_\ell^{-2}(rg_\ell(ih,jh)-\hat{g}_\ell(k))(y_\ell(k+1)-y_\ell(k)-\hat{g}_\ell(k)\Delta) \end{cases} \tag{5.58}$$

where

$$\hat{g}_\ell(k) = \sum_{i=-L_1}^{L_1} \sum_{j=-L_2}^{L_2} g_\ell(ih,jh) P^y_{(i,j),1}(k), \quad \ell=1,\dots,p \tag{5.59}$$

and

$$\begin{cases} \lambda((i,j),r;(i,j),r) = -[\sum_{\nu=1}^{2} (\sigma_\nu^2 + h|f_\nu(ih,jh)|) + qh^2]/h^2 \\ r=0,1 \ ; \quad -L_1 \leq i \leq L_1 \ , \quad -L_2 \leq j \leq L_2 \end{cases} \tag{5.60}$$

$$\begin{cases} \lambda((i-1,j),r;(i,j),r) = (\sigma_1^2/2 + hf_1^+(ih-h,jh))/h^2 \\ r=0,1 \ ; \quad -L_1+1 \leq i \leq L_1 \ , \quad -L_2 \leq j \leq L_2 \end{cases} \tag{5.61}$$

$$\begin{cases} \lambda((i+1,j),r;(i,j),r) = (\sigma_1^2/2 + hf_1^-(ih+h,jh))/h^2 \\ r=0,1 \ ; \quad -L_1 \leq i \leq L_1-1 \ , \quad -L_2 \leq j \leq L_2 \end{cases} \tag{5.62}$$

$$\begin{cases} \lambda((i,j+1),r;(i,j),r) = (\sigma_2^2/2 + hf_2^-(ih,jh+h))/h^2 \\ r=0,1 \ ; \quad -L_1 \leq i \leq L_1 \ , \quad -L_2 \leq j \leq L_2-1 \end{cases} \tag{5.63}$$

$$\begin{cases} \lambda((i,j-1),r;(i,j),r) = (\sigma_2^2/2 + hf_2^+(ih,jh-h))/h^2 \\ \\ r=0,1 \;\; ; \quad -L_1 \le i \le L_1 \quad , \quad -L_2+1 \le j \le L_2 \end{cases} \tag{5.64}$$

$$\lambda((i,j),r;(i,j),\bar{r})=q, \; r=0,1 \; ; \; -L_1 \le i \le L_1 \; , -L_2 \le j \le L_2$$

where $\bar{0} = 1$ and $\bar{1} = 0$,

and $\lambda(\alpha,i;\beta,j) = 0$ for the rest of $\alpha,\beta \in \mathbb{R}_h^2$, $i,j=0,1$.

The following set of systems were considered:

(a)
$$\begin{cases} dx_1 = x_2 dt \\ \qquad\qquad\qquad t > 0 \\ dx_2 = [-2x_2 + 10\text{sat}(x_1)]dt + 0.01dW_2 \end{cases} \tag{5.65}$$

$$dy_i = \theta x_i dt + 0.005dv_i \quad , \quad t > 0 \quad , \quad i=1,2 \quad , \tag{5.66}$$

where

$$\text{sat}(\lambda) \triangleq \begin{cases} 1 & \lambda > 1 \\ \lambda & |\lambda| \le 1 \\ -1 & \lambda < -1 \end{cases} \tag{5.67}$$

In this case the set D was taken to be

$$D \triangleq \{x \in \mathbb{R}^2 : \; |x_i| < 2.4 + \delta \; , \; i=1,2\} \; ,\delta < 0.2 \tag{5.68}$$

and the following set of parameters was used: $\zeta_{xi}(0)=0$, $i=1,2$; $h=0.2$; $L=12$; $q=5$; $\theta(0)=0,1$; $\Delta = 10^{-3}$ and $N=10^4$. Typical extracts from the results are given in the following tables and figures.

TABLE 5.1: $\theta(k)$ and $\hat{\theta}^{h,y}(k)$ as functions of k for the system given by equations (5.65)-(5.66), where q=5 and $\theta(0)=0$.

k	$\theta(k)$	$\hat{\theta}^{h,y}(k)$
50	0.0	.0212
250	0.0	.0503
500	1.0	.0688
750	0.0	.2254
1000	0.0	.0377
1250	1.0	.9372
1500	1.0	.3339
1750	1.0	.3465
2000	0.0	.3493
2250	0.0	.4485
2500	1.0	1.0000
2750	0.0	.0741
3000	0.0	.0153
3250	0.0	.0000

TABLE 5.2: $\theta(k)$ and $\hat{\theta}^{h,y}(k)$ as functions of k for the system given by equations (5.65)-(5.66), where q=5 and $\theta(0)=1$

k	$\theta(k)$	$\hat{\theta}^{h,y}(k)$
50	1.0	.0203
250	1.0	.0555
500	0.0	.0229
750	1.0	.1401
1000	1.0	.0258
1250	0.0	.0651
1500	0.0	.2237
1750	0.0	.0442
2000	1.0	.6679
2250	1.0	.9950
2500	0.0	.0079
2750	1.0	.9950
3000	1.0	.9950
3250	1.0	.9072

(b)
$$\begin{cases} dx_1 = [\text{sign}(x_2) + 10\, x_1(0.5 - |x_1| - |x_2|)]dt + 0.005dW_1 \\ dx_2 = [-\text{sign}(x_1) + 10\, x_2(0.5 - |x_1| - |x_2|)]dt + 0.005dW_2 \end{cases} \quad t > 0 \qquad (5.69)$$

$$dy_i = \theta x_i dt + 0.005dv_i \quad , \quad t > 0 \quad , \quad i=1,2 \ . \qquad (5.70)$$

(See Section 2.5 for more details on equations (5.69). Here also, $|x_i|$ and $\text{sign}(x_i)$ are expressions used for $x_i\tanh(ax_i)$ and $\tanh(ax_i)$, respec= tively, for some $a \gg 1$.)

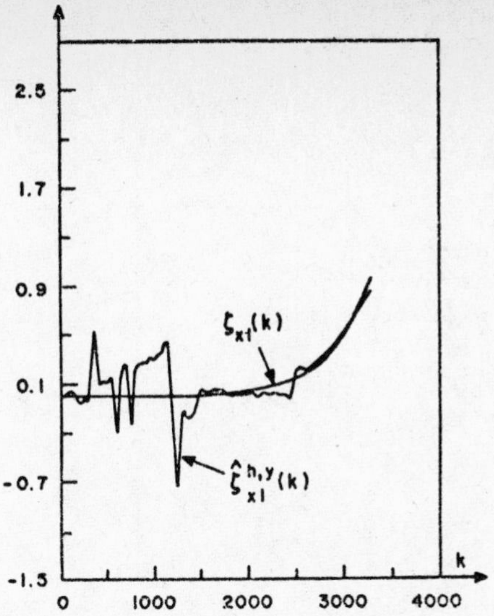

<u>Fig.5.4a</u>: $\zeta_{x1}(k)$ and $\hat{\zeta}^{h,y}_{x1}(k)$ as functions of k for the system given by equations (5.65)-(5.66), where q=5 and $\theta(0)=0$.

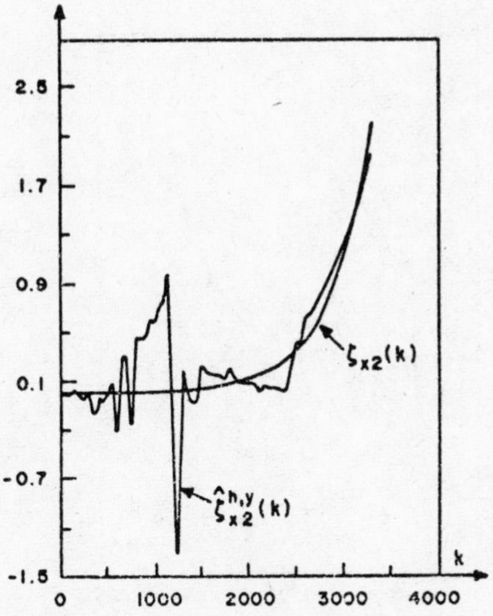

<u>Fig.5.4b</u>: $\zeta_{x2}(k)$ and $\hat{\zeta}^{h,y}_{x2}(k)$ as functions of k for the system given by equations (5.65)-(5.66), where q=5 and $\theta(0)=0$.

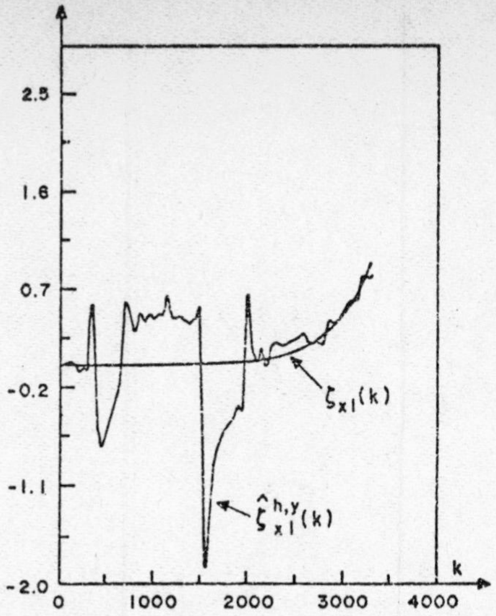

<u>Fig.5.4c</u>: $\zeta_{x1}(k)$ and $\hat{\zeta}^{h,y}_{x1}(k)$ as functions of k for the system given by equations (5.65)-(5.66), where q=5 and $\theta(0)=1$.

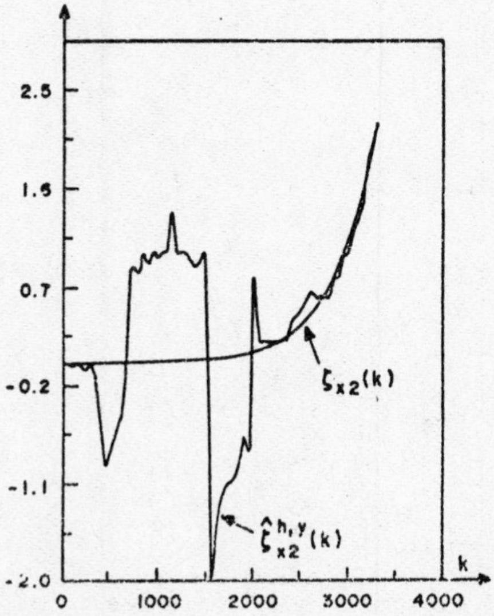

<u>Fig.5.4d:</u> $\zeta_{x2}(k)$ and $\hat{\zeta}^{h,y}_{x2}(k)$ as functions of k for the system given by equations (5.65)-(5.66), where q=5 and $\theta(0)=1$.

In this case the set D was taken to be ·

$$D \triangleq \{x \in \mathbb{R}^2 : |x_i| < 1 + \delta,\ i=1,2\}\ ,\ \delta < 1/12 \qquad (5.71)$$

and the following set of parameters was used: $\zeta_{xi}(0)=0$, $i=1,2$; $h=1/12$; $L=12$; $q=1,5$; $\theta(0)=0,1$; $\Delta = 10^{-3}$ and $N=10^4$. Typical extracts from the results are given in the following tables and figures.

TABLE 5.3: $\theta(k)$ and $\hat{\theta}^{h,y}(k)$ as functions of k for the system given by equations (5.69)-(5.70), where q=1 and $\theta(0)=0$

k	$\theta(k)$	$\hat{\theta}^{h,y}(k)$	k	$\theta(k)$	$\hat{\theta}^{h,y}(k)$	k	$\theta(k)$	$\hat{\theta}^{h,y}(k)$
50	0.0	.0134	4500	1.0	.9922	9000	1.0	.9966
250	1.0	.0047	4750	1.0	1.0000	9250	0.0	.0138
500	0.0	.0103	5000	0.0	.0003	9500	0.0	.0000
750	0.0	.0226	5250	0.0	.0058	9750	0.0	.0478
1000	0.0	.0168	5500	1.0	1.0000	10000	1.0	.9990
1250	0.0	.1296	5750	0.0	.0058			
1500	0.0	.0021	6000	0.0	.0037			
1750	0.0	.0150	6250	1.0	.9989			
2000	0.0	.0109	6500	0.0	.0009			
2250	1.0	.9968	6750	1.0	.9990			
2500	1.0	1.0000	7000	1.0	.9990			
2750	1.0	1.0000	7250	1.0	.9902			
3000	1.0	.9999	7500	1.0	1.0000			
3250	1.0	.9962	7750	1.0	.9961			
3500	1.0	1.0000	8000	1.0	.9988			
3750	1.0	.9875	8250	1.0	1.0000			
4000	1.0	.9942	8500	1.0	1.0000			
4250	1.0	.9999	8750	1.0	.9990			

TABLE 5.4: $\theta(k)$ and $\hat{\theta}^{h,y}(k)$ as functions of k for the system given by equations (5.69)-(5.70), where q=1 and $\theta(0)=1$

k	$\theta(k)$	$\hat{\theta}^{h,y}(k)$	k	$\theta(k)$	$\hat{\theta}^{h,y}(k)$
50	1.0	.0048	5250	0.0	.0034
250	1.0	.0436	5500	0.0	.0195
500	0.0	.0246	5750	0.0	.0444
750	0.0	.1322	6000	0.0	.0592
1000	0.0	.0023	6250	0.0	.0013
1250	1.0	.9990	6500	1.0	1.0000
1500	1.0	.9898	6750	0.0	.0057
1750	1.0	.9945	7000	0.0	.0135
2000	1.0	.9991	7250	1.0	.9982
2250	1.0	.9990	7500	1.0	.9938
2500	1.0	.9973	7750	1.0	1.0000
2750	1.0	.8092	8000	0.0	.0001
3000	1.0	.9706	8250	0.0	.0015
3250	1.0	.9929	8500	0.0	.0026
3500	1.0	.9961	8750	0.0	.0185
3750	1.0	.9963	9000	0.0	.1462
4000	1.0	.9876	9250	0.0	.0002
4250	1.0	.9973	9500	0.0	.0020
4500	1.0	.9167	9750	0.0	.0107
4750	1.0	1.0000	10000	0.0	.0033
5000	0.0	.0033			

TABLE 5.5: $\theta(k)$ and $\hat{\theta}^{h,y}(k)$ as functions of k for the system given by equations (5.69)-(5.70), where q=5 and $\theta(0)=0$.

k	$\theta(k)$	$\hat{\theta}^{h,y}(k)$	k	$\theta(k)$	$\hat{\theta}^{h,y}(k)$
50	0.0	.2205	5250	1.0	1.0000
250	0.0	.0528	5500	1.0	1.0000
500	0.0	.0180	5750	1.0	.9944
750	1.0	.6919	6000	1.0	1.0000
1000	0.0	.0346	6250	1.0	.9518
1250	1.0	.9844	6500	1.0	.9950
1500	0.0	.4543	6750	1.0	.9999
1750	1.0	.9950	7000	0.0	.0268
2000	1.0	1.0000	7250	0.0	.0103
2250	0.0	.0007	7500	0.0	.0081
2500	1.0	1.0000	7750	1.0	.9967
2750	1.0	.9950	8000	1.0	.9892
3000	0.0	.0013	8250	0.0	.0004
3250	0.0	.0124	8500	0.0	.0491
3500	0.0	.0499	8750	1.0	.9998
3750	0.0	.0064	9000	0.0	.0000
4000	0.0	.0536	9250	0.0	.0117
4250	1.0	.9851	9500	0.0	.0061
4500	0.0	.0071	9750	0.0	.0346
4750	1.0	1.0000	10000	0.0	.0079
5000	0.0	.0058			

TABLE 5.6: $\theta(k)$ and $\hat{\theta}^{h,y}(k)$ as functions of k for the system given by equations (5.69)-(5.70), where q=5 and $\theta(0)=1$

k	$\theta(k)$	$\hat{\theta}^{h,y}(k)$	k	$\theta(k)$	$\hat{\theta}^{h,y}(k)$
50	1.0	.5863	5250	1.0	.9888
250	1.0	.0188	5500	0.0	.0124
500	1.0	.5213	5750	1.0	.9950
750	1.0	.9058	6000	1.0	1.0000
1000	1.0	.9700	6250	1.0	.9813
1250	0.0	.0068	6500	1.0	.9867
1500	1.0	1.0000	6750	1.0	.9949
1750	0.0	.0001	7000	1.0	.9948
2000	1.0	.9934	7250	1.0	1.0000
2250	1.0	.9950	7500	1.0	.9925
2500	0.0	.0012	7750	0.0	.0442
2750	1.0	.9826	8000	1.0	.9918
3000	1.0	.9869	8250	1.0	.9950
3250	0.0	.0045	8500	1.0	.9909
3500	0.0	.0325	8750	0.0	.0196
3750	0.0	.0992	9000	0.0	.0235
4000	1.0	1.0000	9250	0.0	.0108
4250	1.0	.9864	9500	1.0	.9877
4500	0.0	.3056	9750	1.0	.9328
4750	1.0	.9878	10000	0.0	.0136
5000	1.0	.6844			

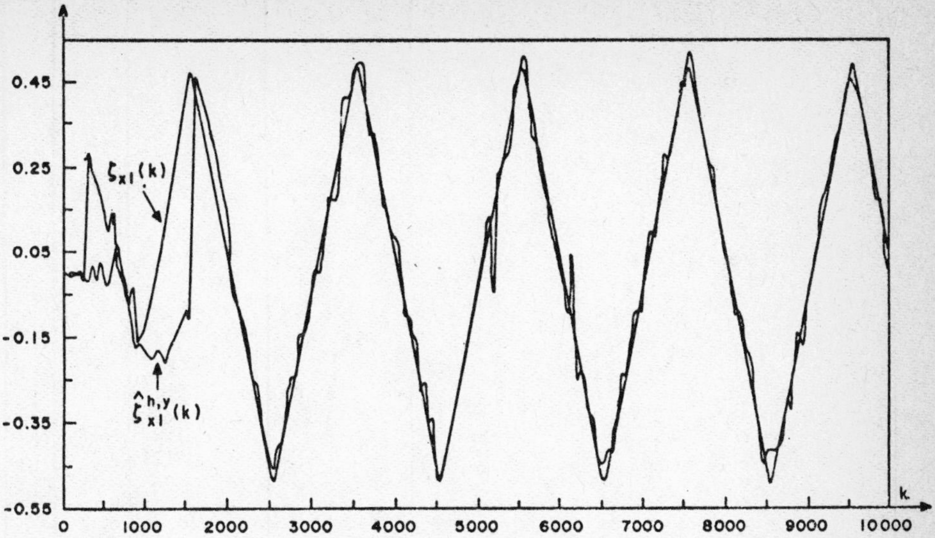

Fig.5.5a: $\zeta_{x1}(k)$ and $\hat{\zeta}^{h,y}_{x1}(k)$ as functions of k for the system given by equations (5.69)-(5.70), where q=1 and $\theta(0)=0$.

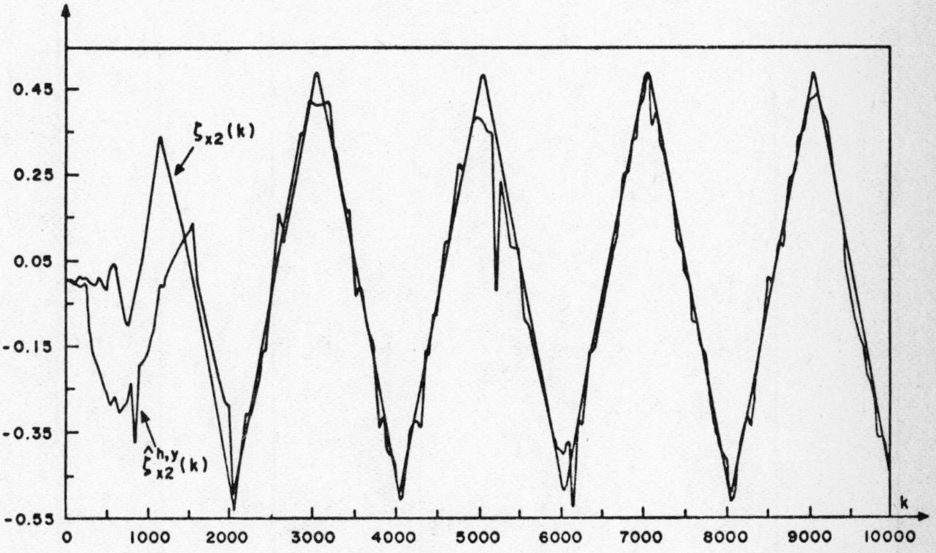

Fig.5.5b: $\zeta_{x2}(k)$ and $\hat{\zeta}^{h,y}_{x2}(k)$ as functions of k for the system given by equations (5.69)-(5.70), where q=1 and $\theta(0)=0$.

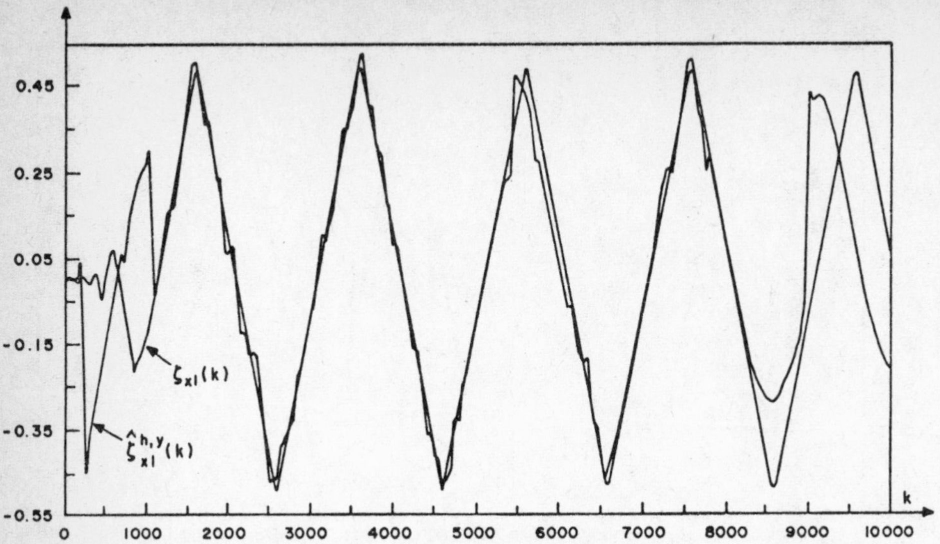

Fig.5.5c: $\zeta_{x1}(k)$ and $\hat{\zeta}_{x1}^{h,y}(k)$ as functions of k for the system given by equations (5.69)-(5.70), where q=1 and $\theta(0)=1$.

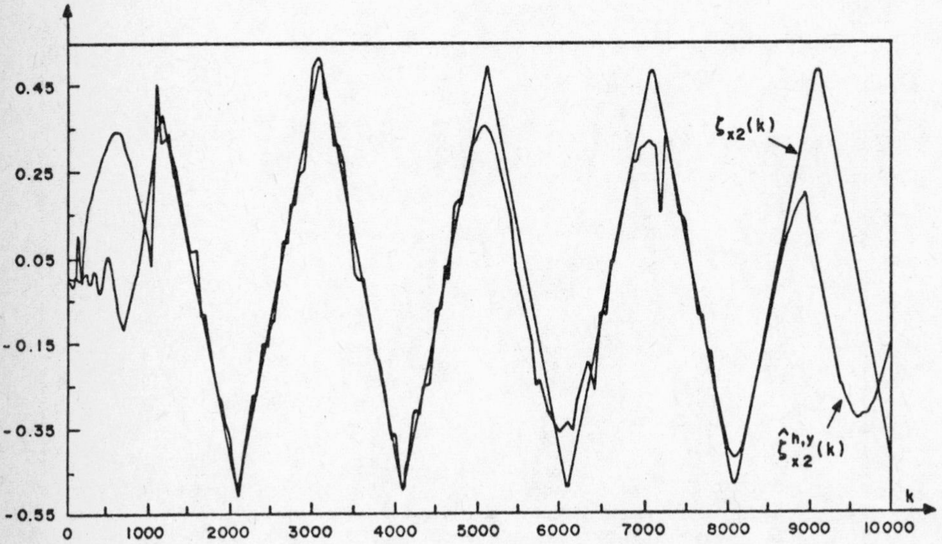

Fig.5.5d: $\zeta_{x2}(k)$ and $\hat{\zeta}_{x2}^{h,y}(k)$ as functions of k for the system given by equations (5.69)-(5.70), where q=1 and $\theta(0)=1$.

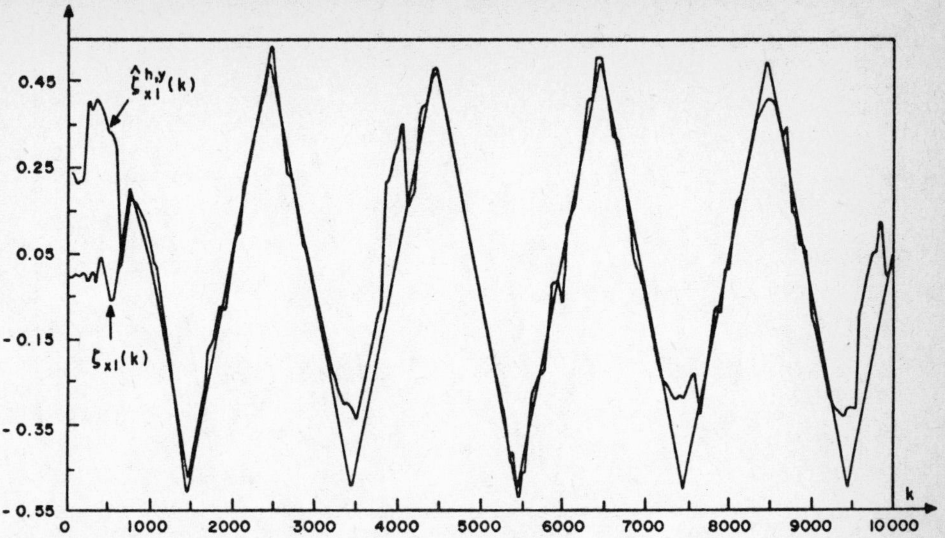

Fig.5.5e: $\zeta_{x1}(k)$ and $\hat{\zeta}_{x1}^{h,y}(k)$ as functions of k for the system given by equations (5.69)-(5.70), where q=5 and $\theta(0)=0$.

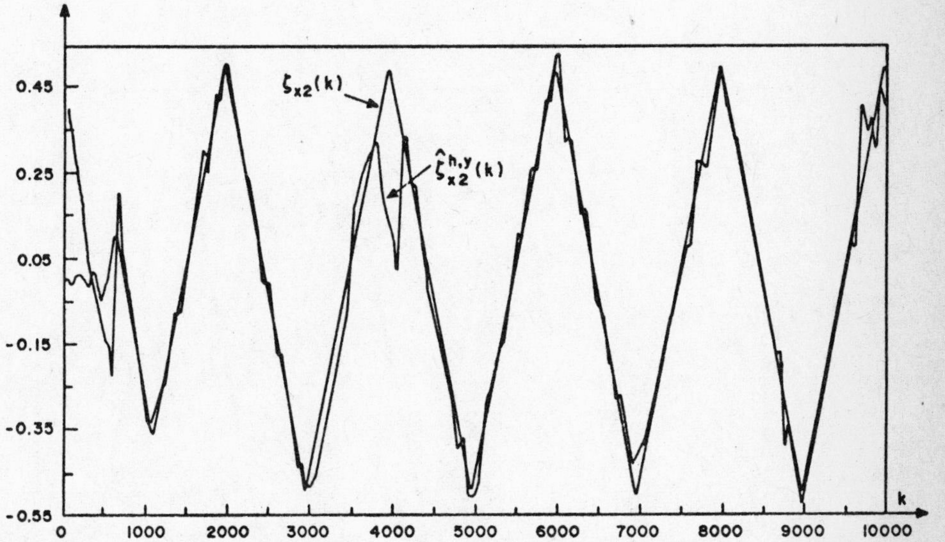

Fig.5.5f: $\zeta_{x2}(k)$ and $\hat{\zeta}_{x2}^{h,y}(k)$ as functions of k for the system given by equations (5.69)-(5.70), where q=5 and $\theta(0)=0$.

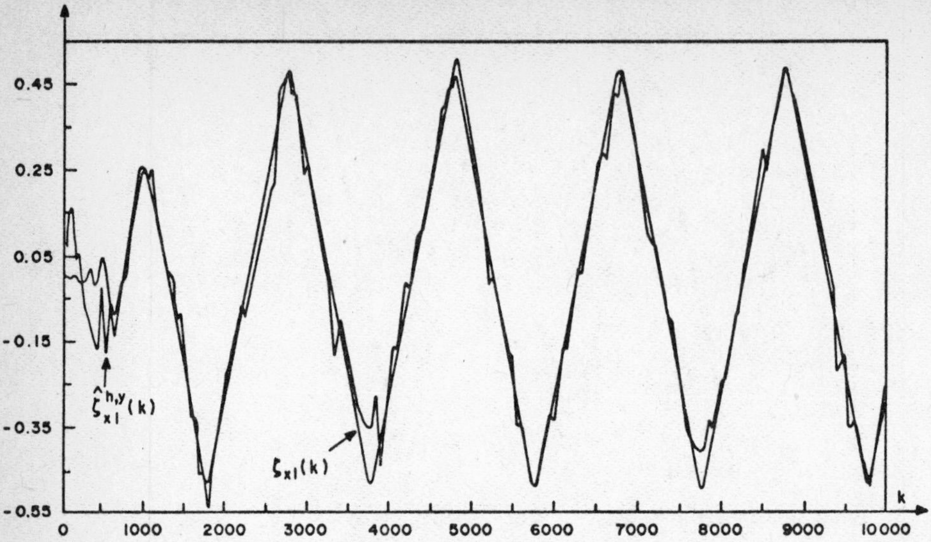

Fig.5.5g: $\zeta_{X1}(k)$ and $\hat{\zeta}^{h,y}_{X1}(k)$ as functions of k for the system given by equations (5.69)-(5.70), where q=5 and $\theta(0)=1$.

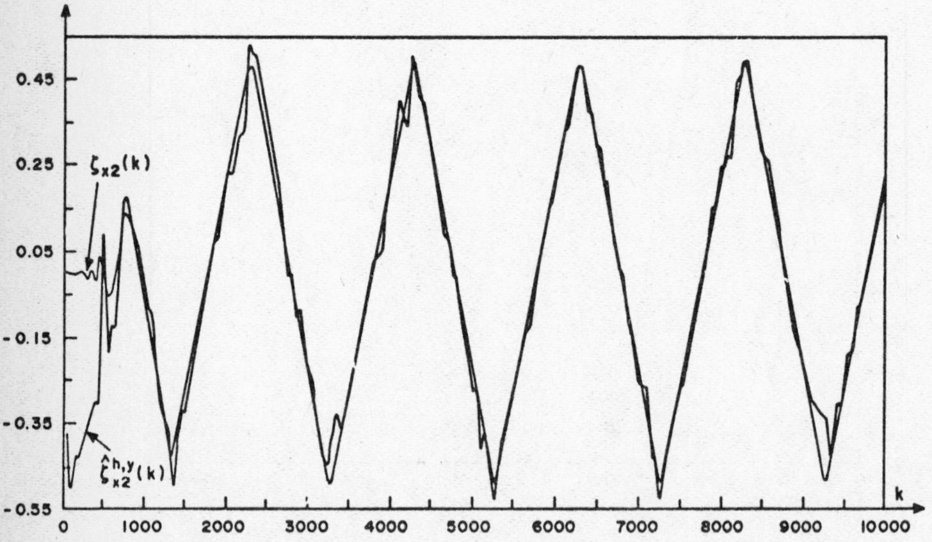

Fig.5.5h: $\zeta_{X2}(k)$ and $\hat{\zeta}^{h,y}_{X2}(k)$ as functions of k for the system given by equations (5.69)-(5.70), where q=5 and $\theta(0)=1$.

(c) $\{\zeta_x(t), t \geq 0\}$ is determined by equations (5.69) and the measure= ment process is given by

$$dy_i = sat(x_i/0.6)dt + 0.005dv_i \quad , \quad t > 0 \quad , \quad i=1,2 \tag{5.72}$$

where the function $sat(\lambda)$, $\lambda \in \mathbb{R}$, is given by equation (5.67).

In this case the set D was taken to be the same as given by equation (5.71), and the same set of parameters as in (b), was used. Some of the results are given in the following figures.

All the graphs in this section were plotted using the set of points $\{t'_k = 50k\Delta : k=0,1,\ldots,200\}$.

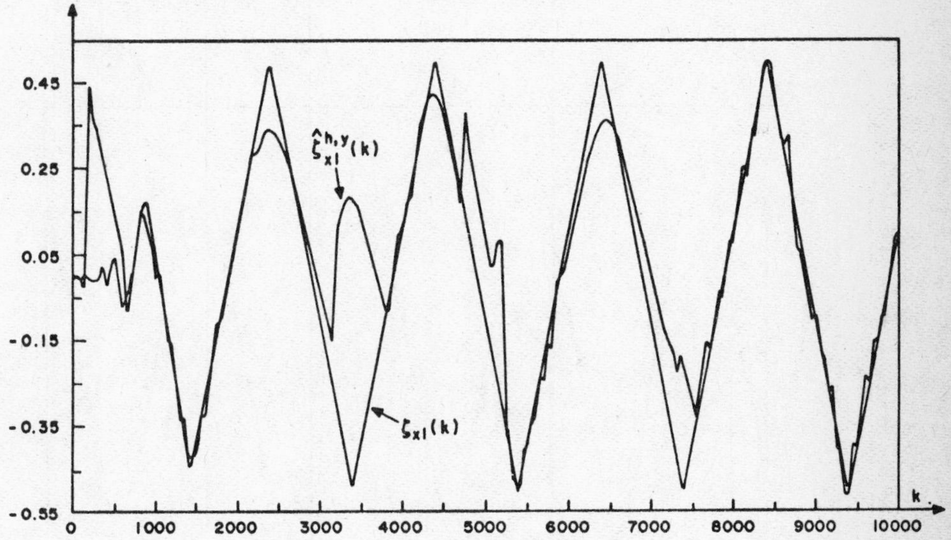

Fig.5.6a: $\zeta_{x1}(k)$ and $\hat{\zeta}^{h,y}_{x1}(k)$ as functions of k for the system given by equations (5.69) and (5.72), where q=1 and $\theta(0)=0$.

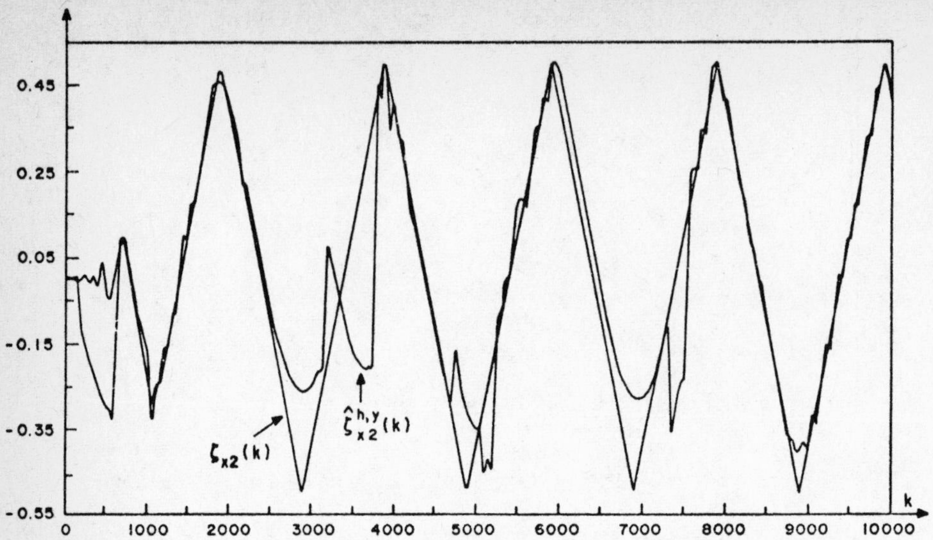

Fig.5.6b: $\zeta_{X2}(k)$ and $\hat{\zeta}^{h,y}_{X2}(k)$ as functions of k for the system given by equations (5.69) and (5.72), where q=1 and $\theta(0)=0$.

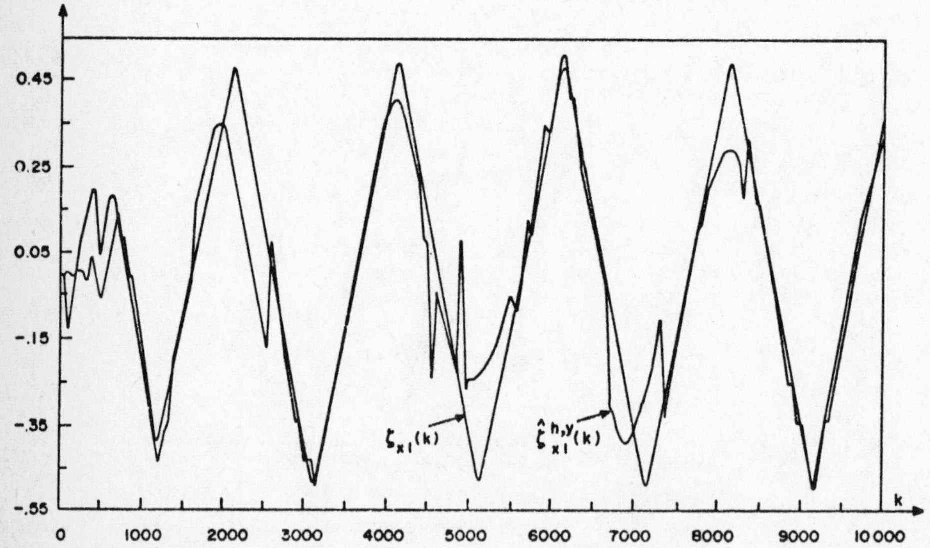

Fig.5.6c: $\zeta_{X1}(k)$ and $\hat{\zeta}^{h,y}_{X1}(k)$ as functions of k for the system given by equations (5.69) and (5.72), where q=1 and $\theta(0)=1$.

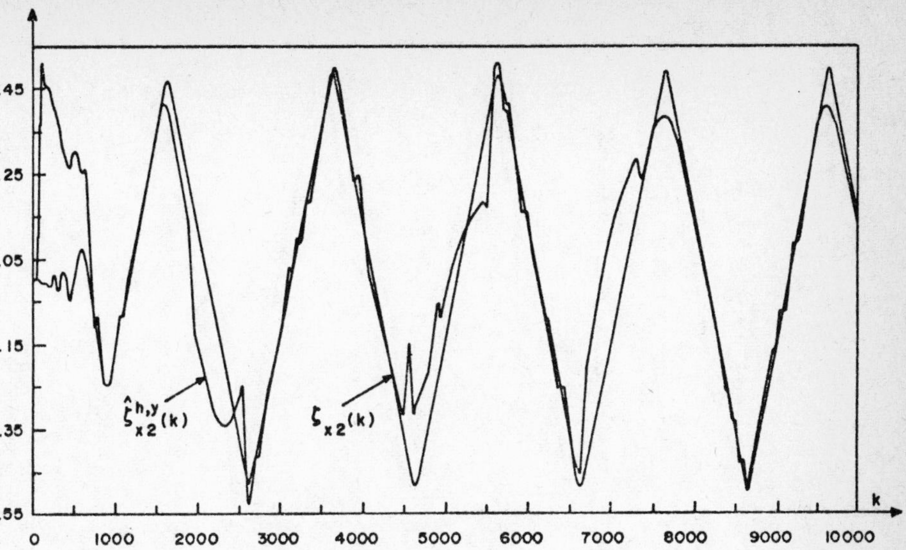

Fig.5.6d: $\zeta_{X2}(k)$ and $\hat{\zeta}^{h,y}_{X2}(k)$ as functions of k for the system given by equations (5.69) and (5.72), where q=1 and $\theta(0)=1$.

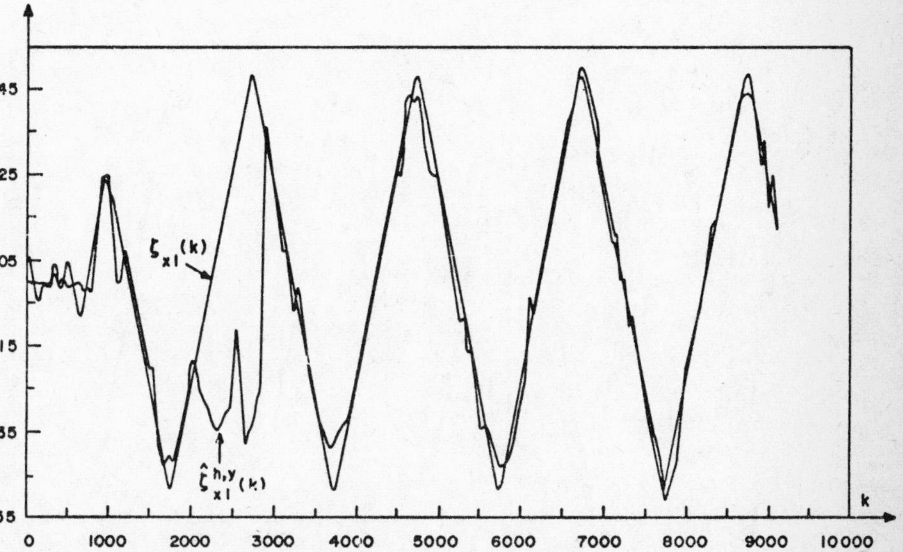

Fig.5.6e: $\zeta_{X1}(k)$ and $\hat{\zeta}^{h,y}_{X1}(k)$ as functions of k for the system given by equations (5.69) and (5.72), where q=5 and $\theta(0)=0$.

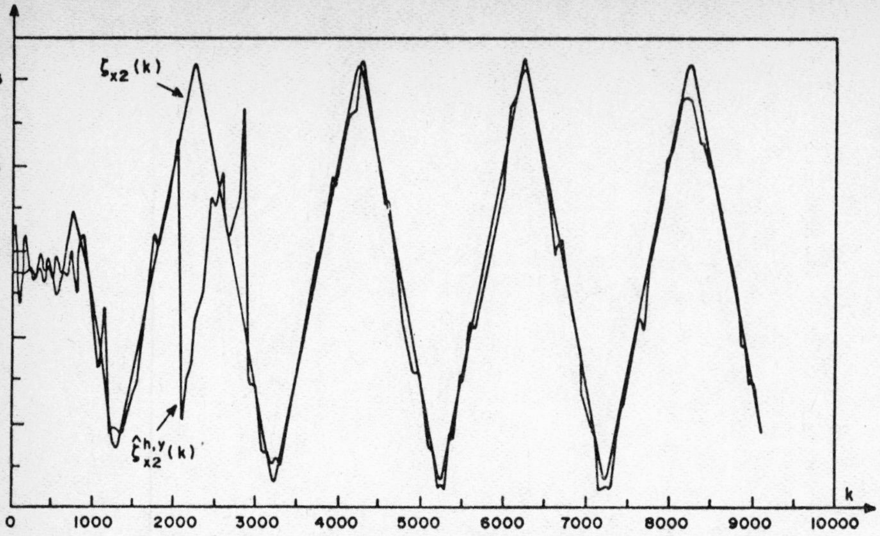

Fig.5.6f: $\zeta_{x2}(k)$ and $\hat{\zeta}_{x2}^{h,y}(k)$ as functions of k for the system given by equations (5.69) and (5.72), where q=5 and $\theta(0)=0$.

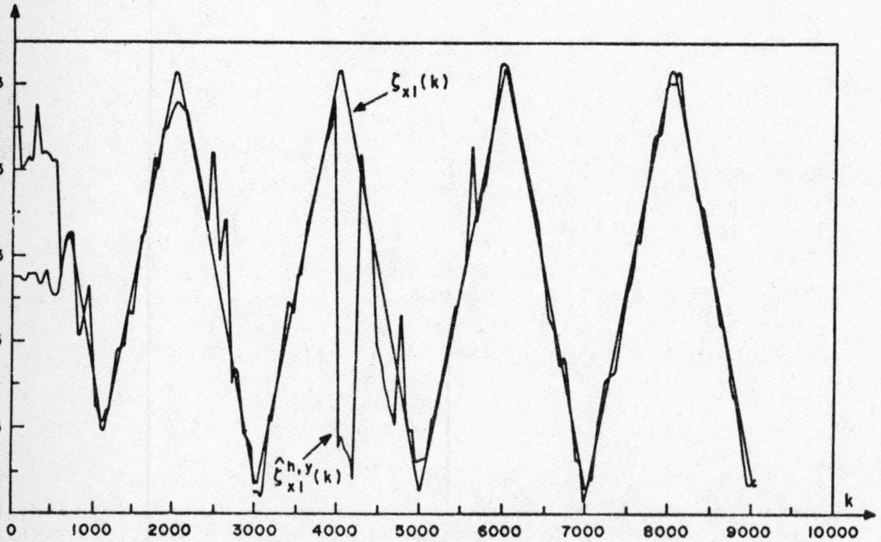

Fig.5.6g: $\zeta_{x1}(k)$ and $\hat{\zeta}_{x1}^{h,y}(k)$ as functions of k for the system given by equations (5.69) and (5.72), where q=5 and $\theta(0)=1$.

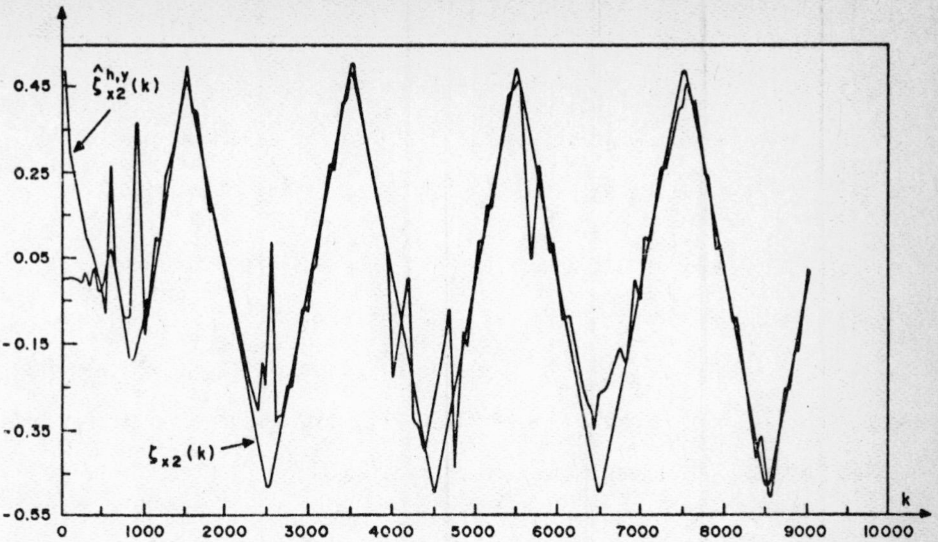

Fig.5.6h: $\zeta_{x2}(k)$ and $\hat{\zeta}^{h,y}_{x2}(k)$ as functions of k for the system given by equations (5.69) and (5.72), where q=5 and $\theta(0)=1$.

5.7 REMARKS

State estimation problems using interrupted observations (i.e.,when the measurement process is given by (5.2)) arise in the context of communication theory where the signal process (ζ_x) is subjected to random attenuation or fading; or in control theory in cases where intermittent faults in the sen= sors lead to a degradation in the measurement quality in such a manner that the measurement process can be modelled by (5.2); or in the estimation of reflection coefficients for seismic data deconvolution [5.2]; etc.

The problem of state estimation using interrupted observations has been treated by several authors, various models and approaches having been used. For more details see, for example, Refs. [4.7], [4.9], [5.1] and [5.3-5.9]. Note that some of the discrete-in-time versions of the problem (Refs. [5.4] and [5.6])are closely related to the problem dealt with in Section 4.6. However, the approaches used in this work are entirely different from those used in the above-mentioned references.

The examples dealt with in Section 5.5 treat some one-dimensional diffu= sion processes ζ_x where the stochastic differential equation, describing the dynamical system, contains a basic nonlinearity as x^2 or sign (x).

The systems given by (5.65) and (5.69), which are dealt with in Section 5.6, describe: a 'classical' closed-loop position servo [5.10] having a nonlinearity - sat (•) (see Fig. 8.5) and dW2/dt acting as a 'white-noise' input (eqns. (5.65)); and a noise-driven triangular-wave generator (eqns. (5.69)), a model which has already been dealt with in Chapters 2 and 3.

5.8 REFERENCES

5.1 Y. Sawaragi, T. Katayama and S. Fujishige, State estimation for con= tinuous-time system with interrupted observation, *IEEE Trans. on Automatic Control*, AC-19, pp 307-314, 1974.

5.2 A.K. Mahalanabis, S. Prasad and K.P. Mohandas, Recursive decision directed estimation of reflection coefficients for seismic data deconvolution, *Automatica*, 18, pp 721-726, 1982.

5.3 B.B. Madan and A.K. Mahalanabis, Continuous time smoothing for sys= tems with interrupted observations, *IEEE Trans. on Autmat. Contr.*, Vol. AC-21, pp 428-430, 1976.

5.4 J.K. Tugnait, Truncated ML estimation of transition probabilities for systems with interrupted observations, *Proceeding of the 19th IEEE Conference on Dicision & Control*, 1, pp 576-577, Albuquerque, New Mexico, December 10-12, 1980.

5.5 J.K. Tugnait, Detection and estimation for abruptly changing systems, *Automatica*, 18, pp 607-615, 1982.

5.6 J.K. Tugnait, Parameter estimation and linear system identification with randomly interrupted observations, *IEEE Trans. on Information Theory*, Vol. IT-29, pp 164-168, 1983.

5.7 J.K. Tugnait, Control of stochastic systems with Markov interrupted observations, *IEEE Trans. on Aerospace and Electronic Systems*, Vol. AES-19, pp 232-238, 1983.

5.8 J.K. Tugnait, On identification and adaptive estimation for systems with interrupted observations, *Automatica*, 19, pp 61-73, 1983.

5.9 D.N.P. Murthy, Filtering with unreliable sensors, Int. J. Systems Sci., 13, pp 437-451, 1982.

5.10 C.J. Savant, Jr., *Control System Design*, McGraw-Hill Book Company, New York, 1964.

CHAPTER 6

ESTIMATION IN A MULTITARGET ENVIRONMENT

6.1 INTRODUCTION

The state estimation of a number of targets using measurements that could not be associated with certainty with the various targets under consideration has received a great deal of attention recently. See for example Bar-Shalom and Tse [6.1], Bar-Shalom [6.2], Fortmann et al. [6.3], Bar-Shalom and Marcus [6.4], Athans et al. [6.5] and the references cited there. In this chapter a particular situation of estimation in a multitarget environment is treated.

It is assumed that there are at most L targets in the (x_1,x_2)-plane each located in a redtangular region

$$R_j = \{x \in \mathbb{R}^2 : |x_1-c_{1j}| < a_{1j},\ |x_2-c_{2j}| < a_{2j}\}\ ,\ j=1,\dots,L \quad (6.1)$$

where c_{ij} and $a_{ij} > 0$, $i=1,2$, $j=1,\dots,L$ are given numbers such that $R_\ell \cap R_m = \phi$, $\ell,m=1,\dots,L$, $\ell \neq m$ (ϕ denotes the empty set).

Define the following random variables

$$\theta_j \triangleq \begin{cases} 1 & \text{if there is a target in } R_j \\ 0 & \text{otherwise} \end{cases} \quad (6.2)$$

$j=1,\dots,L$

and let $\theta \triangleq (\theta_1,\dots,\theta_L)$. Hence θ is a (fixed in time) random element belonging to the discrete sample space $\Omega_\theta = \{\theta^1,\theta^2,\dots,\theta^{2^L}\}$ where

$\theta^1 = (0,0,\dots,0,0)$, $\theta^2 = (0,0,\dots,0,1)$, $\theta^3 = (0,0,\dots,1,0)$, $\theta^4 = (0,0,\dots,1,1),\dots,\theta^{2^L} = (1,1,\dots,1,1)$. Thus, for example, the sample point θ^4 represents the event where there are targets only in R_{L-1} and R_L.

We suppose that the measurement process is given by

$$y_i(t) = \int_0^t c_{in(u)}\theta_{n(u)}du + \gamma_i v_i(t) \quad , \quad t \in [0,T], \quad i=1,2 \tag{6.3}$$

where, on a probability space (Ω,F,P) : $\{v(t) = (v_1(t),v_2(t)),\ t \geq 0\}$ is an $\mathbb{R}^2$-valued standard Wiener process; $\{n(t),\ t \geq 0\}$ is a continuous-time Markov chain with state space $S_n = \{1,\dots,L\}$ and transition probabilities satisfying equations (2.29)-(2.31), and θ is a random element with values in Ω_θ. It is assumed that θ, $\{v(t),\ t \geq 0\}$ and $\{n(t),\ t \geq 0\}$ are mutually independent.

Since (as is tacitly assumed) the processes θ, $\{n(t),\ t \geq 0\}$ and $\{v(t),\ t \geq 0\}$ cannot be observed, it follows that the measurements as given by equations (6.3), cannot be associated with certainty with the various targets under consideration.

Denote by F_t^y the smallest σ-field generated by the family of random elements $Y^t = \{y(s);\ 0 \leq s \leq t\}$. The problem dealt with in this chapter is to find an estimate $\hat{\theta}(t)$ to θ such that

$$\hat{\theta}(t) = E[\theta \mid F_t^y] \ , \quad t \in [0,T] \ . \tag{6.4}$$

In the next section a filter is constructed for the calculation of $\{\hat{\theta}(t),\ t \in [0,T]\}$.

6.2 THE EQUATIONS OF THE OPTIMAL FILTER

Since θ and $\{n(t),\ t \geq 0\}$ are mutually independent, and the parameter θ is fixed in time, it follows that

$$\begin{cases} P((\theta(t+\Delta),n(t+\Delta)) = (\theta^{\ell},j) \mid (\theta(t),n(t)) = (\theta^{k},i)) \\ \\ = \delta_{k\ell}\ P(n(t+\Delta) = j \mid n(t) = i) = \delta_{k\ell}(\lambda_{ij}\Delta + 0(\Delta^2)) \quad i \neq j \end{cases} \tag{6.5}$$

$$\begin{cases} P((\theta(t+\Delta),n(t+\Delta)) = (\theta^{\ell},i) \mid (\theta(t),n(t)) = (\theta^{k},i)) \\ \\ = \delta_{k\ell}\ P(n(t+\Delta) = i \mid n(t) = i) = \delta_{k\ell}(1 + \lambda_{ii}\Delta + 0(\Delta^2)) \end{cases} \tag{6.6}$$

where $\{\lambda_{ij}\}$ satisfy equations (2.31).

Denote

$$G_t \stackrel{\Delta}{=} \sigma(\theta,n(s),v(s);\ 0 \le s \le t) \quad , t \in [0,T] \tag{6.7}$$

$$h_{ti} \stackrel{\Delta}{=} \gamma_i^{-1} c_{in(t)}\theta_{n(t)} \quad , \quad t \in [0,T] \ , \quad i=1,2, \tag{6.8}$$

$$z_i(t) \stackrel{\Delta}{=} \int_0^t h_{ui}du + v_i(t) \quad , \quad t \in [0,T] \ , \quad i=1,2 \tag{6.9}$$

$$\nu_i(t) \stackrel{\Delta}{=} z_i(t) - \int_0^t E[h_{ui} \mid F_u^y]du \quad , \quad t \in [0,T] \ , \ i=1,2 \tag{6.10}$$

$$P_{\ell i}(t) \stackrel{\Delta}{=} P(\theta = \theta^{\ell},\ n(t) = i \mid F_t^y),\ t \in [0,T] \ , \quad \ell=1,\dots,2^L, \quad i=1,\dots,L. \tag{6.11}$$

(Note that $z_i(t) = \gamma_i^{-1}y_i(t)$, $t \in [0,T]$, $i=1,2$).

For each $t \in [0,T]$, the σ-fields G_t and $\sigma(v(t_2)-v(t_1);\ t < t_1 < t_2 \le T)$ are independent, and h_t is G_t-measurable. Thus, the results of Theorem 1.1 (Section 1.2) can be applied to our problem. Let $F : \Omega_\theta \times S_n \to \mathbb{R}$ be a bounded and measurable function. Then, equation (1.25),

$$\begin{aligned} dE[F(\theta,n(t)) \mid F_t^y] &= E[A_tF(\theta,n(t)) \mid F_t^y]dt \\ &+ \sum_{i=1}^{2} (E[F(\theta,n(t))h_{ti} \mid F_t^y] - E[F(\theta,n(t)) \mid F_t^y]E[h_{ti} \mid F_t^y])d\nu_i(t) \end{aligned} \tag{6.12}$$

$t \in (0,T)$,

where $A_t F$ is such that

$$\int_0^T E|A_t F(\theta,n(t))|^2 dt < \infty \tag{6.13}$$

and

$$M_t(F) \triangleq F(\theta,n(t)) - E[F(\theta,n(0))|F_0^y] - \int_0^t A_u F(\theta,n(u))du \tag{6.14}$$

is a (G_t,P)-martingale.

Let $F(\theta,n(t)) = \chi_{\ell j}(t)$, where

$$\chi_{\ell j}(t) \triangleq \begin{cases} 1 & \text{if } \theta = \theta^{\ell} \text{ and } n(t) = j \\ 0 & \text{otherwise} \end{cases} \tag{6.15}$$

and suppose that

$$\int_0^T E|\sum_{i=1}^{L} \lambda_{ij}\, \chi_{\ell i}(t)|^2 dt < \infty \quad . \tag{6.16}$$

Then, it can be shown, in the same manner as in Section 2.7, that

$$\begin{aligned} E[A_t \chi_{\ell j}(t)|F_t^y] &= E[\sum_{i=1}^{L} \lambda_{ij}\chi_{\ell i}(t)|F_t^y] \\ &= \sum_{i=1}^{L} \lambda_{ij}\, E[\chi_{\ell i}(t)|F_t^y] \\ &= \sum_{i=1}^{L} \lambda_{ij}\, P_{\ell i}(t) \end{aligned} \tag{6.17}$$

$$t \in [0,T]\ , \quad j=1,\dots,L, \quad \ell=1,\dots,2^L.$$

Using

$$E[\chi_{\ell j}(t)h_{ti}|F_t^y] = E[\chi_{\ell j}(t)\gamma_i^{-1}\sum_{r=1}^{L} c_{ir}\,\delta(n(t),r)\theta_r|F_t^y]$$

$$= \gamma_i^{-1}\sum_{r=1}^{L} c_{ir}\; E[\chi_{\ell j}(t)\delta(n(t),r)\theta_r|F_t^y]$$

$$= \gamma_i^{-1}\sum_{r=1}^{L} c_{ir}\;\delta(j,r)\theta_r^{\ell}P_{\ell j}(t) \tag{6.18}$$

$$= \gamma_i^{-1} c_{ij}\;\theta_j^{\ell}\,P_{\ell j}(t),\; t\in[0,T]\;,\; i=1,2,\quad j=1,\dots,L$$

$$\ell=1,\dots,2^L$$

(where $\delta(j,r)=1$ if $j=r$ and $\delta(j,r)=0$ if $j\neq r$) and

$$E[h_{ti}|F_t^y] = E[\gamma_i^{-1}\sum_{r=1}^{L} c_{ir}\;\delta(n(t),r)\theta_r|F_t^y]$$

$$= \gamma_i^{-1}\sum_{r=1}^{L} c_{ir}\sum_{\ell=1}^{2^L}\sum_{j=1}^{L}\delta(j,r)\theta_r^{\ell}\,P_{\ell j}(t)$$

$$= \gamma_i^{-1}\sum_{r=1}^{L} c_{ir}\sum_{\ell=1}^{2^L}\theta_r^{\ell}\,P_{\ell r}(t) \tag{6.19}$$

$$t\in[0,T]\;,\quad i=1,2$$

equation (6.12) reduce to the following:

For $\ell=1,\dots,2^L$

$$\begin{cases} dP_{\ell j}(t) = \sum_{i=1}^{L}\lambda_{ij}\,P_{\ell i}(t)dt + P_{\ell j}(t)\sum_{i=1}^{2}\gamma_i^{-2}(c_{ij}\theta_j^{\ell} - \hat{g}_i(t))(dy_i(t)-\hat{g}_i(t)dt), \\ t\in(0,T)\;,\quad j=1,\dots,L \end{cases} \tag{6.20}$$

where

$$\hat{g}_i(t) = \sum_{\ell=1}^{2^L}\sum_{r=1}^{L} c_{ir}\,\theta_r^{\ell}\,P_{\ell r}(t)\quad,\quad t\in[0,T]\;,\quad i=1,2\quad. \tag{6.21}$$

Thus, the optimal (minimum variance) estimate of θ given Y^t, viz. $\hat{\theta}(t) = E[\theta|F_t^y]$, is given by

$$\hat{\theta}_j(t) = \sum_{\ell=1}^{2^L}\theta_j^{\ell}\sum_{r=1}^{L}P_{\ell r}(t)\quad,\quad t\in[0,T]\;,\quad j=1,\dots,L \tag{6.22}$$

and $\hat{n}(t) = E[n(t)|F_t^y]$ is given by

$$\hat{n}(t) = \sum_{j=1}^{L} j \sum_{\ell=1}^{2^L} P_{\ell j}(t) \quad , \quad t \in [0,T] \; . \tag{6.23}$$

Equations (6.20)-(6.21) cosntitute the filter equations for computing $\hat{\theta}(t)$ and $\hat{n}(t)$.

Given $\ell \in \{1,\ldots,2^L\}$ and $\{\hat{g}(t),\ t \in [0,T]\}$. Then equations (6.20) determine a filter, whose input is $\{dy(t),\ \hat{g}(t);\ \ t \in [\ 0,T]\}$ and whose output is $\{P_{\ell j}(t),\ j=1,\ldots,L,\ t \in [0,T]\}$. Denote this filter by F_ℓ. Fig. 6.1 shows the block diagram of the optimal estimator, where $\underline{P}_\ell(t) \triangleq (P_{\ell 1}(t),\ldots,P_{\ell L}(t))$, $\ell=1,\ldots,2^L$.

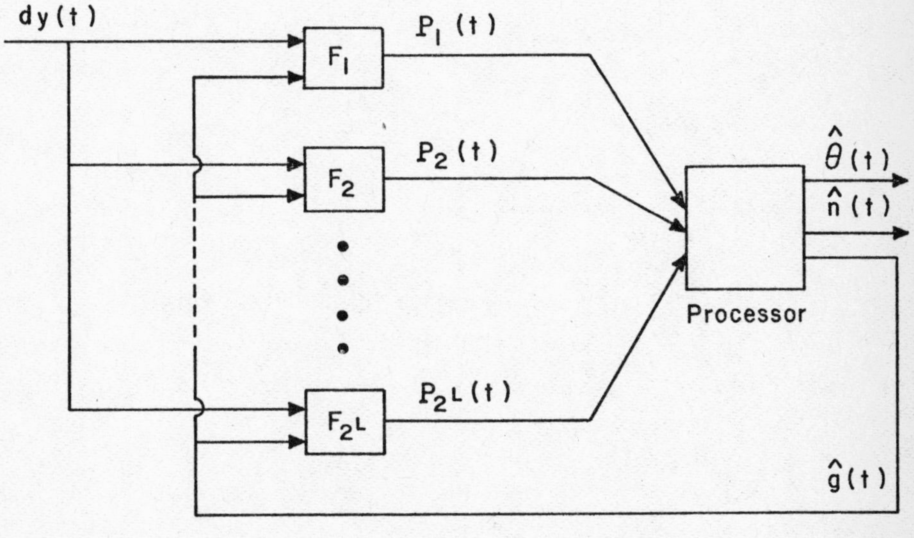

Fig.6.1: The block diagram of the optimal filter

The Processor in Fig.6.1 computes $\hat{g}(t)$, $\hat{\theta}(t)$ and $\hat{n}(t)$ by using equations (6.21), (6.22) and (6.23), respectively. Fig. 6.1 illustrates the parallel processing features of the optimal filter. In general, the numbers $\{P_{\ell j}(0)\}$ are unknown. In the next section an algorithm for computing $\{(\hat{\theta}(k\Delta),\hat{n}(k\Delta))\}_{k=1}^{N}$ is suggested.

6.3 AN ALGORITHM FOR COMPUTING $(\hat{\theta},\hat{n})$

In the sequel the following notation is used: $y(k) \triangleq y(k\Delta)$, $P_{\ell j}(k) \triangleq P_{\ell j}(k\Delta)$, $\hat{g}_i(k) \triangleq \hat{g}_i(k\Delta)$, $\hat{\theta}(k) \triangleq \hat{\theta}(k\Delta)$ and $\hat{n}(k) \triangleq \hat{n}(k\Delta)$, $k=0,1,\ldots,N$; $\ell=1,\ldots,2^L$; $j=1,\ldots,L$, $i=1,2$.

We choose $P_{\ell j}(0) \triangleq 2^{-L} L^{-1}$, $\ell=1,\ldots,2^L$, $j=1,\ldots,L$. Then,

$$\hat{g}_i(0) = 2^{-L} L^{-1} \sum_{\ell=1}^{2^L} \sum_{r=1}^{L} c_{ir} \theta_r^\ell \quad , \quad i=1,2,.$$

Let ε be a given positive number.

Algorithm 6.3

1. $k:=0$

2. For $\ell=1,\ldots,2^L$

   for $j=1,\ldots,L$ calculate

$$P_{\ell j}(k+1):=P_{\ell j}(k) + \sum_{i=1}^{L} \lambda_{ij} P_{\ell i}(k)\Delta + P_{\ell j}(k) \sum_{i=1}^{2} \gamma_i^{-2}(c_{ij}\theta_j^\ell - \hat{g}_i(k))(y_i(k+1)-y_i(k)-\hat{g}_i(k)\Delta) \tag{6.24}$$

$$P_{\ell j}(k+1):=\max(0,P_{\ell j}(k+1)) \tag{6.25}$$

3. $$Z(k+1):= \sum_{\ell=1}^{2^L} \sum_{j=1}^{L} P_{\ell j}(k+1) \tag{6.26}$$

4. If $Z(k+1) \geq \varepsilon$ then: for $\ell=1,\ldots,2^L$; $j=1,\ldots,L$,

$$P_{\ell j}(k+1):=P_{\ell j}(k+1)/Z(k+1) \tag{6.27}$$

   Otherwise : stop.

5. For $i=1,2$

$$\hat{g}_i(k+1):= \sum_{\ell=1}^{2^L} \sum_{r=1}^{L} c_{ir} \theta_r^\ell P_{\ell r}(k+1) \tag{6.28}$$

6. For $j=1,\ldots,L$

$$\hat{\theta}_j(k+1) := \sum_{\ell=1}^{2^L} \theta_j^{\ell} \sum_{r=1}^{L} P_{\ell r}(k+1) \tag{6.29}$$

7. $$\hat{n}(k+1) := \sum_{j=1}^{L} j \sum_{\ell=1}^{2^L} P_{\ell j}(k+1) \tag{6.30}$$

8. If k=N stop. Otherwise k:=k+1 and go to 2.

6.4 EXAMPLES

Suppose that $\{\lambda_{ij}\}$ satisfy the following equations

$$\lambda_{j-1,j} = q/2 \qquad 3 \le j \le L \tag{6.31}$$

$$\lambda_{j+1,j} = q/2 \qquad 1 \le j \le L-2 \tag{6.32}$$

$$\lambda_{j,j} = -q \qquad 1 \le j \le L \tag{6.33}$$

$$\lambda_{1,2} = \lambda_{L,L-1} = q \tag{6.34}$$

$$\lambda_{o,1} = \lambda_{L+1,L} = 0.$$

The numerical study of Algorithm 6.3 was carried out via numerical ex= perimentation. Each experiment (run) consisted of the following stages:

(i) A sample path of $\{n(t), t \in [0,T]\}$ is constructed by applying the following procedure:

1. Construct the sequence $\{T_i\}_{i=0}^{M+1}$ in the same manner as in Section 5.4
2. Take $n(0)=j$, $j \in \{1,2,\ldots,L\}$, then

$$n(k) = j \qquad 0 \le k\Delta < T_1 .$$

For $i=0,1,\ldots,M-1$, $T_M \le T < T_{M+1}$,

If $n(k) = \ell$, $T_i \le k\Delta < T_{i+1}$, $\ell \in \{2,3,\ldots,L-1\}$,

then

$$n(k) = \begin{cases} \ell-1 & \text{w.p. } 0.5 \\ \ell+1 & \text{w.p. } 0.5 \end{cases} \qquad T_{i+1} \le k\Delta < T_{i+2}$$

If $n(k) \in \{1,L\}$, $T_i \le k\Delta < T_{i+1}$,

then $n(k)=2$ or $n(k)=L-1$, respectively, for $T_{i+1} \le k\Delta < T_{i+2}$.

(ii) Simulate equations (6.3) by applying the following procedure:

$y_i(0)=0$, $i=1,2$. For $k=0,1,\ldots,N$

1. For $i=1,2$, calculate

$$y_i(k+1) = y_i(k) + c_{in(k)}\theta_{n(k)}\Delta + \sqrt{\Delta}\,\gamma_i\, v_i(k) \tag{6.36}$$

where $\{v(k)\}_{k=0}^{N}$ is a sequence of independent $\mathbb{R}^2$-valued Gaussian random elements with

$$Ev(k)=0 \text{ and } E[v(k)v'(\ell)] = \delta_{k\ell}\, I_2, \quad k,\ell=0,\ldots,N. \tag{6.37}$$

(iii) Apply Algorithm 6.3, where $\{y(k+1)-y(k)\}_{k=0}^{N-1}$ act as an input to the filter and $\{\hat{\theta}(k),\hat{n}(k)\}_{k=1}^{N}$ act as the filter's output.

6.4.1 Example 1

In this example the following parameters were used: $L=8$,

$$\{(c_{1j},c_{2j})\}_{j=1}^{8} = \{(0.5,0.5),(1.0,0.5),(1.5,0.5),(2.0,0.5),(0.5,1.0),$$
$$(1.0,1.0),(1.5,1.0),(2.0,1.0)\}$$

$a_{1j} = a_{2j} = 0.4999$, $j=1,\ldots,8$; $\gamma_1 = \gamma_2 = 0.05$; $\Delta = 10^{-3}$; $n(0)=3$; $q=3,6,9$. Some of the numerical results are given below.

(a) $\theta = (1,0,1,0,1,0,1,0)$

$q=3$: $\hat{\theta}(9999) = (1.0000,.0000,1.0000,.0005,1.0000,.5064,.5000,.5000)$

q=6: $\hat{\theta}(9999)$ = (1.0000,.0000,1.0000,.0000,1.0000,.0000,1.0000,.0359)

q=9: $\hat{\theta}(9999)$ = (1.0000,.0000,1.0000,.0000,1.0000,.0011,1.0000,.0449)

(b) θ = (0,1,1,0,1,1,0,0)

q=3: $\hat{\theta}(9999)$ = (.0000,1.0000,1.0000,.0000,1.0000,.5067,.4999,.5000)

q=6: $\hat{\theta}(9999)$ = (.0000,1.0000, 1.0000,.0000,1.0000,1.0000,.0000,.0047)

q=9: $\hat{\theta}(9999)$ = (.0000,1.0000,1.0000,.0000,1.0000,1.0000,.0001,.0130)

(c) θ = (0,1,0,1,0,1,0,1)

q=3: $\hat{\theta}(9999)$ = (.0000,1.0000,.0001,1.0000,.1339,.3952,.4823,.4977)

q=6: $\hat{\theta}(9999)$ = (.0395,1.0000,.0000,1.0000,.0011,1.0000,.0000,1.0000)

q=9: $\hat{\theta}(9999)$ = (.0000,1.0000,.0000,1.0000,.0000,1.0000,.0007,.9951)

(d) θ = (0,0,0,0,0,0,0,0)

q=3: $\hat{\theta}(9999)$ = (.5581,.7205,.1717,.0054,.0084,.0314,.0975,.1893)

q=6: $\hat{\theta}(9999)$ = (.5549,.7402,.1342,.0013,.0006,.0016,.0062,.0241)

q=9: $\hat{\theta}(9999)$ = (.5891,.6781,.0829,.0006,.0002,.0006,.0012,.0054)

(e) θ = (1,1,1,1,1,1,1,1)

q=3: $\hat{\theta}(9999)$ = (1.0000,1.0000,1.0000,1.0000,1.0000,.5076,.4999,.5000)

q=6: $\hat{\theta}(9999)$ = (1.0000,1.0000,1.0000,1.0000,1.0000,1.0000,1.0000,1.0000)

q=9: $\hat{\theta}(9999)$ = (1.0000,1.0000,1.0000,1.0000,1.0000,1.0000,1.0000,.7650)

6.4.2 Example 2

In this example the following parameters were used: L=8,

$$\{(c_{1j},c_{2j})\}_{j=1}^{8} = \{(1.0,1.0),(2.0,2.0),(3.0,3.0),(4.0,4.0),(-1.0,1.0),(-2.0,2.0),(-3.0,3.0),(-4.0,4.0)\}$$

$a_{1j} = a_{2j} = 0.9999$, $j=1,\ldots,8$; $\gamma_1 = \gamma_2 = 0.5$; $\Delta = 10^{-3}$; $n(0)=3$; $q=3,6,9$. Some of the numerical results are given below.

(a) $\theta = (1,0,1,0,1,0,1,0)$

q=3: $\hat{\theta}(9999) = (.4017,.3558,.6558,.4419,.3846,.6329,.5034,.4766)$

q=6: $\hat{\theta}(9999) = (.3360,.3417,.9697,.1409,.3618,.4676,.7990,.2351)$

q=9: $\hat{\theta}(9999) = (.5685,.0784,.6060,.4196,.6844,.5981,.8679,.5897)$

(b) $\theta = (0,1,0,1,0,1,0,1)$

q=3: $\hat{\theta}(9999) = (.4805,.9394,.2916,.3632,.5200,.4944,.4959,.4985)$

q=6: $\hat{\theta}(9999) = (.3425,.8668,.2167,.8113,.2140,.5681,.4546,.5241)$

q=9: $\hat{\theta}(9999) = (.4861,.9256,.2005,.7851,.2314,.7163,.3211,.3135)$

(c) $\theta = (0,1,1,0,1,1,0,0)$

q=3: $\hat{\theta}(9999) = (.5104,.9967,.9638,.0694,.6571,.7152,.5390,.5014)$

q=6: $\hat{\theta}(9999) = (.3066,.9605,.9559,.2549,.2558,.8827,.1288,.1343)$

q=9: $\hat{\theta}(9999) = (.3309,.9596,.7696,.4020,.5016,.8744,.0665,.0952)$

(d) $\theta = (0,0,0,0,0,0,0,0)$

q=3: $\hat{\theta}(9999) = (.3908,.2147,.1458,.1131,.3378,.2105,.2367,.2806)$

q=6: $\hat{\theta}(9999) = (.3905,.1819,.0849,.0375,.3230,.1411,.1552,.2097)$

q=9: $\hat{\theta}(9999) = (.3908,.1560,.0523,.0134,.3212,.0998,.0974,.1553)$

(e) $\theta = (1,1,1,1,1,1,1,1)$

q=3: $\hat{\theta}(9999) = (.7433,.9999,.9985,.3931,.7083,.7304,.5414,.5018)$

q=6: $\hat{\theta}(9999) = (.4002,.8534,.9979,.9959,.5446,.7619,.9993,.9703)$

q=9: $\hat{\theta}(9999)$ = (.6110,.9984,.9975,.9713,.2990,.9214,.9967,.8916)

6.4.3 Example 3

In this example the same set of parameters as in Example 2 was used, except for $\gamma_1 = \gamma_2 = 0.1$. Some of the numerical results are given below.

(a) θ = (1,0,1,0,1,0,1,0)

q=3: $\hat{\theta}(9999)$ = (1.0000,.0000,1.0000,.0000,1.0000,.5291,.4993,.5000)

q=6: $\hat{\theta}(9999)$ = (1.0000,.0000,1.0000,.0000,1.0000,.0000,1.0000,.0033)

q=9: $\hat{\theta}(9999)$ = (1.0000,.0000,1.0000,.0000,1.0000,.0000,1.0000,.0549)

(b) θ = (0,1,1,0,1,1,0,0)

q=3: $\hat{\theta}(9999)$ = (.0000,1.0000,1.0000,.0000,1.0000,.5297,.4993,.5000)

q=6: $\hat{\theta}(9999)$ = (.0002,1.0000,1.0000,.0000,1.0000,1.0000,.0000,.0008)

q=9: $\hat{\theta}(9999)$ = (.0000,1.0000,1.0000,.0000,1.0000,1.0000,.0005,.0528)

(c) θ = (0,1,0,1,0,1,0,1)

q=3: $\hat{\theta}(9999)$ = (.0001,1.0000,.0000,1.0000,.0905,.3700,.4868,.4986)

q=6: $\hat{\theta}(9999)$ = (.8503,1.0000,.0000,1.0000,.0001,1.0000,.0000,1.0000)

q=9: $\hat{\theta}(9999)$ = (.0000,1.0000,.0000,1.0000,.0000,1.0000,.0000,1.0000)

(d) θ = (0,0,0,0,0,0,0,0)

q=3: $\hat{\theta}(9999)$ = (.2408,.1646,.1065,.0691,.3747,.1813,.3004,.5795)

q=6: $\hat{\theta}(9999)$ = (.0939,.0277,.0161,.0127,.2259,.1577,.3424,.6813)

q=9: $\hat{\theta}(9999)$ = (.0465,.0051,.0030,.0023,.1358,.0945,.2966,.7516)

(e) θ = (1,1,1,1,1,1,1,1)

q=3: $\hat{\theta}(9999)$ = (1.000,1.0000,1.0000,1.0000,1.0000,.5387,.4992,.5000) q=6: $\hat{\theta}(9999)$ = (1.0000,1.0000,1.0000,1.0000,1.0000,1.0000,1.0000,1.0000)

q=9: $\hat{\theta}(9999)$ = (1.0000,1.0000,1.0000,1.0000,1.0000,1.0000,1.0000,1.0000)

TABLE 4.1: n(k) and $\hat{n}(k)$ as functions of k for the case where θ = (0,1,0,1,0,1,0,1), and q=9

k	n(k)	$\hat{n}(k)$	k	n(k)	$\hat{n}(k)$
99	4	3.9374	5299	6	7.6169
299	2	2.0015	5499	6	6.0026
499	2	1.9808	5699	5	5.9298
699	3	2.4333	5899	3	3.9694
899	4	3.9835	6099	3	2.0363
1099	4	4.0000	6299	4	3.9998
1299	2	1.9777	6499	2	2.0481
1499	2	1.9508	6699	1	2.0902
1699	3	2.2095	6899	1	2.6696
1899	3	2.2426	7099	1	2.0006
2099	2	1.9932	7299	1	3.2968
2299	3	3.9836	7499	1	2.0240
2499	3	4.5781	7699	3	2.1377
2699	2	2.0373	7899	5	4.0003
2899	1	1.7727	8099	6	6.0077
3099	2	2.0537	8299	5	6.0017
3299	3	2.0113	8499	5	5.5591
3499	2	2.0004	8699	4	4.0000
3699	1	2.1748	8899	5	4.0001
3899	3	2.3396	9099	2	2.0003
4099	5	4.0422	9299	1	2.0234
4299	8	6.1307	9499	1	2.0045
4499	7	6.3612	9699	2	2.0006
4699	7	5.9584	9899	1	2.0596
4899	7	6.7046	9999	2	2.0015
5099	7	6.5966			

6.5 REFERENCES

6.1 Y.Bar-Shalom and E.Tse, Tracking in a cluttered environment with probabilistic data association, *Automatica*, 11, pp 451-460, 1975.

6.2 Y. Bar-Shalom, Tracking methods in a multitarget environment, *IEEE Trans. on Automatic Control*, AC-23, pp 618-626, 1978.

6.3 T.E. Fortmann, Y. Bar-Shalom and M. Scheffe, Multitarget tracking using probabilistic data association, Proceedings of the 19th IEEE Conference on Decision & Control, Vol.2, pp 807-811, Albuquerque, New Mexico, December 1980.

6.4 Y. Bar-Shalom and G.D.Marcus, Tracking with measurements of uncertain origin and random arrival times, *IEEE Trans. on Automatic Control*, AC-25, pp 802-807, 1980.

6.5 M. Athans, W.B. Davenport, E.R. Ducot and R.R. Tenney, Editors, *Surveillance and Target Tracking*, Proceeding of the Fourth MIT/ONR Workshop on Distributed Information and Decision Systems Motivated by Command-Control-Communications (C^3) Problems, Vol.1, LIDS-R-1156, Laboratory for Information and Decision Systems, MIT, Cambridge, Mass., 1981.

CHAPTER 7

STATE AND PARAMETER ESTIMATION

7.1 INTRODUCTION

Consider the $\mathbb{R}^m$-valued stochastic process $\zeta_x = \{\zeta_x(t),\ t \geq 0\}$ satisfying the equation

$$\zeta_x(t) = x + \int_0^t f(\zeta_x(s),\theta)ds + B(\theta)W(t),\ t \geq 0,\quad x \in \mathbb{R}^m \tag{7.1}$$

with the measurement process given by

$$y(t) = \int_0^t g(\zeta_x(s),\theta)ds + \Gamma v(t)\ ,\quad t \geq 0,\ \ y(t) \in \mathbb{R}^p \tag{7.2}$$

where on a probability space (Ω,F,P), $\mathbf{W} = \{W(t),\ t \geq 0\}$ and $\mathbf{V} = \{v(t),\ t \geq 0\}$ are $\mathbb{R}^m$-valued are $\mathbb{R}^p$-valued standard Wiener processes respectively and θ is a random element with values in a compact set Ω_θ, $\Omega_\theta \subset \mathbb{R}^r$. It is assumed that $\mathbf{W}$, $\mathbf{V}$ and θ are mutually independent. $f : \mathbb{R}^m \times \Omega_\theta \to \mathbb{R}^m$ and $g : \mathbb{R}^m \times \Omega_\theta \to \mathbb{R}^p$ are given functions satisfying:

(i) For any $\theta \in \Omega_\theta$, $f(\cdot,\theta)$ and $g(\cdot,\theta)$ are continuously differentiable with respect to x_i, $i=1,\ldots,m$, on $\mathbb{R}^m$.

(ii) For any $\theta \in \Omega_\theta$, $|f(x,\theta)|_m^2 \leq M(1 + |x|_m^2)$ and $|g(x,\theta)|_p^2 \leq M(1 + |x|_m^2)$, $x \in \mathbb{R}^m$, for some $0 < M < \infty$.

$B(\theta) \in \mathbb{R}^{m\times m}$ and $\Gamma \in \mathbb{R}^{p\times p}$ are matrices satisfying $B_{ij}(\theta) = \delta_{ij}\ \sigma_i(\theta)$, $i,j=1,\ldots,m$ and $\Gamma_{ij} = \delta_{ij}\ \gamma_i$, $i,j=1,\ldots,p$, where for any $\Theta \in \Omega_\theta$, $\{\sigma_i(\theta)\}$ and $\{\gamma_i\}$ are given positive numbers.

Under the assumptions on $f, B(\theta)$, θ and $\mathbf{W}$ it can be shown that equations (7.1) have a unique solution and that (ζ_x,θ) is a Markov process on (Ω,F,P).

Also, using the assumptions on g, Γ and **V** it follows that (ζ_x,θ,Y), where $Y = \{y(t), t \geq 0\}$ is a Markov process on (Ω,F,P).

Denote by F_t^y the smallest σ-field generated by the family of random elements $Y^t = \{y(s) ; 0 \leq s \leq t\}$. The problem dealt with in this chap= ter is to find approximations $\hat{\zeta}_x^{h,y}(k)$ and $\hat{\theta}^{h_1,y}(k)$ to $\hat{\zeta}_x(t)$ and $\hat{\theta}(t)$ respectively, at the instants $t_k = k\Delta$, $k\Delta \in [0,T]$, where

$$\hat{\zeta}_x(t) = E[\zeta_x(t\wedge\tau_T-) \mid F^y_{t\wedge\tau_T-}] , \quad t \in [0,T] \tag{7.3}$$

$$\hat{\theta}(t) = E[\theta \mid F^y_{t\wedge\tau_T-}] , \quad t \in [0,T] \tag{7.4}$$

where $\tau_T = \tau_T(x)$ is the first exit time of $\zeta_x(t)$ from an open and bounded domain $D \subset \mathbb{R}^m$ (see equation (3.20) for the definition of τ_T).

It is well-known that $E[\zeta_x(t) \mid F_t^y]$ and $E[\theta \mid F_t^y]$ are the minimum variance estimates of $\zeta_x(t)$ and θ respectively given Y^t. The problem of finding $(E[\zeta_x(t) \mid F_t^y], E[\theta \mid F_t^y])$, is called *state and parameter estimation*. Ex= tensive work on state and parameter estimation has been done, various approaches being used. For more details see, for example, McGarty [7.1], Anderson and Moore [7.2], Ljung [7.3], Hazewinkel and Willems (the chap= ter on Identification) [7.4], Le Gland in [7.4] and Maybeck [7.5].

In this chapter, the problem of state and parameter estimation is treated by using methods different from those used in the above-mentioned references. Here, methods, similar to those used in Chapters 3 and 5, are applied. Given an open and bounded domain $D \subset \mathbb{R}^m$, let $\tau_T = \tau_T(x)$ be the first exit time of $\zeta_x(t)$ from D, during the time interval $[0,T]$. First, the process $\{(\zeta_x(t\wedge\tau_T),\theta), t \in [0,T]\}$ is approximated by a con= tinuous-time Markov chain $\{(\zeta_x^h(t\wedge\tau_T^h),\theta^{h_1}(t)), t \in [0,T]\}$ with a finite state space $S = D_h \times \{\theta^1,\ldots,\theta^L\}$, where $D_h \triangleq \mathbb{R}_h^m \cap D$ ($\mathbb{R}_h^m$ is given by equation (3.4)). Second, an optimal least-squares filter is derived for the on-line computation of $(\hat{\zeta}_x^h(t), \hat{\theta}^{h_1}(t)) \triangleq (E[\zeta_x^h(t\wedge\tau_T^h-) \mid F^{y,h}_{t\wedge\tau_T^h-}]$,

$E[\theta^{h_1}(t)|F^{y,h}_{t\wedge\tau^h_T-}])$; ($\tau^h_T$ and $F^{y,h}_t$ are defined in a similar manner as in Section 3.2). Third, an estimator $\{(\hat{\zeta}^{h,y}_x(k), \hat{\theta}^{h_1,y}(k)), k\Delta \in [0,T]\}$ is constructed as an approximation to $\{(\hat{\zeta}_x(k\Delta),\hat{\theta}(k\Delta)), k\Delta \in [0,T]\}$ (equations (7.3)-(7.4)) and this estimator is simulated for a variety of examples.

7.2 CONSTRUCTION OF THE MARKOV CHAIN

Let $\mathbb{R}^m_h$ be a grid on $\mathbb{R}^m$ with a constant mesh size h along all axes as de= fined by equation (3.4), and denote by e^i the unit vector along the i-th axis, i=1,...,m. Also, let $\Omega^{h_1}_\theta \triangleq \Omega_\theta \cap \mathbb{R}^r_{h_1}$, where $\mathbb{R}^r_{h_1} \triangleq \{(i_1h_1,i_2h_2,\ldots,i_rh_r): i_k = 0, \pm 1, \pm 2,\ldots; k=1,\ldots,r\}$ and $\{h_i\}^r_{i=1}$ are given positive numbers. Then, $\Omega^{h_1}_\theta$ can be written as $\Omega^{h_1}_\theta = \{\theta^1,\ldots,\theta^L\}$ for some integer $0 < L < \infty$.

Define the following function $\lambda : (\mathbb{R}^m_h \times \Omega^{h_1}_\theta) \times (\mathbb{R}^m_h \times \Omega^{h_1}_\theta) \to \mathbb{R}$ by

$$\begin{cases} \lambda(x,\theta^j;x,\theta^j) \triangleq -\sum_{i=1}^{m} (\sigma^2_i(\theta^j) + h|f_i(x,\theta^j)|)/h^2 , \\ x \in \mathbb{R}^m_h , \quad j=1,\ldots,L \end{cases} \tag{7.6}$$

$$\begin{cases} \lambda(x,\theta^j;x+e^ih,\theta^j) \triangleq (\sigma^2_i(\theta^j)/2 + h\, f^+_i(x,\theta^j))/h^2, \quad i=1,\ldots,m \\ x \in \mathbb{R}^m_h , \quad j=1,\ldots,L \end{cases} \tag{7.7}$$

$$\begin{cases} \lambda(x,\theta^j;x-e^ih,\theta^j) \triangleq (\sigma^2_i(\theta^j)/2 + h\, f^-_i(x,\theta^j))/h^2, \quad i=1,\ldots,m \\ x \in \mathbb{R}^m_h , \quad j=1,\ldots,L \end{cases} \tag{7.8}$$

$$\lambda(x,\theta^j;z,\theta^\ell) = 0 , \quad x,z \in \mathbb{R}^m_h , \quad j \neq \ell \tag{7.9}$$

$$\lambda(x,\theta^j;z,\theta^j) = 0 , \quad x \in \mathbb{R}^m_h , \quad z \in U_x, \; j=1,\ldots,L \tag{7.10}$$

where for any $\alpha \in \mathbb{R}$, $\alpha^+ = \max(0,\alpha)$, $\alpha^- = -\min(0,\alpha)$ and

$$U_x \triangleq \{z \in \mathbb{R}_h^m : z \neq x \text{ and } z \neq x \pm e^i h \quad , \quad i=1,\ldots,m\}. \tag{7.11}$$

Note that $\lambda(x,\theta^j;z,\theta^\ell) \geq 0$ for $x,z \in \mathbb{R}_h^m$, $x \neq z$, and $j,\ell=1,\ldots,L$, and

$$\sum_{\substack{z \in \mathbb{R}_h^m \\ z \neq x}} \lambda(x,\theta^j;z,\theta^j) = 0 \quad , \quad x \in \mathbb{R}_h^m \quad , \quad j=1,\ldots,L \; . \tag{7.12}$$

Hence, given $(x,\theta) \in \mathbb{R}_h^m \times \Omega_\theta^{h_1}$, we can construct a continuous-time Markov chain $\{(\zeta_x^h(t), \theta^{h_1}(t)),\ t \in [0,T]\}$ with state space $\tilde{S} = \mathbb{R}_h^m \times \Omega_\theta^{h_1}$, by defining the following set of tranisition probabilities

$$\begin{cases} P((\zeta_x^h(t+\Delta),\theta^{h_1}(t+\Delta))=(z,\theta^j) \mid (\zeta_x^h(t),\theta^{h_1}(t))=(z,\theta^j)) \triangleq 1+\lambda(z,\theta^j;z,\theta^j)\Delta+O(\Delta^2) \\ (z,\theta^j) \in \mathbb{R}_h^m \times \Omega_\theta^{h_1} \end{cases} \tag{7.13}$$

$$\begin{cases} P((\zeta_x^h(t+\Delta),\theta^{h_1}(t+\Delta))=(z\pm e^i h,\theta^j) \mid (\zeta_x^h(t),\theta^{h_1}(t))=(z,\theta^j)) \triangleq \lambda(z,\theta^j;z\pm e^i h,\theta^j)\Delta + O(\Delta^2) \\ (z,\theta^j) \in \mathbb{R}_h^m \times \Omega_\theta^{h_1} \ , \ i=1,\ldots,m \end{cases} \tag{7.14}$$

$$\begin{cases} \sum_{y \in U_z} P((\zeta_x^h(t+\Delta),\theta^{h_1}(t+\Delta))=(y,\theta^j) \mid (\zeta_x^h(t),\theta^{h_1}(t))=(z,\theta^j)) \triangleq O(\Delta^2), \\ (z,\theta^j) \in \mathbb{R}_h^m \times \Omega_\theta^{h_1} \end{cases} \tag{7.15}$$

$$\begin{cases} \sum_{y \in \mathbb{R}_h^m} P((\zeta_x^h(t+\Delta),\theta^{h_1}(t+\Delta))=(y,\theta^\ell) \mid (\zeta_x^h(t),\theta^{h_1}(t))=(z,\theta^j)) = \delta_{j\ell} \\ (z,\theta^j) \in \mathbb{R}_h^m \times \Omega_\theta^{h_1} \ , \ \theta^\ell \in \Omega_\theta^{h_1} \end{cases} \tag{7.16}$$

$$P(\zeta_x^h(0) = x) = 1, \ x \in \mathbb{R}_h^m; \ P(\theta^{h_1}(0) = \theta^j) = \pi_j \ , \ j=1,\ldots,L \; . \tag{7.17}$$

Thus, using equations (7.13)-(7.16) it follows that

$$\begin{cases} E[\zeta_{xi}^h(t+\Delta) - \zeta_{xi}^h(t) \mid (\zeta_x^h(t),\theta^{h_1}(t)) = (z,\theta^j)] = f_i(z,\theta^j)\Delta + hO(\Delta^2) \\ i=1,\ldots,m \quad , \quad (z,\theta^j) \in \mathbb{R}_h^m \times \Omega_\theta^{h_1} \end{cases} \tag{7.18}$$

$$\begin{cases} E[(\zeta^h_{xi}(t+\Delta) - \zeta^h_{xi}(t))(\zeta^h_{xj}(t+\Delta) - \zeta^h_{xj}(t)) | (\zeta^h_x(t),\theta^{h_1}(t)) = (z,\theta^\ell)] \\ \qquad = \delta_{ij}(\sigma_i^2(\theta^\ell) + h|f_i(z,\theta^\ell)|)\Delta + (\delta_{ij}+1)h^2 O(\Delta^2) \\ (z,\theta^\ell) \in \mathbb{R}^m_h \times \Omega^{h_1}_\theta , \quad i,j=1,\ldots,m . \end{cases} \tag{7.19}$$

Equations (7.16) and (7.18)-(7.19) illustrate the relations between the Markov chain $\{(\zeta^h_x(t),\theta^{h_1}(t)),\ t \in [0,T]\}$ and the Markov process $\{(\zeta_x(t),\theta),\ t \in [0,T]\}$.

Let set D and the stopping times τ_T and τ^h_T be defined in the same manner as in equations (3.19),(3.20) and (3.21) respectively, where $D_h \triangleq \mathbb{R}^m_h \cap D$. Define

$$y^h(t) \triangleq \int_0^t g(\zeta^h_x(u),\theta^{h_1}(u))du + \Gamma v(t) \ , \ t \in [0,T] \tag{7.20}$$

and denote by $F^{y,h}_t$ the σ-field generated by $\{y^h(u) \ ; \ 0 \le u \le t\}$.

In the next section an optimal least-squares filter is constructed for the computation of $(E[\zeta^h_x(t\wedge\tau^h_T-) | F^{y,h}_{t\wedge\tau^h_T-}],\ E[\theta^{h_1}(t\wedge\tau^h_T-) | F^{y,h}_{t\wedge\tau^h_T-}])$.

7.3 THE EQUATIONS OF THE OPTIMAL FILTER

Assume that $\sup_{t \in [0,T]} E|\zeta^h_x(t)|^2 < \infty$, $x \in \mathbb{R}^m_h$ and denote

$$G_t \triangleq \sigma((\zeta^h_x(u),\theta^{h_1}(u)),\ v(u);\ \ 0 \le u \le t) \ , \ t \in [0,T] \tag{7.21}$$

$$h_t \triangleq \Gamma^{-1} g(\zeta^h_x(t),\theta^{h_1}(t)) \quad , \quad t \in [0,T] \ , \tag{7.22}$$

$$z^h(t) \triangleq \int_0^t h_u \, du + v(t) \quad , \quad t \in [0,T] \ , \tag{7.23}$$

$$\nu^h(t) \triangleq z^h(t) - \int_0^t E[h_u | F^{y,h}_u]du \quad , \quad t \in [0,T], \tag{7.24}$$

$$\begin{cases} \tilde{p}_{ai}(t) \triangleq P((\zeta^h_x(t),\theta^{h_1}(t)) = (a,\theta^i) | F^{y,h}_t) \\ t \in [0,T] \ , \ (a,\theta^i) \in \mathbb{R}^m_h \times \Omega^{h_1}_\theta \end{cases} \tag{7.25}$$

$$\begin{cases} P_{ai}(t) \triangleq P((\zeta_x^h(t\wedge\tau_T^h-),\theta^{h_1}(t\wedge\tau_T^h-)) = (a,\theta^i)\,|\,F^{y,h}_{t\wedge\tau_T^h-}) \\ t \in [0,T] \quad , \quad (a,\theta^i) \in \mathbb{R}_h^m \times \Omega_\theta^{h_1} \end{cases} \tag{7.26}$$

We further assume that $\int_0^T E|h_t|^2 dt < \infty$.

For each $t \in [0,T]$, the σ-fields G_t and $\sigma(v(s_2)-v(s_1);\ t < s_1 < s_2 \leq T)$ are independent and h_t is G_t-measurable. Thus, by following the same development given in Section 2.7 we obtain:

For j=1,...,L

$$\begin{cases} d\tilde{P}_{aj}(t) = \sum_{c \in \mathbb{R}_h^m} \lambda(c,\theta^j;a,\theta^j)\tilde{P}_{cj}(t)dt \\ \quad + \tilde{P}_{aj}(t) \sum_{i=1}^{p} \gamma_i^{-2}(g_i(a,\theta^j) - \hat{\hat{g}}_i(t))(dy_i^h(t) - \hat{\hat{g}}_i(t)dt) \\ t \in (0,T), \quad (a,\theta^j) \in \mathbb{R}_h^m \times \Omega_\theta^{h_1} \end{cases} \tag{7.27}$$

and

$$\hat{\hat{g}}_i(t) \triangleq \sum_{b \in \mathbb{R}_h^m} \sum_{\ell=1}^{L} g_i(b,\theta^\ell)\tilde{P}_{b\ell}(t), \; t \in [0,T], \; i=1,\ldots,p, \tag{7.28}$$

where $\{\lambda(c,\theta^j;a,\theta^j)\}$ are given by equations (7.6)-(7.10).

Since

$$\begin{cases} P(\zeta_x^h(t) = a|F_t^{y,h}) = P(\bigcup_{j=1}^{L}\{(\zeta_x^h(t),\theta^{h_1}(t)) = (a,\theta^j)\}|F_t^{y,h}) \\ \qquad = \sum_{j=1}^{L} P((\zeta_x^h(t),\theta^{h_1}(t)) = (a,\theta^j)|F_t^{y,h}) \\ \qquad = \sum_{j=1}^{L} \tilde{P}_{aj}(t) \end{cases} \tag{7.29}$$

and

$$\begin{cases} P(\theta^{h_1}(t) = \theta^i|F_t^{y,h}) = P(\bigcup_{a \in \mathbb{R}_h^m}\{(\zeta_x^h(t),\theta^{h_1}(t)) = (a,\theta^i)\}|F_t^{y,h}) \\ \qquad = \sum_{a \in \mathbb{R}_h^m} P((\zeta_x^h(t),\theta^{h_1}(t)) = (a,\theta^i)|F_t^{y,h}) \\ \qquad = \sum_{a \in \mathbb{R}_h^m} \tilde{P}_{ai}(t) \end{cases} \tag{7.30}$$

it follows that

$$E[\zeta_x^h(t)|\mathcal{F}_t^{y,h}] = \sum_{a\in\mathbb{R}_h^m} a \sum_{j=1}^{L} \tilde{P}_{aj}(t) \quad , \quad t\in[0,T] \tag{7.31}$$

and

$$E[\theta^{h_1}(t)|\mathcal{F}_t^{y,h}] = \sum_{j=1}^{L} \theta^j \sum_{a\in\mathbb{R}_h^m} \tilde{P}_{aj}(t) \quad , \quad t\in[0,T]. \tag{7.32}$$

In order to obtain the filter equations for computing $\{P_{ai}(t), (a,\theta^i)\in D_h\times\Omega_\theta^{h_1}, t\in[0,T]\}$ we follow the discussion given in Section 3.3.

Let $\{\lambda(c,\theta^i;a,\theta^j), a,c\in\mathbb{R}_h^m;\ \theta^i,\theta^j\in\Omega_\theta^{h_1}\}$ be defined by equations (7.6)-(7.10) together with the additional condition

$$\lambda(a,\theta^j;z,\theta^j) = 0,\ (a,\theta^j)\in(\mathbb{R}_h^m-D_h)\times\Omega_\theta^{h_1},\ z\in\mathbb{R}_h^m \tag{7.33}$$

Then, by following the development given in Section 3.3 (see also Section 5.3) the following equations are obtained:

For j=1,...,L

$$\begin{cases} dP_{aj}(t) = \sum_{c\in D_h} \lambda(c,\theta^j;a,\theta^j)P_{cj}(t)dt \\ \qquad + P_{aj}(t)\sum_{i=1}^{p}\gamma_i^{-2}(g_i(a,\theta^j)-\hat{g}_i(t))(dy_i^h(t)-\hat{g}_i(t)dt) \\ t\in(0,T),\quad a\in D_h \end{cases} \tag{7.34}$$

$$\hat{g}_i(t) = \sum_{b\in D_h}\sum_{\ell=1}^{L} g_i(b,\theta^\ell)P_{b\ell}(t) \ ,\ t\in[0,T],\ i=1,\dots,p \tag{7.35}$$

and

$$\hat{\zeta}_x^h(t) = E[\zeta_x^h(t\wedge\tau_T^h-)|\mathcal{F}_{t\wedge\tau_T^h-}^{y,h}] = \sum_{a\in D_h} a\sum_{j=1}^{L}P_{aj}(t) \ ,\ t\in[0,T], \tag{7.36}$$

$$\hat{\theta}^{h_1}(t) = E[\theta^{h_1}(t\wedge\tau_T^h-)|\mathcal{F}_{t\wedge\tau_T^h-}^{y,h}] = \sum_{j=1}^{L}\theta^j\sum_{a\in D_h}P_{aj}(t) \ ,\ t\in[0,T] \tag{7.37}$$

where in equation (3.19), $a_i = L_i h$, i=1,...,m; $\{L_i\}$ are given positive integers, and $\{\lambda(c,\theta^j;a,\theta^\ell)\}$ are given by equations (7.6)-(7.10) and (7.33).

Given $\theta^j \in \Omega_\theta^{h_1}$ and $\{\hat{g}(t),\ t \in [0,T]\}$, equations (7.34) determine a filter, whose input is $\{dy^h(t),\ \hat{g}(t),\ t \in [0,T]\}$ and whose output is $\{P_{aj}(t),\ a \in D_h;\ t \in [0,T]\}$. Denote this filter by $F(\theta^j)$. Fig. 7.1 shows the block diagram of the optimal estimator (equations (7.34)-(7.37)), where $\underline{P}_j(t) \triangleq \{P_{aj}(t),\ a \in D_h\}$, j=1,...,L.

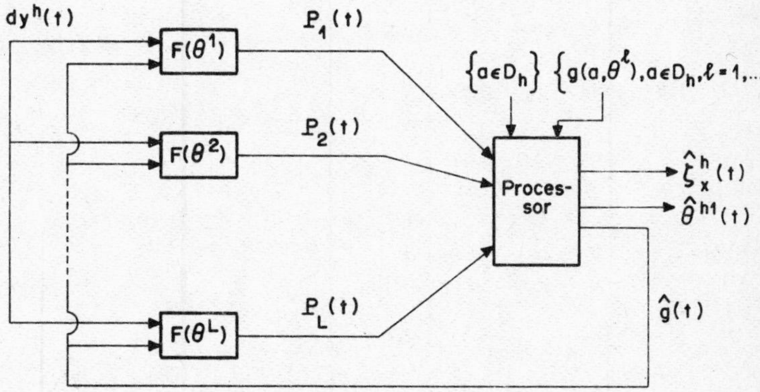

Fig. 7.1: Block diagram of the optimal filter

The Processor in Fig. 7.1 computes $\hat{g}(t)$, $\hat{\zeta}_x^h(t)$ and $\hat{\theta}^{h_1}(t)$ by using equations (7.35), (7.36) and (7.37) respectively. Fig. 7.1 illustrates the parallel processing properties of the optimal filter.

Given $Y^t = \{y(s);\ 0 \le s \le t\}$, $t \in [0,T]$. Then, in order to compute an approximation to $(\hat{\zeta}_x(t),\ \hat{\theta}(t))$ (equations (7.3)-(7.4)), equations (7.34)-(7.35) are solved, where in equations (7.34) the increment dy^h is replaced by dy. Let $\{P^y_{ai}(t), (a,\theta^i) \in D_h \times \Omega_\theta^{h_1}, t \in [0,T]\}$, denote the solution to equations (7.34)-(7.35) (where dy replaces dy^h in (7.34)). Then, a process $(\hat{\zeta}_x^{h,y}, \hat{\theta}^{h_1,y}) = \{(\hat{\zeta}_x^{h,y}(t), \hat{\theta}^{h_1,y}(t)), t \in [0,T]\}$ is defined by

$$\begin{cases} \hat{\zeta}_x^{h,y}(t) \triangleq \sum\limits_{a \in D_h} a \sum\limits_{j=1}^{L} P^y_{aj}(t) \quad , \quad t \in [0,T] \quad , \\ \\ \hat{\theta}^{h_1,y}(t) \triangleq \sum\limits_{j=1}^{L} \theta^j \sum\limits_{a \in D_h} P^y_{aj}(t), \quad t \in [0,T]. \end{cases} \tag{7.38}$$

$(\hat{\zeta}_x^{h,y}, \hat{\theta}^{h_1,y})$ serves here as an approximation to $(\hat{\zeta}_x, \hat{\theta})$ (equations (7.3)-(7.4)). In the next section a procedure for computing $\{(\hat{\zeta}_x^{h,y}(k\Delta), \hat{\theta}^{h_1,y}(k\Delta)), k\Delta \in [0,T]\}$ is suggested. We assume that $\Pi_{ai} = P^y_{ai}(0)$, $(a,\theta^i) \in D_h \times \Omega_\theta^{h_1}$, are unknown.

7.4 AN ALGORITHM FOR COMPUTING $(\hat{\zeta}_x^{h,y}, \hat{\theta}^{h_1,y})$

In the sequel the following notations are used: $y(k) \triangleq y(k\Delta)$, $P^y_{ai}(k) \triangleq P^y_{ai}(k\Delta)$, $\hat{g}_j(k) \triangleq \hat{g}_j(k\Delta)$, $\hat{\zeta}_x^{h,y}(k) \triangleq \hat{\zeta}_x^{h,y}(k\Delta)$, $\hat{\theta}^{h_1,y}(k) \triangleq \hat{\theta}^{h_1,y}(k\Delta)$; $k=0,1,\ldots,N$; $(a,\theta^i) \in D_h \times \Omega_\theta^{h_1}$, $j=1,\ldots,p$.

We choose $P^y_{ai}(0) \triangleq \prod\limits_{i=1}^{m} (2L_i + 1)^{-1} \cdot L^{-1}$, $(a,\theta^i) \in D_h \times \Omega_\theta^{h_1}$. Then,

$$\hat{g}_\ell(0) = L^{-1} \prod_{i=1}^{m} (2L_i + 1)^{-1} \sum_{b \in D_h} \sum_{j=1}^{L} g_\ell(b,\theta^j), \quad \ell=1,\ldots,p.$$

Let ε be a given positive number.

Algorithm 7.4

1. $k:=0$
2. For $j=1,\ldots,L$

   For $a \in D_h$, calculate

$$\begin{aligned} P^y_{aj}(k+1) := P^y_{aj}(k) &+ \sum_{c \in D_h} \lambda(c,\theta^j;a,\theta^j) P^y_{cj}(k)\Delta \\ &+ P^y_{aj}(k) \sum_{i=1}^{p} \gamma_i^{-2}(g_i(a,\theta^j)-\hat{g}_i(k))(y_i(k+1)-y_i(k)-\hat{g}_i(k)\Delta) \end{aligned} \tag{7.39}$$

$$P^y_{aj}(k+1) := \max(0, P^y_{aj}(k+1)) \tag{7.40}$$

3. $$Z(k+1) := \sum_{a \in D_h} \sum_{j=1}^{L} P^y_{aj}(k+1) \tag{7.41}$$

4. If $Z(k+1) \geq \varepsilon$ then : for $(a,\theta^j) \in D_h \times \Omega_\theta^{h_1}$, $P^y_{aj}(k+1) := P^y_{aj}(k+1)/Z(k+1)$ (7.42)

   Otherwise: stop.

5. For $\ell=1,\ldots,p$ calculate

$$\hat{g}_\ell(k+1) := \sum_{b \in D_h} \sum_{j=1}^{L} g_\ell(b,\theta^j) P^y_{bj}(k+1) \tag{7.43}$$

6. $$\hat{\zeta}^{h,y}_x(k+1) := \sum_{a \in D_h} a \sum_{j=1}^{L} P^y_{aj}(k+1) \tag{7.44}$$

$$\hat{\theta}^{h_1,y}(k+1) := \sum_{j=1}^{L} \theta^j \sum_{a \in D_h} P^y_{aj}(k+1) \tag{7.45}$$

7. If $k=N$ or if $\hat{\zeta}^{h,y}_x(k+1) \notin D$ then stop. Otherwise : $k := k+1$ and go to 2.

The problem of establishing conditions for the weak convergence of $(\hat{\zeta}^{h,y}_x, \hat{\theta}^{h_1,y})$ to $(\hat{\zeta}_x, \hat{\theta})$(equations (7.3)-(7.4)) as $h \downarrow 0$ and $\max_i h_i \downarrow 0$ is beyond the scope of this work. Instead, the role of $(\hat{\zeta}^{h,y}_x, \hat{\theta}^{h_1,y})$, as an approxi= mation to $(\hat{\zeta}_x, \hat{\theta})$, is demonstrated in the sequel by means of numerical experimentation.

The numerical study of Algorithm 7.4 was carried out via numerical ex= perimentation. Each experiment (run) was executed in the same manner as in Section 3.4.

7.5 EXAMPLES : THE CASE m=1

In this section cases where $\{\zeta_x(t), t \geq 0\}$ (equation (7.1)) is an $\mathbb{R}$-valued stochastic process are considered. In these cases

$$D_h \triangleq \{ih : i=0,\pm1,\ldots,\pm L_1\} \tag{7.46}$$

and equations (7.39) and (7.43) of Algorithm 7.4 reduce to

$$P^y_{\nu,j}(k+1) := P^y_{\nu,j}(k) + [\lambda(\nu+1,\theta^j;\nu,\theta^j)P^y_{\nu+1,j}(k) + \lambda(\nu,\theta^j;\nu,\theta^j)P^y_{\nu,j}(k)$$

$$+ \lambda(\nu-1,\theta^j;\nu,\theta^j)P^y_{\nu-1,j}(k)]\Delta \qquad (7.47)$$

$$+ P^y_{\nu,j}(k)\gamma^{-2}(g(\nu h,\theta^j) - \hat{g}(k))(y(k+1)-y(k) - \hat{g}(k)\Delta)$$

$j=1,\ldots,L$, $\nu = -L_1,-L_1+1,\ldots,L_1$

and

$$\hat{g}(k+1) = \sum_{\nu=-L_1}^{L_1} \sum_{j=1}^{L} g(\nu h,\theta^j)P^y_{\nu,j}(k+1) \qquad (7.48)$$

respectively, where

$$\lambda(i,\theta^j;\ell,\theta^j) \triangleq \lambda(ih,\theta^j;\ell h,\theta^j) \ , \ j=1,\ldots,L \ , \ i,\ell=0,\pm1,\ldots,\pm L_1 . \qquad (7.49)$$

The following set of systems was considered:

(a) $dx = -\theta x^2 dt + 7dW; \ dy = xdt + 3dv \ , \ t > 0, \ x,y \in \mathbb{R}$ (7.50)

In this case set D was taken to be

$$D \triangleq \{x \in \mathbb{R} : |x| < 250 + \delta\} \ , \ \delta < 1.25 \qquad (7.51)$$

and the following set of parameters was used : $h = 1.25$, $L_1 = 200$, $\zeta_x(0) = 200$, $\Delta = 10^{-3}$ and $N = 10^4$. The following cases were numerically experimented with:

(a-1) $\theta = 5\cdot10^{-5}$; $\theta^1 = 2\cdot10^{-5}$, $\theta^2 = 4\cdot10^{-5}$, $\theta^3 = 6\cdot10^{-5}$, $\theta^4 = 8\cdot10^{-5}$, $\theta^5 = 10^{-4}$

(a-2) $\theta = 5\cdot10^{-5}$; $\theta^1 = 10^{-5}$, $\theta^2 = 3\cdot10^{-5}$, $\theta^3 = 5\cdot10^{-5}$, $\theta^4 = 7\cdot10^{-5}$, $\theta^5 = 9\cdot10^{-5}$.

(b) $dx = -5\cdot10^{-5}x^2 dt + \theta dW; \ dy = xdt + 3dv \ , \ t > 0, \ x,y \in \mathbb{R}$ (7.52)

In this case set D is given by equation (7.51), $h=1.25$, $L_1=200$, $\zeta_x(0)=200$, $\Delta = 10^{-3}$ and $N = 10^4$. The following cases were numerically experimented with:

(b-1) $\theta = 7;\ \theta^1 = 4,\ \theta^2 = 5.5,\ \theta^3 = 7,\ \theta^4 = 8.5,\ \theta^5 = 10.$

(b-2) $\theta = 7;\ \theta^1 = 4,\ \theta^2 = 6,\ \theta^3 = 8,\ \theta^4 = 10.$

(c) $dx = -5\cdot 10^{-5}\, x^2 dt + 7dW;\quad dy = \theta x dt + 3dv,\ t > 0,\quad x,y \in \mathbb{R}$ (7.53)

In this case set D is given by equation (7.51), $h=1.25$, $L_1=200$, $\zeta_x(0)=200$, $\Delta = 10^{-3}$ and $N = 10^4$. The following cases were numerically experimented with:

(c-1) $\theta = 1;\ \theta^1 = 0.8,\ \theta^2 = 0.9,\ \theta^3 = 1,\ \theta^4 = 1.1,\ \theta^5 = 1.2$

(c-2) $\theta = 1;\ \theta^1 = 0.7,\ \theta^2 = 0.9,\ \theta^3 = 1.1,\ \theta^4 = 1.3.$

(d) $dx = -\theta_1 x^2 dt + \theta_2 dW;\ dy = x dt + 3dv\ ,\ t > 0\ ,\quad x,y \in \mathbb{R}$ (7.54)

In this case set D is given by equation (7.51), $h=1.25$, $L_1=200$, $\zeta_x(0)=200$, $\Delta = 10^{-3}$ and $N = 10^4$. The following cases were numerically experimented with:

(d-1) $\theta = (5\cdot 10^{-5},7);\theta^1 = (2\cdot 10^{-5},5),\ \theta^2 = (4\cdot 10^{-5},7),\ \theta^3 = (6\cdot 10^{-5},9),\ \theta^4 = (8\cdot 10^{-5},11)$

(d-2) $\theta = (5\cdot 10^{-5},7);\ \theta^1 = (3\cdot 10^{-5},4),\ \theta^2 = (5\cdot 10^{-5},6),\ \theta^3 = (7\cdot 10^{-5},8),\ \theta^4 = (9\cdot 10^{-5},10)$

(d-3) $\theta = (5\cdot 10^{-5},7);\ \{\theta^i\}_{i=1}^9 = \{(2\cdot 10^{-5},5),(2\cdot 10^{-5},7),(2\cdot 10^{-5},9),(4\cdot 10^{-5},5)$ $(4\cdot 10^{-5},7),(4\cdot 10^{-5},9),(6\cdot 10^{-5},5),(6\cdot 10^{-5},7),(6\cdot 10^{-5},9)\}$

(d-4) $\theta = (5\cdot 10^{-5},7);\{\theta^i\}_{i=1}^9 = \{(3\cdot 10^{-5},4),(3\cdot 10^{-5},6),(3\cdot 10^{-5},8),(5\cdot 10^{-5},4),$ $(5\cdot 10^{-5},6),(5\cdot 10^{-5},8),(7\cdot 10^{-5},4),(7\cdot 10^{-5},6),(7\cdot 10^{-5},8)\}.$

Some of the results of the corresponding runs are given in the following figures. All the graphs in this section were plotted using the set of points $\{t'_k = 50k\Delta : k=0,1,\dots,200\}$.

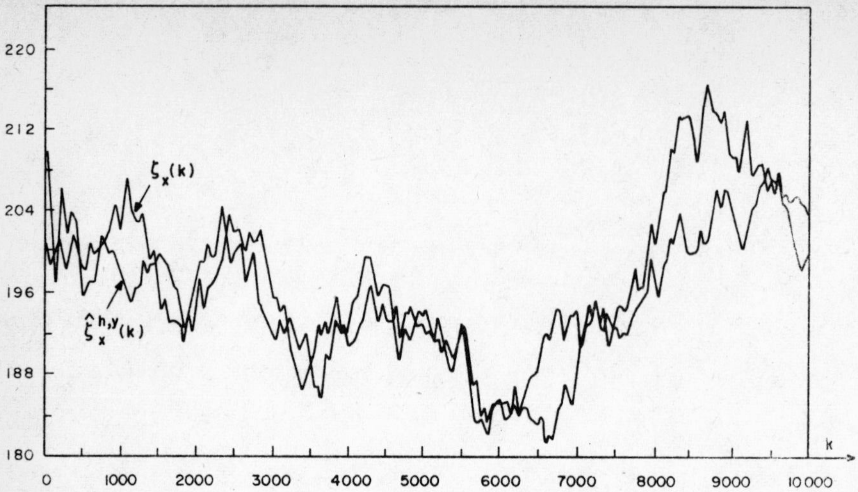

<u>Fig.7.2a:</u> $\zeta_x(k)$ and $\hat{\zeta}_x^{h,y}(k)$ as functions of k for the system given by equations (7.50), case (a-1).

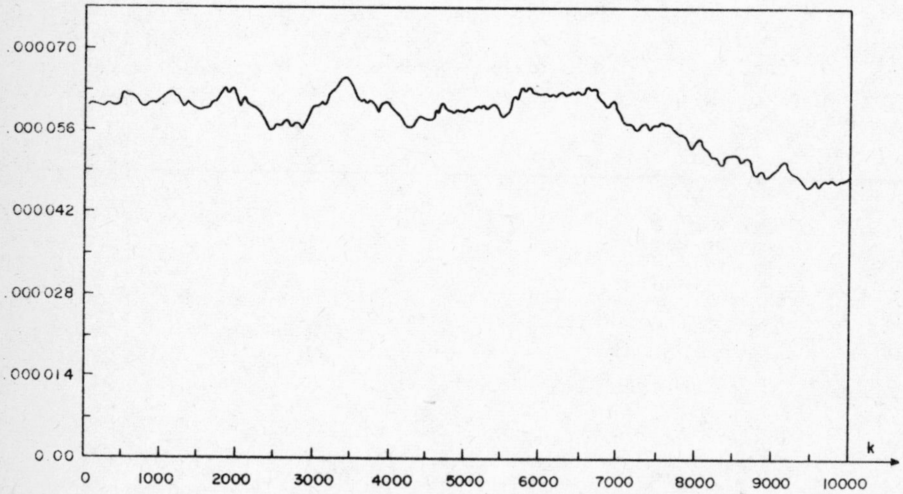

<u>Fig.7.2b:</u> $\hat{\theta}^{h_1,y}(k)$ as a function of k for the system given by equations (7.50), case (a-1).

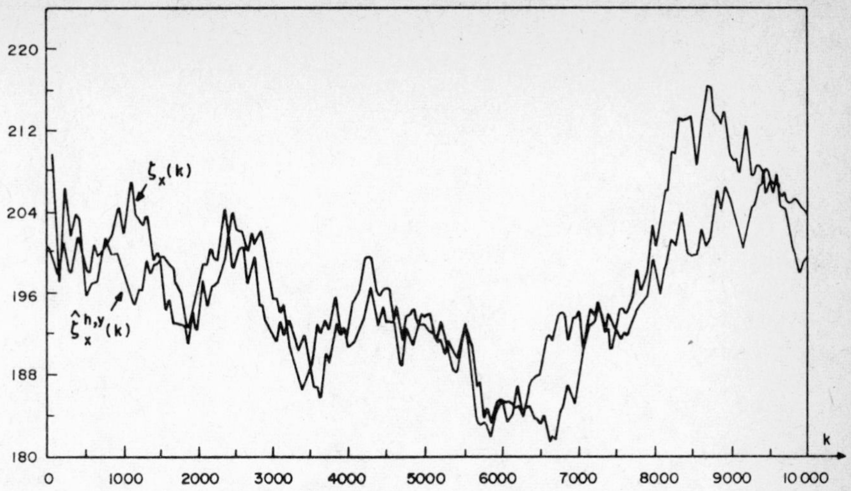

<u>Fig.7.2c:</u> $\zeta_x(k)$ and $\hat{\zeta}_x^{h,y}(k)$ as functions of k for the system given by equations (7.50), case (a-2).

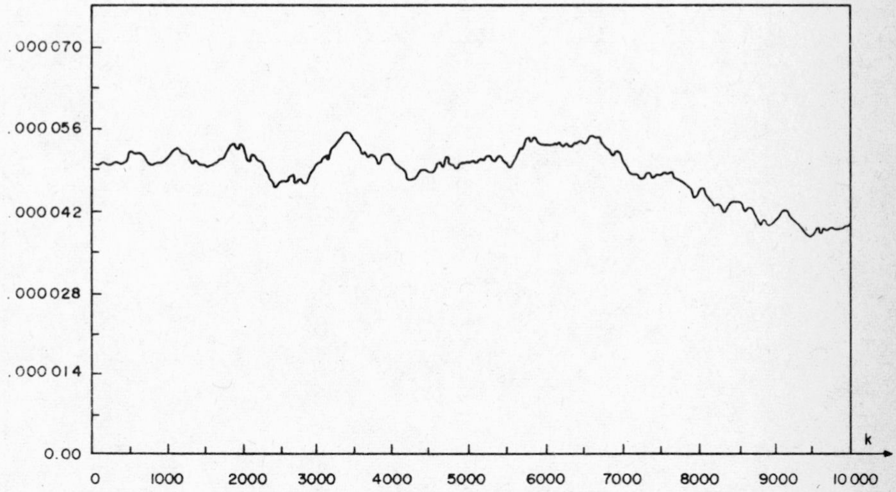

<u>Fig.7.2d:</u> $\hat{\theta}^{h_1,y}(k)$ as a function of k for the system given by equations (7.50), case (a-2).

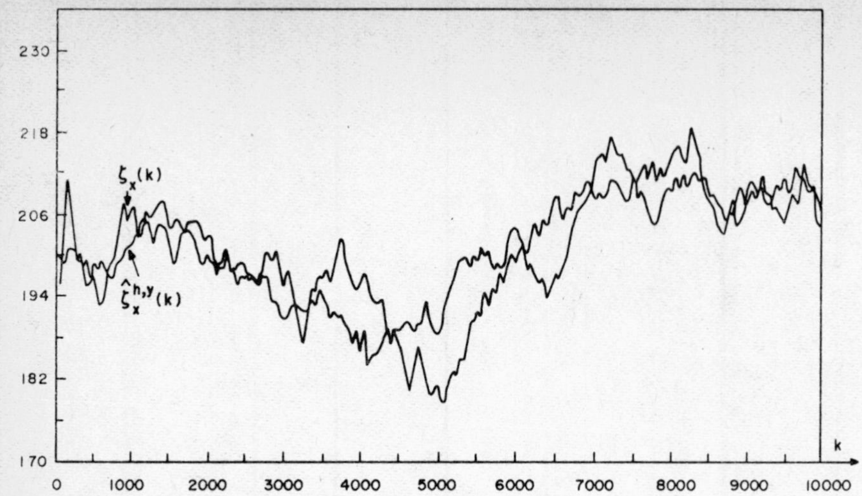

Fig.7.3a: $\zeta_x(k)$ and $\hat{\zeta}_x^{h,y}(k)$ as functions of k for the system given by equations (7.52), case (b-1).

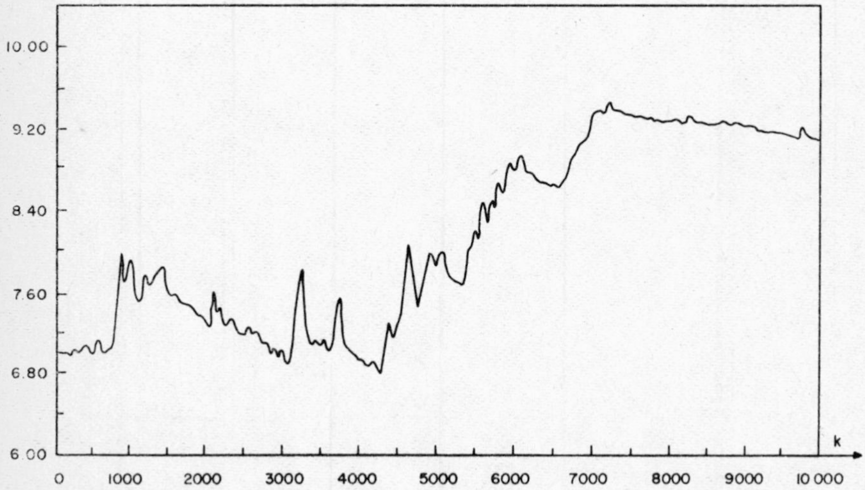

Fig.7.3b: $\hat{\theta}^{h_1,y}(k)$ as a function of k for the system given by equations (7.52), case (b-1).

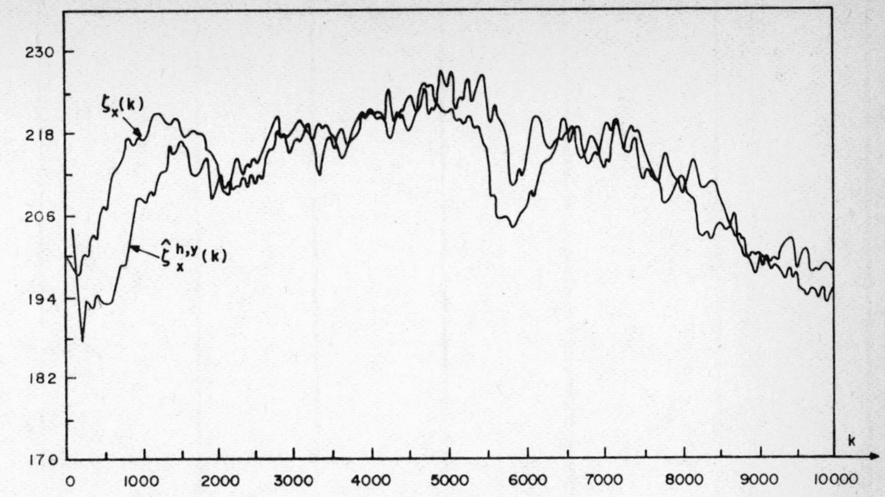

Fig.7.3c: $\zeta_x(k)$ and $\hat{\zeta}_x^{h,y}(k)$ as functions of k for the system given by equations (7.52), case (b-2).

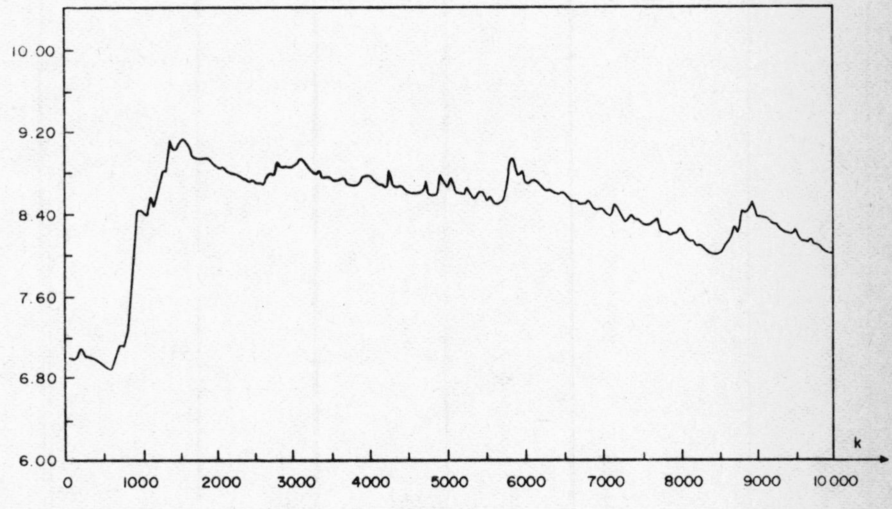

Fig.7.3d: $\hat{\theta}^{h_1,y}(k)$ as a function of k for the system given by equations (7.52), case (b-2).

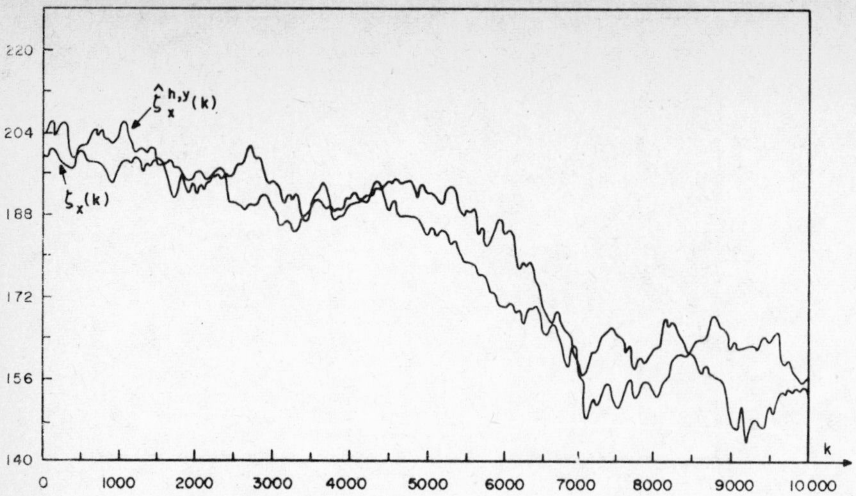

Fig.7.4a: $\zeta_x(k)$ and $\hat{\zeta}_x^{h,y}(k)$ as functions of k for the system given by equations (7.53), case (c-1).

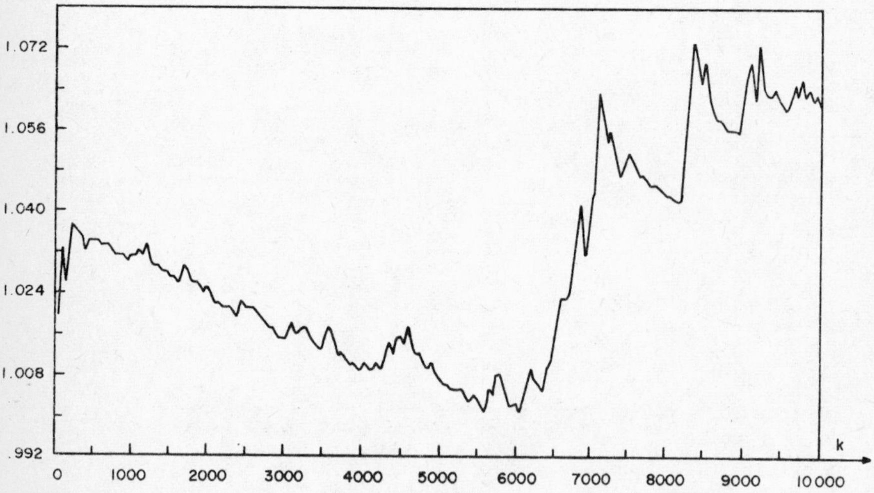

Fig.7.4b: $\hat{\theta}^{h_1,y}(k)$ as a function of k for the system given by equations (7.53), case (c-1).

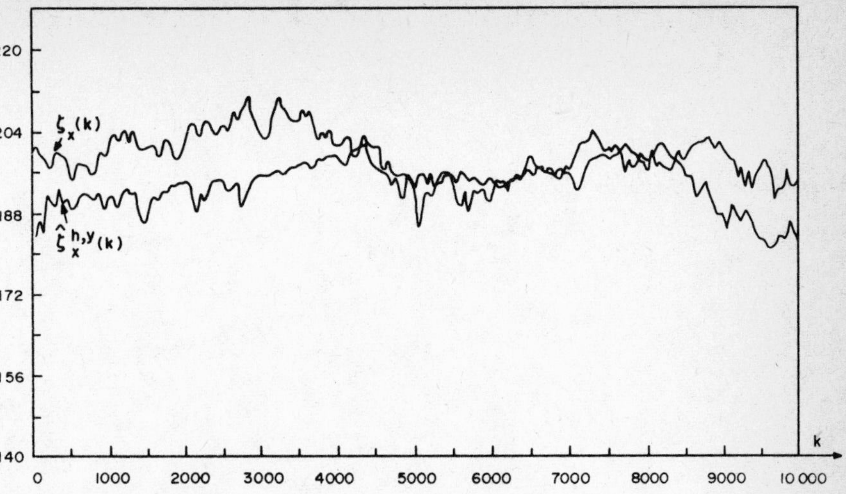

Fig.7.4c: $\zeta_x(k)$ and $\hat{\zeta}_x^{h,y}(k)$ as functions of k for the system given by equations (7.53), case (c-2).

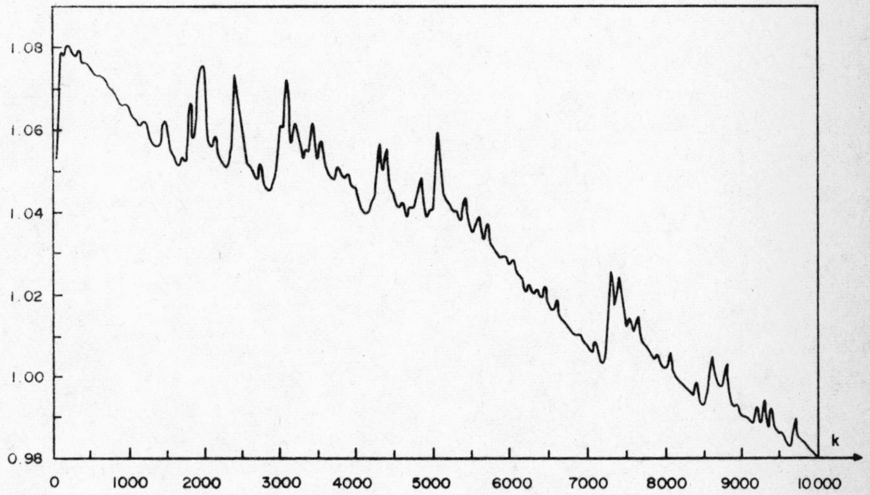

Fig.7.4d: $\hat{\theta}^{h_1,y}(k)$ as a function of k for the system given by equations (7.53), case (c-2).

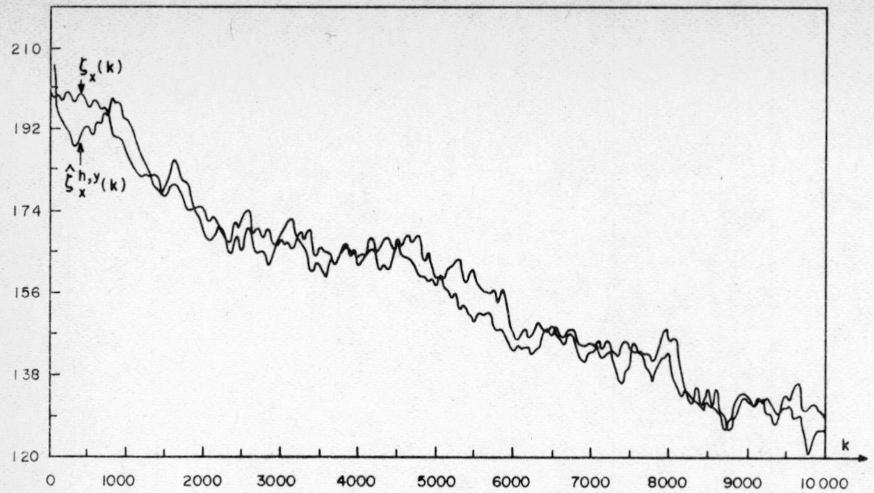

Fig.7.5a: $\zeta_x(k)$ and $\hat{\zeta}_x^{h,y}(k)$ as functions of k for the system given by equations (7.54), case (d-1).

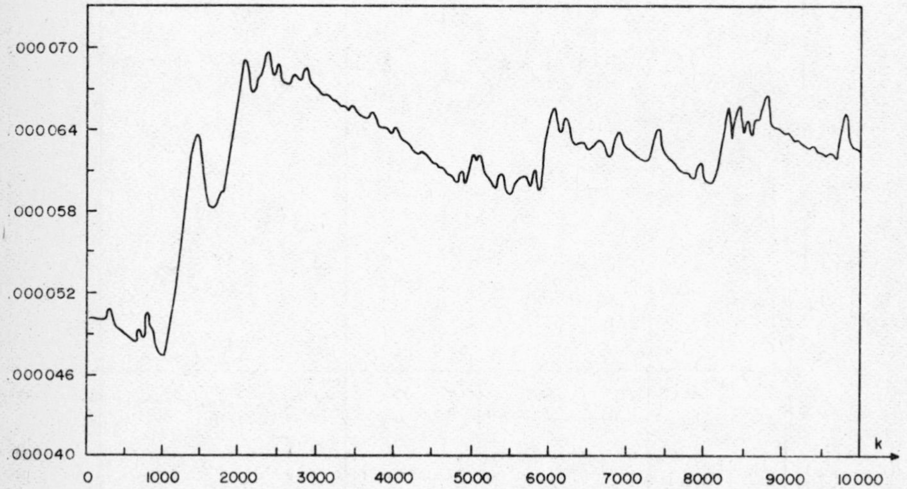

Fig.7.5b: $\hat{\theta}_1^{h_1,y}(k)$ as a function of k for the system given by equations (7.54), case (d-1).

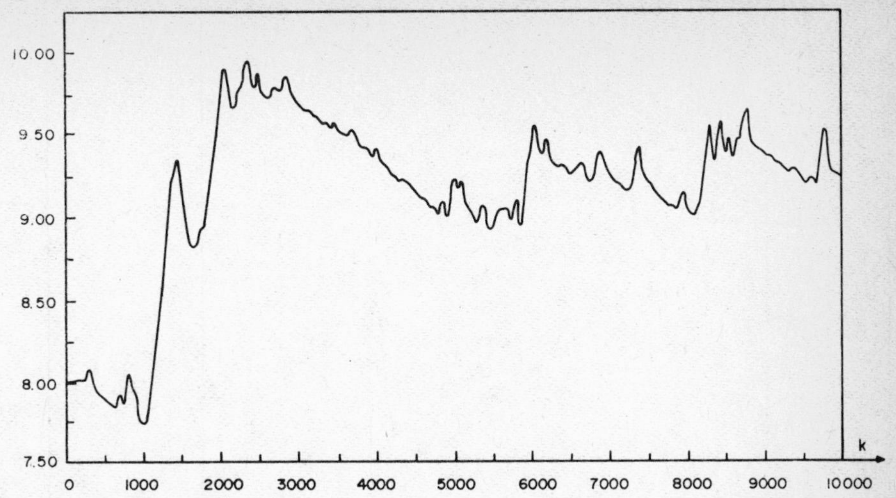

<u>Fig.7.5c:</u> $\hat{\theta}_2^{h_1,y}(k)$ as a function of k for the system given by equations (7.54), case (d-1).

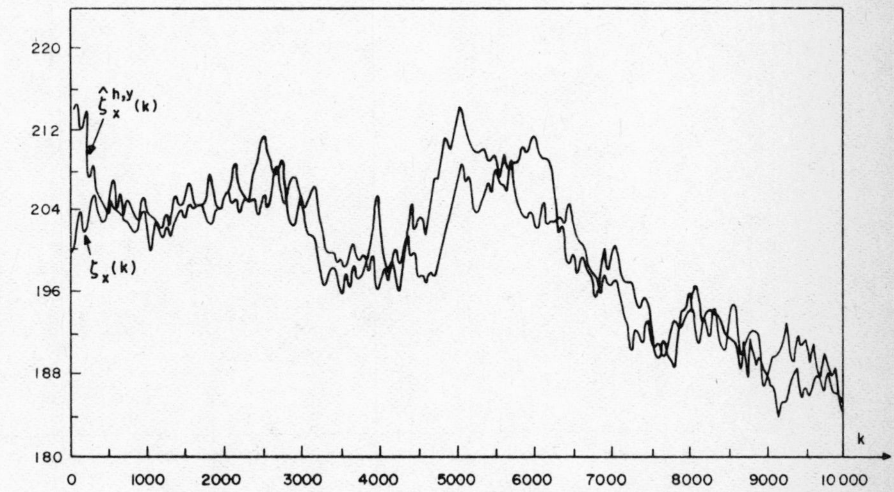

<u>Fig.7.5d:</u> $\zeta_x(k)$ and $\hat{\zeta}_x^{h,y}(k)$ as functions of k for the system given by equations (7.54), case (d-2).

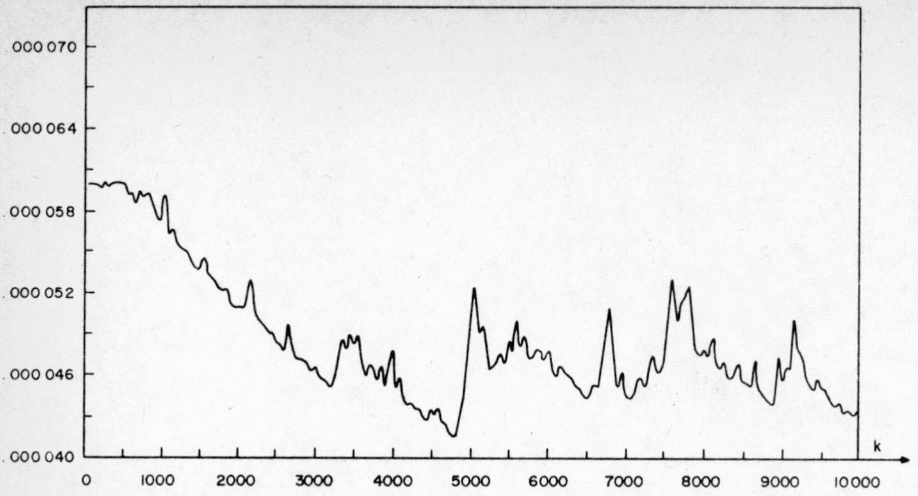

Fig.7.5e: $\hat{\theta}_1^{h_1,y}(k)$ as a function of k for the system given by equations (7.54), case (d-2).

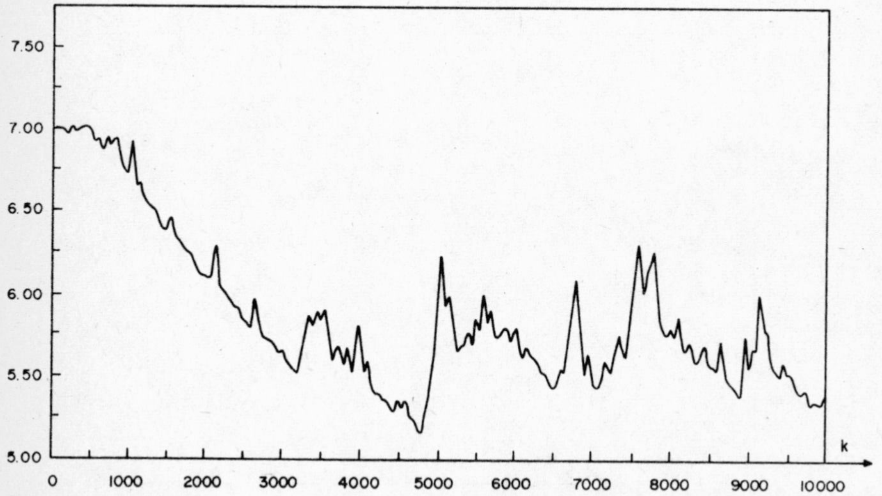

Fig.7.5f: $\hat{\theta}_2^{h_1,y}(k)$ as a function of k for the system given by equations (7.54), case (d-2).

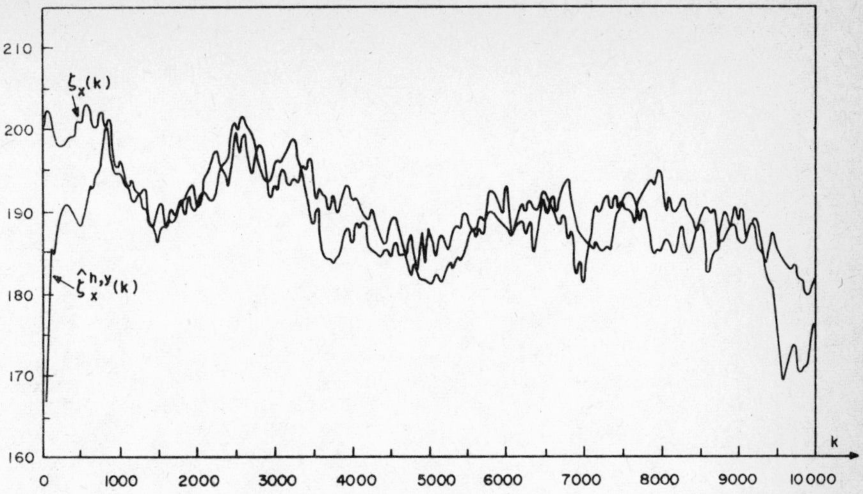

Fig.7.5g: $\zeta_x(k)$ and $\hat{\zeta}_x^{h,y}(k)$ as functions of k for the system given by equations (7.54), case (d-3).

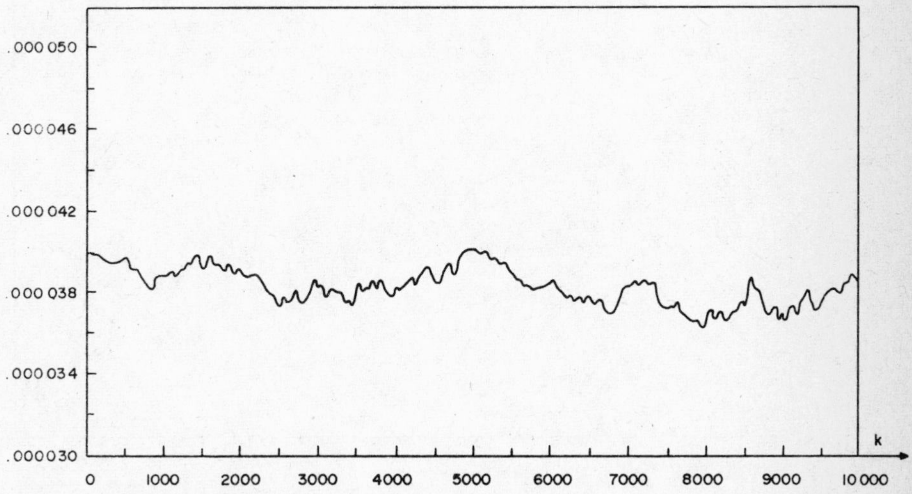

Fig.7.5h: $\hat{\theta}_1^{h_1,y}(k)$ as a function of k for the system given by equations (7.54), case (d-3).

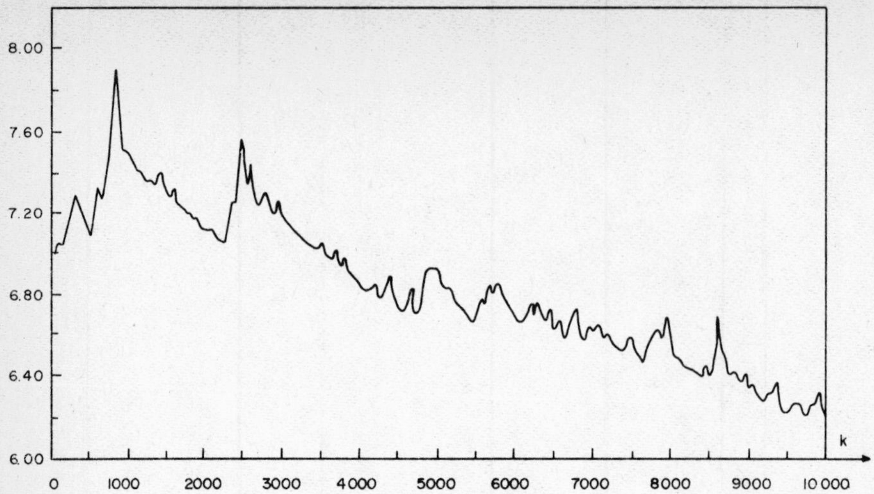

Fig.7.5i: $\hat{\theta}_2^{h_1,y}(k)$ as a function of k for the system given by equations (7.54), case (d-3).

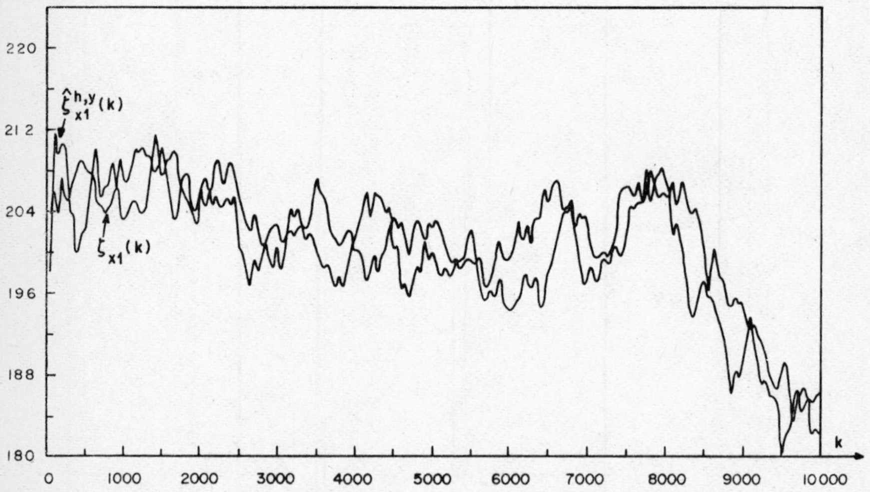

Fig.7.5j: $\zeta_x(k)$ and $\hat{\zeta}_x^{h,y}(k)$ as functions of k for the system given by equations (7.54), case (d-4).

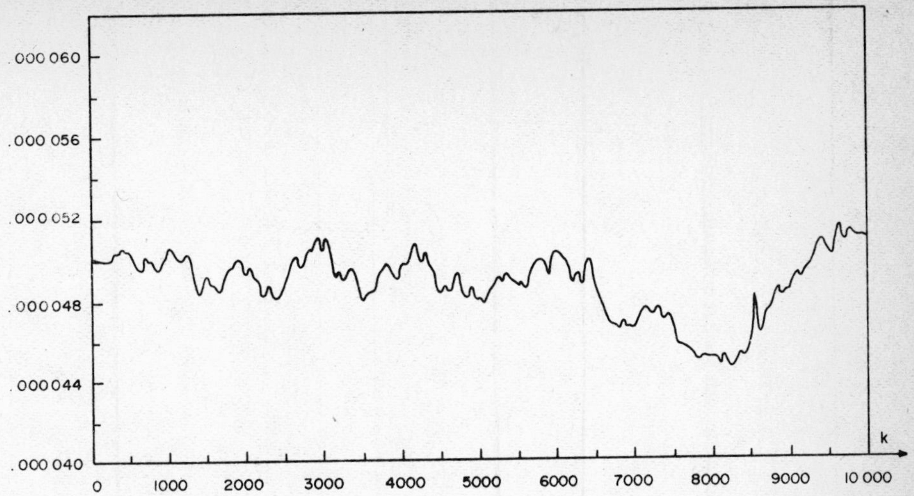

Fig.7.5k: $\hat{\theta}_1^{h_1,y}(k)$ as a function of k for the system given by equations (7.54), case (d-4).

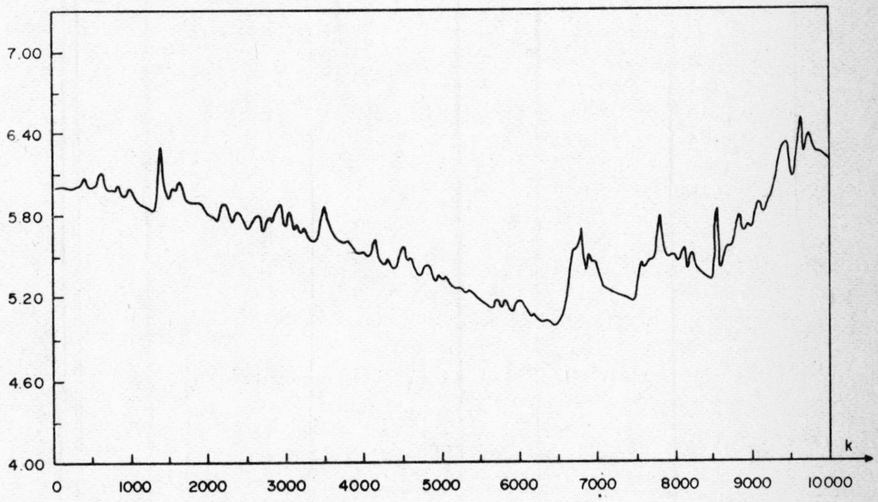

Fig.7.5ℓ: $\hat{\theta}_2^{h_1,y}(k)$ as a function of k for the system given by equations (7.54), case (d-4).

7.6 EXAMPLES : THE CASE m=2

In this section cases where $\zeta_x = \{\zeta_x(t),\ t \geq 0\}$ (equation (7.1)) is an $\mathbb{R}^2$-valued stochastic process are considered. In these cases

$$D_h \triangleq \{(ih,jh) : i=0,\pm1,\dots,\pm L_1,\ j=0,\pm1,\dots,\pm L_2\} \tag{7.55}$$

and equations (7.39) and (7.43) of Algorithm 7.4 reduce to

$$\begin{aligned} P^y_{i,j,\nu}(k+1) := P^y_{i,j,\nu}(k) &+ [\lambda(i-1,j,\theta^\nu;i,j,\theta^\nu)P^y_{i-1,j,\nu}(k) \\ &+ \lambda(i+1,j,\theta^\nu;i,j,\theta^\nu)P^y_{i+1,j,\nu}(k) + \lambda(i,j,\theta^\nu;i,j,\theta^\nu)P^y_{i,j,\nu}(k) \\ &+ \lambda(i,j-1,\theta^\nu;i,j,\theta^\nu)P^y_{i,j-1,\nu}(k) + \lambda(i,j+1,\theta^\nu;i,j,\theta^\nu)P^y_{i,j+1,\nu}(k)]\Delta \\ &+ P^y_{i,j,\nu}(k)\sum_{\ell=1}^{p} \gamma_\ell^{-2}(g_\ell(ih,jh,\theta^\nu) - \hat{g}_\ell(k))(y_\ell(k+1)-y_\ell(k)-\hat{g}_\ell(k)\Delta) \end{aligned} \tag{7.56}$$

$$\nu=1,\dots,L \quad , \quad -L_1 \leq i \leq L_1 \quad , \quad -L_2 \leq j \leq L_2$$

and

$$\hat{g}_\ell(k+1) := \sum_{i=-L_1}^{L_1} \sum_{j=-L_2}^{L_2} \sum_{\nu=1}^{L} g_\ell(ih,jh,\theta^\nu)P^y_{i,j,\nu}(k+1),\ \ell=1,\dots,p. \tag{7.57}$$

respectively, where

$$\begin{cases} \lambda(i,j,\theta^\nu;q,s,\theta^\nu) = \lambda((ih,jh),\theta^\nu;(qh,sh),\theta^\nu) \\ \nu=1,\dots,L \quad , \quad i,q=0,\pm1,\dots,\pm L_1 \quad , \quad j,s=0,\pm1,\dots,\pm L_2. \end{cases} \tag{7.58}$$

The following set of systems was considered:

(a)
$$\begin{cases} dx_1 = [-\theta x_2 + 50x_1(0.36 - x_1^2 - x_2^2)]dt + 0.01dW_1 \\ dx_2 = [\theta x_1 + 50x_2(0.36 - x_1^2 - x_2^2)]dt + 0.01dW_2 \end{cases} \quad t > 0 \tag{7.59}$$

$$dy_i = x_i dt + 0.002dv_i, \quad t > 0, \quad i=1,2 \tag{7.60}$$

In this case the set D is given by

$$D \triangleq \{x \in \mathbb{R}^2 \; : \; |x_i| < 1 + \delta, \; i=1,2\} \; , \; \delta < 1/12 \tag{7.61}$$

and the following set of parameters was used: $h=1/12$, $L_1=L_2=12$, $\zeta_{xi}(0)=0.1$, $i=1,2$; $\theta=1$, $\theta^1=0.75$, $\theta^2=0.85$, $\theta^3=0.95$, $\theta^4=1.05$, $\theta^5=1.15$, $\Delta=10^{-3}$, $N=10^4$.

(b)
$$\begin{cases} dx_1 = [\theta \operatorname{sign}(x_2) + 50x_1(0.6 - |x_1| - |x_2|)]dt + 0.005dW_1 \\ \qquad\qquad\qquad\qquad\qquad\qquad\qquad\qquad\qquad\qquad t > 0 \\ dx_2 = [-\theta \operatorname{sign}(x_1) + 50x_2(0.6 - |x_1| - |x_2|)]dt + 0.005dW_2 \end{cases} \tag{7.62}$$

$$dy_i = x_i dt + 0.005dv_i \quad , \quad t > 0 \quad , \quad i=1,2 \tag{7.63}$$

In this case D is given by

$$D \triangleq \{x \in \mathbb{R}^2 \; : \; |x_i| < 1.2 + \delta, \; i=1,2\} \; , \; \delta < 0.1 \tag{7.64}$$

and the following set of parameters was used: $h=0.1$, $L_1=L_2=12$, $\zeta_{xi}(0)=0.1$, $i=1,2$; $\theta=1$, $\theta^1=0.75$, $\theta^2=0.85$, $\theta^3=0.95$, $\theta^4=1.05$, $\theta^5=1.15$; $\Delta=10^{-3}$, $N=10^4$.

(c)
$$dx_i = \theta dW_i \quad , \quad t > 0 \quad , \quad i=1,2 \tag{7.65}$$

$$dy_i = x_i^3 dt + \gamma_i dv_i \quad , \quad t > 0 \quad , \quad i=1,2. \tag{7.66}$$

In this case D is given by

$$D \triangleq \{x \in \mathbb{R}^2 \; : \; |x_i| < 0.6 + \delta, \; i=1,2\} \; , \; \delta < 0.05 \tag{7.67}$$

and the following set of parameters was used: $h=0.05$, $L_1=L_2=12$, $\zeta_{xi}(0)=0.1$, $i=1,2$; $\gamma_1=\gamma_2=10^{-4}$, $\theta=0.025$, $\Delta=10^{-3}$, $N=10^4$. The following cases were numerically experimented with:

(c-1) $\theta^1 = 0.15$, $\theta^2 = 0.20$, $\theta^3 = 0.25$, $\theta^4 = 0.30$, $\theta^5 = 0.35$.

(c-2) $\theta^1 = 0.015$, $\theta^2 = 0.020$, $\theta^3 = 0.025$, $\theta^4 = 0.030$, $\theta^5 = 0.035$.

Some of the results of the corresponding runs are given in the following figures. All the graphs in this section were plotted using the set of points $\{t_k' = 50k\Delta : k=0,1,\ldots,200\}$.

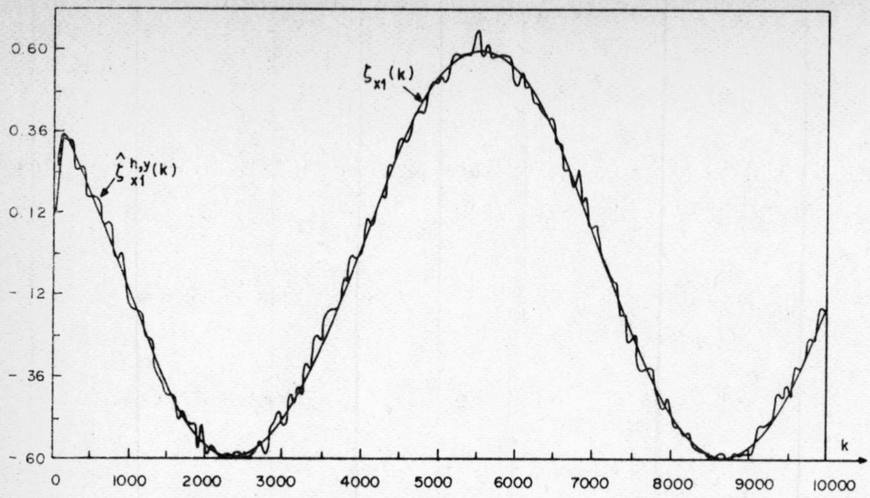

Fig.7.6a: $\zeta_{X1}(k)$ and $\hat{\zeta}^{h,y}_{X1}(k)$ as functions of k for the system given by equations (7.59)-(7.60).

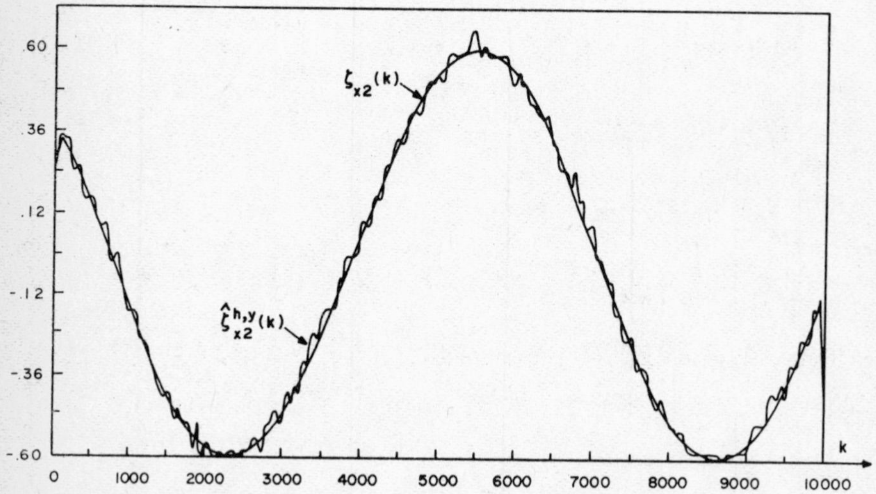

Fig.7.6b: $\zeta_{X2}(k)$ and $\hat{\zeta}^{h,y}_{X2}(k)$ as functions of k for the system given by equations (7.59)-(7.60).

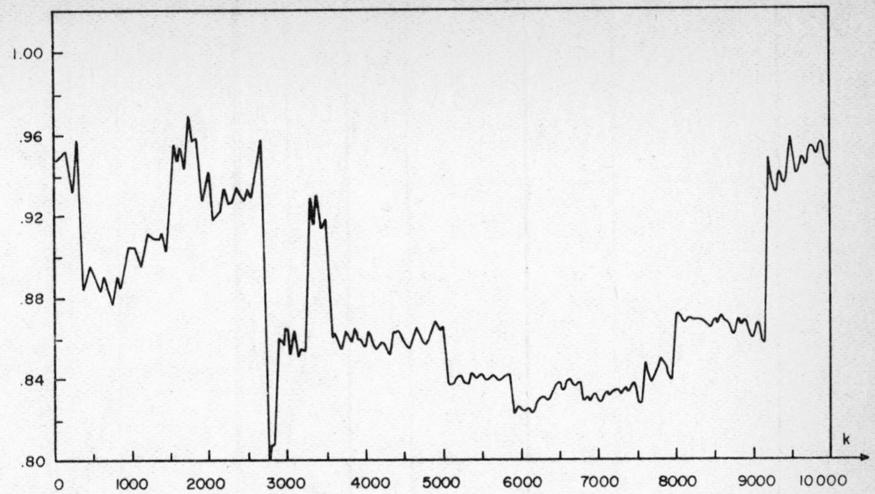

Fig.7.6c: $\hat{\theta}^{h_1,y}(k)$ as a function of k for the system given by equations (7.59)-(7.60).

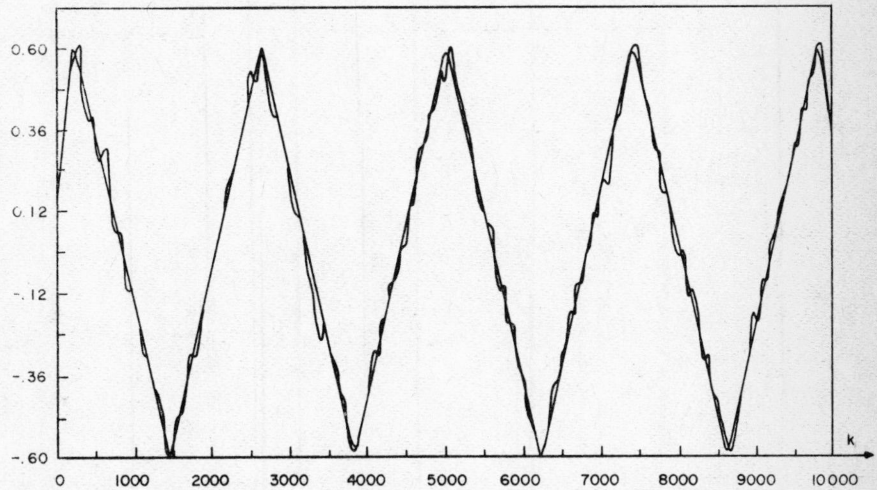

Fig.7.7a: $\zeta_{x1}(k)$ and $\hat{\zeta}^{h,y}_{x1}(k)$ as functions of k for the system given by equations (7.62)-(7.63).

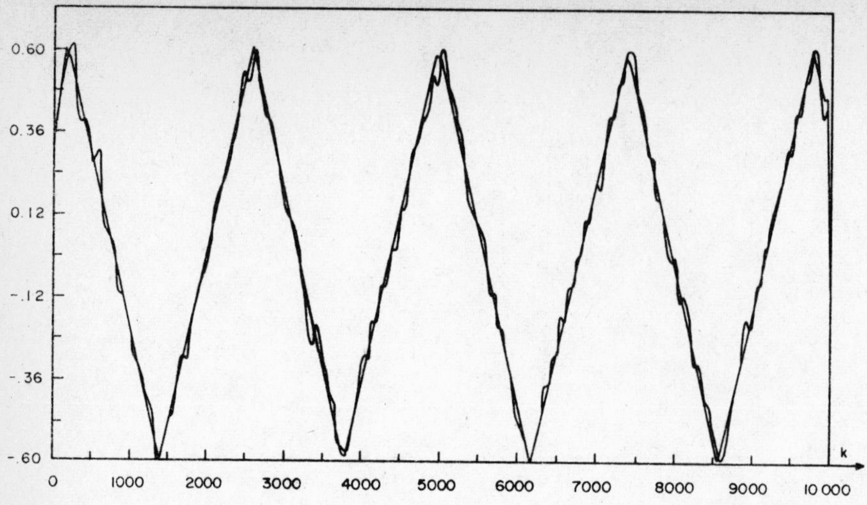

Fig.7.7b: $\zeta_{x2}(k)$ and $\hat{\zeta}_{x2}^{h,y}(k)$ as functions of k for the system given by equations (7.62)-(7.63).

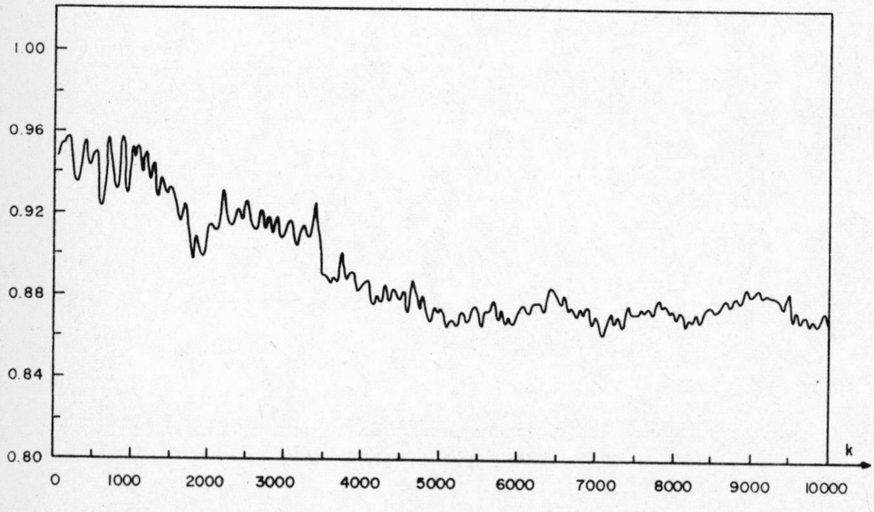

Fig.7.7c: $\hat{\theta}^{h_1,y}(k)$ as a function of k for the system given by equations (7.62)-(7.63).

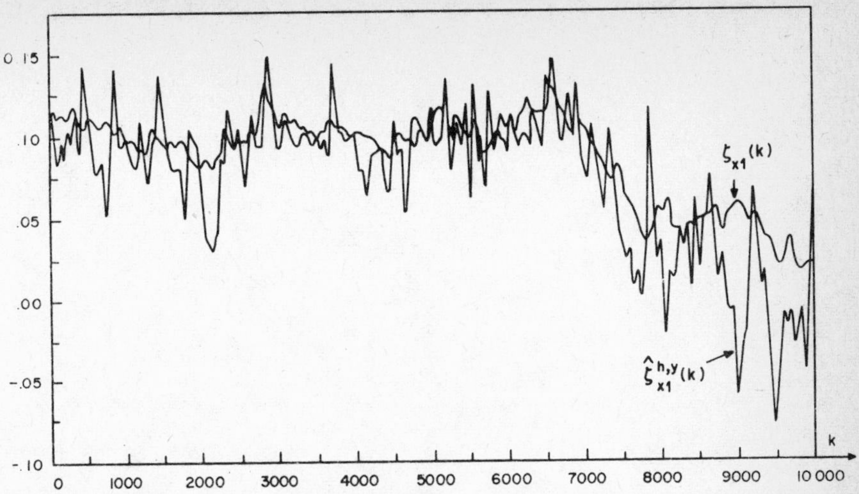

Fig.7.8a: $\zeta_{x1}(k)$ and $\hat{\zeta}^{h,y}_{x1}(k)$ as functions of k for the system given by equations (7.65)-(7.66), case (c-1).

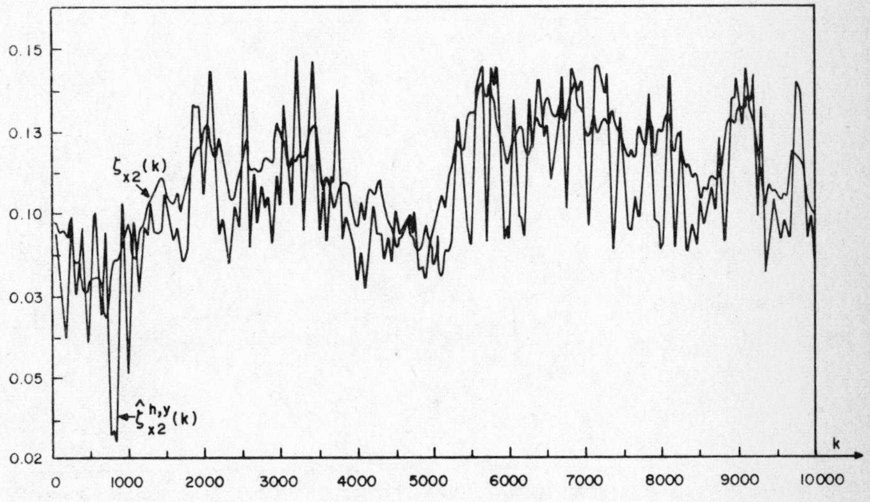

Fig.7.8b: $\zeta_{x2}(k)$ and $\hat{\zeta}^{h,y}_{x2}(k)$ as functions of k for the system given by equations (7.65)-(7.66), case (c-1).

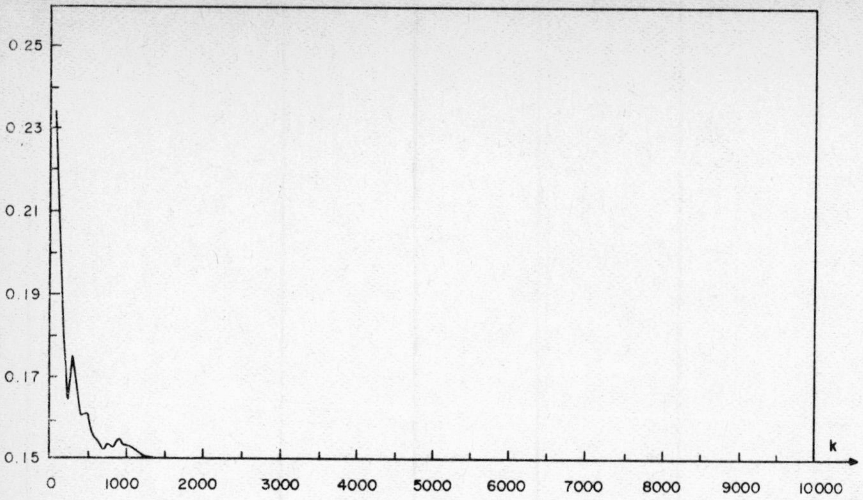

Fig.7.8c: $\hat{\theta}^{h_1,y}(k)$ as a function of k for the system given by equations (7.65)-(7.66), case (c-1).

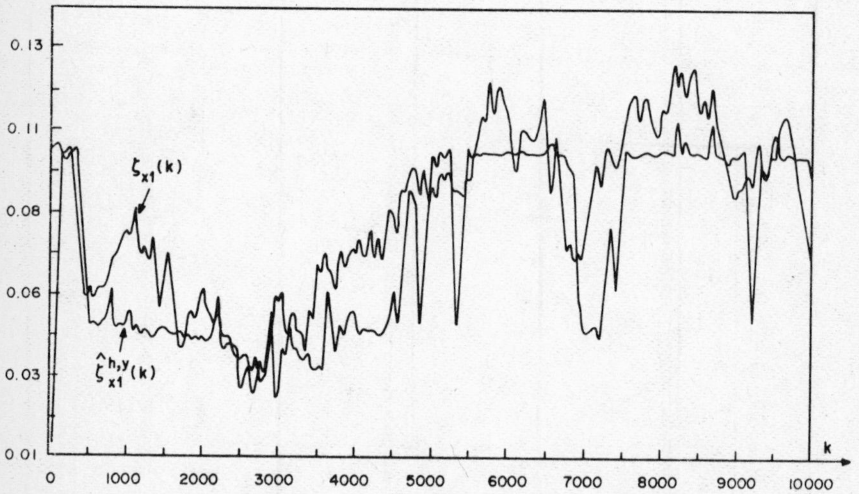

Fig.7.8d: $\zeta_{x1}(k)$ and $\hat{\zeta}_{x1}^{h,y}(k)$ as functions of k for the system given by equations (7.65)-(7.66), case (c-2).

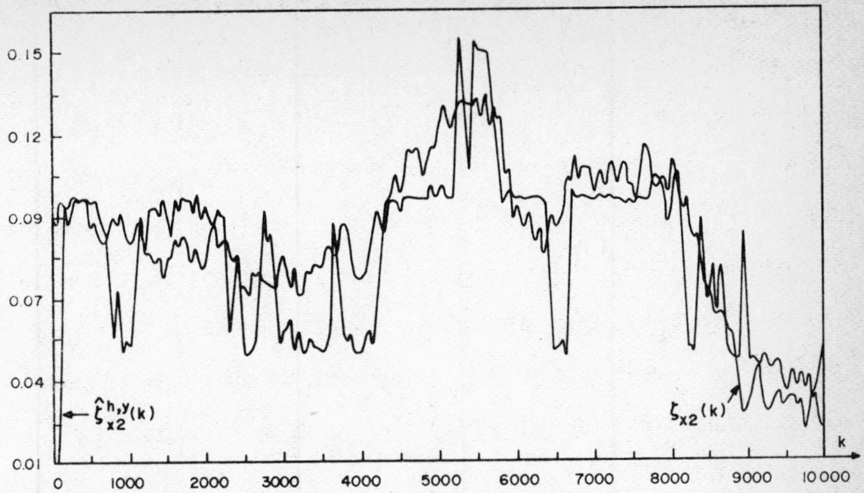

Fig.7.8e: $\zeta_{x2}(k)$ and $\hat{\zeta}^{h,y}_{x2}(k)$ as functions of k for the system given by equations (7.65)-(7.66), case (c-2).

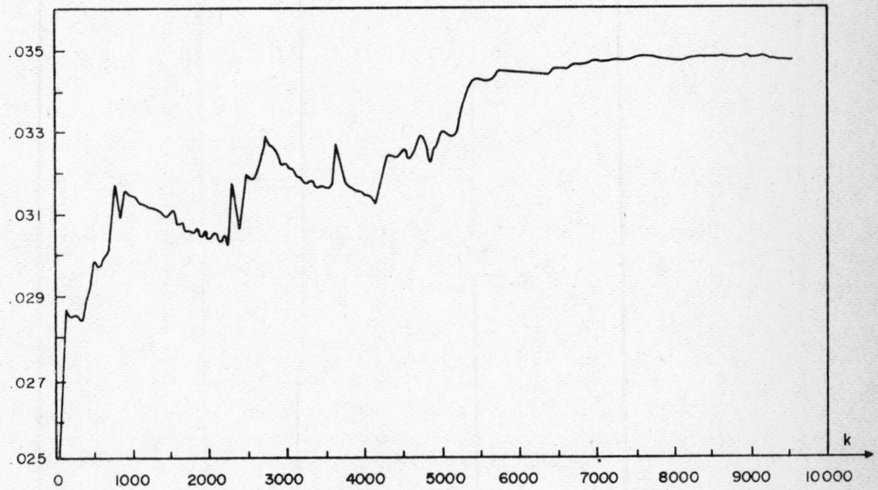

Fig.7.8f: $\hat{\theta}^{h_1,y}(k)$ as a function of k for the system given by equations (7.65)-(7.66), case (c-2).

7.7 REMARKS

The examples given in Section 7.5 deal with state and parameter estimation concerning a one-dimensional motion of a body in a resisting medium. The problem of the estimation of the drag coefficient (the parameter θ in eqns. (7.50)) of a flying object is useful when one wants to distinguish between a threat and a decoy. The examples given in Section 7.6 deal with the state and frequency estimation of a sine-wave oscillator (7.59) and a triangular-wave generator (7.62), using noisy measurements of the state. This state and frequency estimation problem is an extension of the problem dealt with in Sections 2.4-2.5. The last example in Section 7.6 (eqns. (7.65)-(7.66)) is an extension of the 'cubic sensor problem' (see [7.6] and the refenrences cited there).

7.8 REFERENCES

7.1 T.P. McGarty, *Stochastic Systems and State Estimation*, John Wiley & Sons, New York, 1974.

7.2 B.D.O. Anderson and J.B. Moore, *Optimal Filtering*, Prentice-Hall, Englewood Cliffs, 1979.

7.3 L.Ljung, Asymptotic behavior of the extended Kalman filter as a parameter estimator for linear systems, *IEEE Trans. on Automatic Control*, AC-24, pp 36-50, 1979.

7.4 M. Hazewinkel and J.C. Willems, Editors; *Stochastic Systems : The Mathematics of Filtering and Identification and Applications*, D. Reidel Publishing Company, Dordracht, 1981.

7.5 P.S. Maybeck, *Stochastic Models, Estimation, and Control*, Vol.2, Academic Press, New York, 1982.

7.6 M. Hazewinkel and S.I. Marcus, On Lie algebras and finite dimen= sional filtering, *Stochastics*, 7, pp 29-62, 1982.

CHAPTER 8

STATE ESTIMATION FOR SYSTEMS DRIVEN BY WIENER AND POISSON PROCESSES

8.1 INTRODUCTION

Consider the $\mathbb{R}^m$-valued Markov process $\zeta_x = \{\zeta_x(t),\ t \geq 0\}$ satisfying the equation

$$\zeta_x(t) = x + \int_0^t f(\zeta_x(s-))ds + BW(t) + CN(t) \ ,\ t \geq 0 \ ,\ x \in \mathbb{R}^m \tag{8.1}$$

and let the observation process Y be given by

$$y(t) = \int_0^t g(\zeta_x(s))ds + \Gamma v(t) \quad ,\quad y(t) \in \mathbb{R}^p \ ,\quad t \geq 0, \tag{8.2}$$

where on a probability space (Ω,F,P), $\mathbf{W} = \{W(t) = (W_1(t),\ldots,W_m(t)),$ $t \geq 0\}$ and $\mathbf{V} = \{v(t) = (v_1(t),\ldots,v_p(t)),\ t \geq 0\}$ are $\mathbb{R}^m$-valued and $\mathbb{R}^p$-valued standard Wiener processes respectively, and $\mathbf{N} = \{N(t) = (N_1(t),\ldots,N_r(t)),\ t \geq 0\}$ is a vector of mutually independent Poisson processes with parameter $Q = (q_1,\ldots,q_r)$. It is assumed that the processes $\mathbf{W}$, $\mathbf{V}$ and $\mathbf{N}$ are mutually independent. $f : \mathbb{R}^m \to \mathbb{R}^m$ and $g : \mathbb{R}^m \to \mathbb{R}^p$ are given functions satisfying the conditions stated in Section 3.1. $B \in \mathbb{R}^{m\times m}$ and $\Gamma \in \mathbb{R}^{p\times p}$ are matrices satisfying $B_{ij} = \sigma_i\ \delta_{ij}$, $i,j=1,\ldots,m$, and $\Gamma_{ij} = \gamma_i\ \delta_{ij}$, $i,j=1,\ldots,p$, where σ_i, $i=1,\ldots,m$ and γ_i, $i=1,\ldots,p$ are given positive numbers. $C \in \mathbb{R}^{m\times r}$ is a given matrix.

Let $F_t^y = \sigma(y(s);\ \ 0 \leq s \leq t)$. The problem dealt with in this chapter is to find an approximation $\hat{\zeta}_x^{h,y}(k)$ to

$$\hat{\zeta}_x(t) \triangleq E[\zeta_x(t\wedge\tau_T-)\,|\,F^y_{t\wedge\tau_T-}] \ ,\quad t \in [0,T] \ , \tag{8.3}$$

at the instants $t_k = k\Delta$, $k\Delta \in [0,T]$, where $\tau_T = \tau_T(x)$ is the first exit time of $\zeta_x(t)$ from an open and bounded domain $D \subset \mathbb{R}^m$.

The problem of finding $E[\zeta_x(t)|F_t^y]$, where ζ_x and Y are given by equations (8.1)-(8.2), has been treated in the past by several authors, see for example, Fisher [8.1], McGarty [8.2], Kwakernaak [8.3], Au [8.4], Marcus [8.5,8.6] and Au et al. [8.7].

In this chapter, the nonlinear filtering problem (i.e. the problem of finding $E[\zeta_x(t)|F_t^y]$) is treated by using methods different from those used in [8.1] - [8.7]. We here apply the same methods used in Chapters 3,5 and 7 and construct a process $\{\hat{\zeta}_x^{h,y}(k),\ k\Delta \in [0,T]\}$ which serves as an approximation to $\{\hat{\zeta}_x(k\Delta),\ k\Delta \in [0,T]\}$ (equation (8.3)).

8.2 CONSTRUCTION OF THE MARKOV CHAIN

Let $\mathbb{R}_h^m$ be a grid on $\mathbb{R}^m$ with a constant mesh size h along all axes as defined by equation (3.4). Denote by e^i the unit vector in $\mathbb{R}^m$ along the i-th axis, i=1,...,m; and denote by ε^j the unit vector in $\mathbb{R}^r$ along the j-th axis, j=1,...,r.

Suppose that the matrix C has the following form (or that it can be ap= proximated by the following expression)

$$C = h\Lambda = h \begin{pmatrix} \nu_{11} & \nu_{12} & \cdots & \nu_{1r} \\ \nu_{21} & \nu_{22} & \cdots & \nu_{2r} \\ \vdots & & & \\ \nu_{m1} & \nu_{m2} & \cdots & \nu_{mr} \end{pmatrix} \qquad (8.4)$$

where ν_{ij}, i=1,...,m, j=1,...,r, are given integers.

Define the following function $\lambda : \mathbb{R}_h^m \times \mathbb{R}_h^m \to \mathbb{R}$ by

$$\lambda(x,x) \overset{\Delta}{=} -[\sum_{i=1}^{m} (\sigma_i^2 + h|f_i(x)|) + h^2 \sum_{\ell=1}^{r} q_\ell]/h^2,\ x \in \mathbb{R}_h^m \qquad (8.5)$$

$$\lambda(x,x+e^i h) \triangleq (\sigma_i^2/2 + h\, f_i^+(x))/h^2,\ i=1,\ldots,m,\ x\in \mathbb{R}_h^m \quad (8.6)$$

$$\lambda(x,x-e^i h) \triangleq (\sigma_i^2/2 + h\, f_i^-(x))/h^2,\ i=1,\ldots,m,\ x\in \mathbb{R}_h^m \quad (8.7)$$

$$\lambda(x,x+h\Lambda\varepsilon^j) \triangleq q_j,\quad j=1,\ldots,r\ ,\quad x\in \mathbb{R}_h^m \quad (8.8)$$

$$\lambda(x,y) \triangleq 0\ ,\quad x\in \mathbb{R}_h^m\ ,\quad y\in U_x \quad (8.9)$$

where

$$U_x \triangleq \{y\in \mathbb{R}_h^m : y\neq x \text{ and } y\neq x\pm e^i h,\ i=1,\ldots,m, \text{ and } y\neq x+h\Lambda\varepsilon^j, j=1,\ldots,r\}. \quad (8.10)$$

It is assumed here that $h\Lambda\varepsilon^j \neq \pm e^i h$, $i=1,\ldots,m$, $j=1,\ldots,r$.

Note that $\lambda(x,y)\geq 0$ for $x,y\in \mathbb{R}_h^m$, $x\neq y$, and

$$\sum_{y\in \mathbb{R}_h^m} \lambda(x,y) = 0,\quad x\in \mathbb{R}_h^m. \quad (8.11)$$

Hence, given $x\in \mathbb{R}_h^m$, we can construct a continuous-time Markov chain $\{\zeta_x^h(t),\ t\in [0,T]\}$ with state space $S=\mathbb{R}_h^m$ by defining the following set of transition probabilities.

$$P(\zeta_x^h(t+\Delta) = z\,|\,\zeta_x^h(t)=z) \triangleq 1+\lambda(z,z)\Delta + O(\Delta^2),\ z\in \mathbb{R}_h^m\ , \quad (8.12)$$

$$P(\zeta_x^h(t+\Delta) = z\pm e^i h\,|\,\zeta_x^h(t)=z) \triangleq \lambda(z,z\pm e^i h)\Delta + O(\Delta^2),\ z\in \mathbb{R}_h^m,\ i=1,\ldots,m, \quad (8.13)$$

$$P(\zeta_x^h(t+\Delta) = z+h\Lambda\varepsilon^j\,|\,\zeta_x^h(t)=z) \triangleq q_j\Delta + O(\Delta^2),\quad z\in \mathbb{R}_h^m,\quad j=1,\ldots,r, \quad (8.14)$$

$$\sum_{y\in U_z} P(\zeta_x^h(t+\Delta) = y\,|\,\zeta_x^h(t)=z) = O(\Delta^2)\quad ,\quad z\in \mathbb{R}_h^m. \quad (8.15)$$

Thus, using equations (8.12)-(8.15) it follows that

$$\begin{cases} E[\zeta^h_{xi}(t+\Delta) - \zeta^h_{xi}(t) | \zeta^h_x(t) = z] = (f_i(z) + h \sum_{j=1}^{r} \nu_{ij}\, q_j)\Delta + h\, O(\Delta^2) \\ \qquad = (f_i(z) + \sum_{j=1}^{r} c_{ij}\, q_j)\Delta + h\, O(\Delta^2) \\ i=1,\dots,m \quad , z \in \mathbb{R}^m_h \end{cases} \tag{8.16}$$

and

$$\begin{cases} E[(\zeta^h_{xi}(t+\Delta) - \zeta^h_{xi}(t))(\zeta^h_{xj}(t+\Delta) - \zeta^h_{xj}(t)) | \zeta^h_x(t) = z] \\ = \delta_{ij}(\sigma_i^2 + h|f_i(z)|)\Delta + h^2 \sum_{\ell=1}^{r} \nu_{i\ell}\, \nu_{j\ell}\, q_\ell \Delta + h^2 O(\Delta^2) \\ = \delta_{ij}(\sigma_i^2 + h|f_i(z)|)\Delta + \sum_{\ell=1}^{r} c_{i\ell}\, c_{j\ell}\, q_\ell \Delta + h^2 O(\Delta^2),\ i,j=1,\dots,m \quad , z \in \mathbb{R}^m_h \end{cases} \tag{8.17}$$

Remark 8.2.1: Note that equations (8.1) can be written as

$$\zeta_x(t) = x + \int_0^t [f(\zeta_x(s-)) + CQ]ds + BW(t) + CM(t),\ t \geq 0, x \in \mathbb{R}^m \tag{8.18}$$

where $M(t) \overset{\Delta}{=} N(t) - Qt$ is a zero mean $(\sigma(W(s),N(s);\ \ 0 \leq s \leq t),P)$-mar= tingale on $[0,T]$.

Equations (8.4) and (8.16)-(8.18) illustrate the relations between the Markov chain $\{\zeta^h_x(t),\ t \in [0,T]\}$ and the Markov process $\{\zeta_x(t),\ t \in [0,T]\}$ $0 < T < \infty$.

Remark 8.2.2: In Kushner and Yu [8.8] and Kushner and DiMasi [8.9] ζ_x is approximated by a discrete-time Markov chain ζ^h_x (in [8.8] the case where $B = 0$ is considered), and conditions are established for the con= vergence of functionals of ζ^h_x to the corresponding functionals of ζ_x, as $h \downarrow 0$.

Let the set D and the stopping times τ_T and τ^h_T be defined in the same manner as in equations (3.19), (3.20) and (3.21) respectively, where $D_h \overset{\Delta}{=} \mathbb{R}^m_h \cap D$. Define

$$y^h(t) \overset{\Delta}{=} \int_0^t g(\zeta^h_x(s))ds + \Gamma\, v(t),\ t \in [0,T] \tag{8.19}$$

and let $F_t^{y,h} \triangleq \sigma(y^h(s);\ 0 \le s \le t)$, $t \in [0,T]$. In the next section an optimal least-squares filter is constructed for the computation of $E[\zeta_x^h(t \wedge \tau_T^h-) \mid F_{t\wedge\tau_T^h-}^{y,h}]$.

8.3 THE EQUATIONS OF THE OPTIMAL FILTER

Assume that $\sup_{t \in [0,T]} E|\zeta_x^h(t)|^2 < \infty$, $x \in \mathbb{R}_h^m$, and denote

$$G_t \triangleq \sigma(\zeta_x^h(s),\ v(s);\ \ 0 \le s \le t),\ t \in [0,T] \tag{8.20}$$

$$h_t \triangleq \Gamma^{-1}\ g(\zeta_x^h(t)), \quad t \in [0,T], \tag{8.21}$$

$$\tilde{P}_a(t) \triangleq P(\zeta_x^h(t) = a \mid F_t^{y,h})\ ,\ t \in [0,T],\ \ a \in \mathbb{R}_h^m \tag{8.22}$$

$$P_a(t) \triangleq P(\zeta_x^h(t \wedge \tau_T^h-) = a \mid F_{t\wedge\tau_T^h-}^{y,h}]\ ,\ t \in [0,T],\ \ a \in \mathbb{R}_h^m. \tag{8.23}$$

It is further assumed that $\int_0^T E|h_t|^2 dt < \infty$.

For each $t \in [0,T]$ the σ-fields G_t and $\sigma(v(s_2) - v(s_1);\ \ t < s_1 < s_2 \le T)$ are independent and h_t is G_t-measurable. Thus, by following the same development given in Section 2.7 we obtain:

$$\begin{cases} d\tilde{P}_a(t) = \sum_{b \in \mathbb{R}_h^m} \lambda(b,a)\tilde{P}_b(t)dt \\ \qquad + \tilde{P}_a(t) \sum_{\ell=1}^{p} \gamma_\ell^{-2}(g_\ell(a) - \hat{\tilde{g}}_\ell(t))(dy_\ell^h(t) - \hat{\tilde{g}}_\ell(t)dt) \\ a \in \mathbb{R}_h^m\ \ ,\ \ t \in (0,T) \end{cases} \tag{8.24}$$

$$\hat{\tilde{g}}_\ell(t) = \sum_{a \in \mathbb{R}_h^m} g_\ell(a)\tilde{P}_a(t)\quad ,\ \ \ell=1,\dots,p\ ,\ \ t \in [0,T] \tag{8.25}$$

and

$$E[\zeta_x^h(t) \mid F_t^{y,h}] = \sum_{a \in \mathbb{R}_h^m} a\ \tilde{P}_a(t)\quad ,\ \ t \in [0,T] \tag{8.26}$$

where $\{\lambda(b,a),\ \ a,b \in \mathbb{R}_h^m\}$ are given by equations (8.5)-(8.9).

In order to obtain the filter equations for computing $\{P_a(t),\ a \in D_h,$

$t \in [0,T]\}$ we follow the discussion given in Section 3.3. Let $\{\lambda(b,a), a,b \in \mathbb{R}_h^m\}$ be defined by equations (8.5)-(8.9) together with the additional condition

$$\lambda(b,a) = 0 \quad , \quad b \in \mathbb{R}_h^m - D_h, \quad a \in \mathbb{R}_h^m \quad . \tag{8.27}$$

Then, by following the same development given in Section 3.3, the follow= ing equations are obtained:

$$\begin{cases} dP_a(t) = \sum_{b \in D_h} \lambda(b,a) P_b(t) dt \\ \quad + P_a(t) \sum_{\ell=1}^{p} \gamma_\ell^{-2} (g_\ell(a) - \hat{g}_\ell(t))(dy_\ell^h(t) - \hat{g}_\ell(t)dt) \\ a \in D_h, \quad t \in (0,T) \end{cases} \tag{8.28}$$

$$\hat{g}_\ell(t) = \sum_{a \in D_h} g_\ell(a) P_a(t) \quad , \quad \ell=1,\ldots,p \quad , \quad t \in [0,T] \tag{8.29}$$

and

$$E[\zeta_x^h(t \wedge \tau_T^h -) | F_{t \wedge \tau_T^h -}^{y,h}] = \sum_{a \in D_h} a P_a(t) \quad , \quad t \in [0,T] \; , \tag{8.30}$$

where in equation (3.19), $a_i = L_i h$, $i=1,\ldots,m$, and $\{L_i\}$ are given positive integers.

Given $Y^t = \{y(s), 0 \le s \le t\}$, $t \in [0,T]$. Then, in order to approximate $\hat{\zeta}_x(t)$ (equation (8.3)), equations (8.28)-(8.29) are solved, where in equations (8.28) the increment dy^h is replaced by dy. Let $\{P_a^y(t), a \in D_h, t \in [0,T]\}$ denote the solution to equations (8.28)-(8.29) (where dy re= places dy^h in (8.28)). Then, a process $\hat{\zeta}_x^{h,y} = \{\hat{\zeta}_x^{h,y}(t), t \in [0,T]\}$is defined by

$$\hat{\zeta}_x^{h,y}(t) \triangleq \sum_{a \in D_h} a \, P_a^y(t) \quad , \quad t \in [0,T]. \tag{8.31}$$

$\hat{\zeta}_x^{h,y}$ serves here as an approximation to $\{\hat{\zeta}_x(t), t \in [0,T]\}$.

Note that equations (8.24)-(8.25), and (8.28)-(8.29) are of the same form as equations (3.28)-(3.29) and (3.36)-(3.37) respectively. Hence Algorithm 3.4, with $\{\lambda(b,a),\ b,a \in D_h\}$ given by equations (8.5)-(8.9) and (8.27), can be applied for the computation of $\hat{\zeta}_x^{h,y}(k) = \hat{\zeta}_x^{h,y}(k\Delta)$, $k\Delta \in [0,T]$ (where $\hat{\zeta}_x^{h,y}(t)$ is given by (8.31)).

The problem of establishing conditions for the weak convergence of $\{\hat{\zeta}_x^{h,y}(t),\ t \in [0,T]\}$ to $\{\hat{\zeta}_x(t),\ t \in [0,T]\}$ as $h \downarrow 0$ is beyond the scope of this work. Instead, the role of $\hat{\zeta}_x^{h,y}$, as an approximation to $\hat{\zeta}_x$, is demonstrated in the sequel by means of numerical experimentation with several examples.

Let $\mathbf{N} = \{N(t) = (N_1(t),\dots,N_r(t)),\ t \geq 0\}$ be a vector of r mutually in= dependent Poisson processes. Throughout the numerical experimentation equations (8.1) were simulated by applying the following procedure:

$$\zeta_x(0) = x. \quad \text{For } k=0,1,\dots,N$$

1. For i=1,...,m calculate

$$\bar{x}_i(k+1) = \zeta_{xi}(k) + f_i(\zeta_x(k))\Delta + \sqrt{\Delta}\,\sigma_i\,W_i(k) + \sum_{j=1}^{r} c_{ij}(N_j(k+1)-N_j(k))$$

2. For i=1,...,m calculate

$$\zeta_{xi}(k+1) = \zeta_{xi}(k) + [f_i(\zeta_x(k)) + f_i(\bar{x}\,(k+1))]\Delta/2 + \sqrt{\Delta}\,\sigma_i\,W_i(k)$$

$$+ \sum_{j=1}^{r} c_{ij}(N_j(k+1)-N_j(k))$$

where $N_j(k) \triangleq N_j(k\Delta)$, $j=1,\dots,r$, $k=0,1,\dots,N$ and $\{W(k)\}_{k=0}^{N}$ is a se= quence of independent $\mathbb{R}^m$-valued Gaussian elements satisfying equations (3.51). The sequence $\{y(k)\}$ is generated in the same manner as described in Section 3.4.

8.4 EXAMPLES: THE CASE m=1

In this section cases are considered where $\zeta_x = \{\zeta_x(t),\ t \geq 0\}$ (equations (8.1)) is an $\mathbb{R}$-valued Markov process. In these cases the set D_h is given by equation (3.53) and the equations for computing $\hat{\zeta}_x^{h,y}$ (i.e. equations (8.28)-(8.29) and (8.31) where dy replaces dy^h in (8.28)) reduce to

$$\begin{aligned} dP_i^y(t) = [\sum_{\ell=1}^{r} q_\ell\, P_{i-\nu_{1\ell}}^y(t) + \lambda(i-1,i)P_{i-1}^y(t) + \lambda(i,i)P_i^y(t) \\ + \lambda(i+1,i)P_{i+1}^y(t)]dt \\ + P_i^y(t) \sum_{\ell=1}^{p} \gamma_\ell^{-2}(g_\ell(ih)-\hat{g}_\ell(t))(dy_\ell(t)-\hat{g}_\ell(t)dt) \end{aligned} \tag{8.32}$$

$i=-L,-L+1,\ldots,L\ (L=L_1)\quad,\quad t \in (0,T)$

$$\hat{g}_\ell(t) = \sum_{i=-L}^{L} g_\ell(ih)P_i^y(t) \quad,\quad t \in [0,T]\ ,\ \ell=1,\ldots,p \tag{8.33}$$

and

$$\hat{\zeta}_x^{h,y}(t) = \sum_{i=-L}^{L} ih\ P_i^y(t) \quad,\quad t \in [0,T]\ , \tag{8.34}$$

where

$$\lambda(i-1,i) \triangleq (\sigma^2/2 + h\ f^+(ih-h))/h^2,\ -L+1 \leq i \leq L \tag{8.35}$$

$$\lambda(i+1,i) \triangleq (\sigma^2/2 + h\ f^-(ih+h))/h^2,\ -L \leq i \leq L-1 \tag{8.36}$$

$$\lambda(i,i) \triangleq -(\sigma^2 + h|f(ih)| + h^2 \sum_{\ell=1}^{r} q_\ell)/h^2, -L \leq i \leq L \tag{8.37}$$

$$\lambda(-L-1,-L) = \lambda(L+1,L) = 0 \tag{8.38}$$

The following set of systems was considered:

(a) $dx = -4xdt + 0.05dW + cdN;\quad dy = xdt + 0.01dv,\ x,y \in \mathbb{R},\ t > 0$ (8.39)

In this case the set D was taken to be

$$D \triangleq \{x \in \mathbb{R} :\ |x| < 2 + \delta\}\ ,\quad \delta < 0.01 \tag{8.40}$$

and the following set of parameters was used: $c = 50h$, $h = 0.01$, $L = 200$, $q_1 = 1,2,4$; $\Delta = 10^{-3}$, $N = 10^4$.

(b) $dx = -5\cdot 10^{-5}\, x^2 dt + 7dW + cdN; \quad dy = xdt + 3dv, \; x,y \in \mathbb{R}, \; t > 0,$ (8.41)

In this case the set D was taken to be

$$D \triangleq \{x \in \mathbb{R} : |x| < 500 + \delta\}, \; \delta < 1.25 \qquad (8.42)$$

and the following set of parameters was used: $c = 20h$, $h = 1.25$, $L = 400$, $q_1 = 1,3,6$; $\Delta = 10^{-3}$, $N = 10^4$.

(c) $dx = \sigma dW + cdN; \quad dy = xdt + 0.005dv \;, \quad x,y \in \mathbb{R}, \quad t > 0,$ (8.43)

In this case the set D was taken to be

$$D \triangleq \{x \in \mathbb{R} : |x| < 2 + \delta\}, \quad \delta < 0.01 \qquad (8.44)$$

and the following set of parameters was used : $c = 20h$, $h = 0.01$, $\sigma = 0.05, 0.0$; $L = 200$, $q_1 = 1,3,6$; $\Delta = 10^{-3}$, $N = 10^4$.

Typical extracts from the results are given in the following table and figures.

All the graphs in this section were plotted using the set of points $\{t_k' = 99\Delta + 100k\Delta : k=0,1,\ldots,99\}$.

TABLE 8.1: $\zeta_x(k)$ and $\hat{\zeta}_x^{h,y}(k)$ as functions of k for the system given by equations (8.43), where $\sigma = 0.0$ and $q_1 = 3$. Here, the sample path of N has jumps at $\{T_1,...\} = \{.1477,.1693,.2129,.2423,.3103,.44156,.8471,1.1434,1.5848,2.1152,...\}$ and $\tau_T^h \cong 2.1150$

k	$\zeta_x(k)$	$\hat{\zeta}_x^{h,y}(k)$	k	$\zeta_x(k)$	$\hat{\zeta}_x^{h,y}(k)$
99	0.0000	0.0005	1199	1.6000	1.6086
199	0.4000	0.4018	1299	1.6000	1.6155
299	0.8000	0.8053	1399	1.6000	1.6236
399	1.0000	1.0113	1499	1.6000	1.6152
499	1.2000	1.2151	1599	1.8000	1.8086
599	1.2000	1.2167	1699	1.8000	1.8082
699	1.2000	1.2102	1799	1.8000	1.8114
799	1.2000	1.2085	1899	1.8000	1.8081
899	1.4000	1.4025	1999	1.8000	1.8071
999	1.4000	1.4062	2099	1.8000	1.8069
1099	1.4000	1.4067			

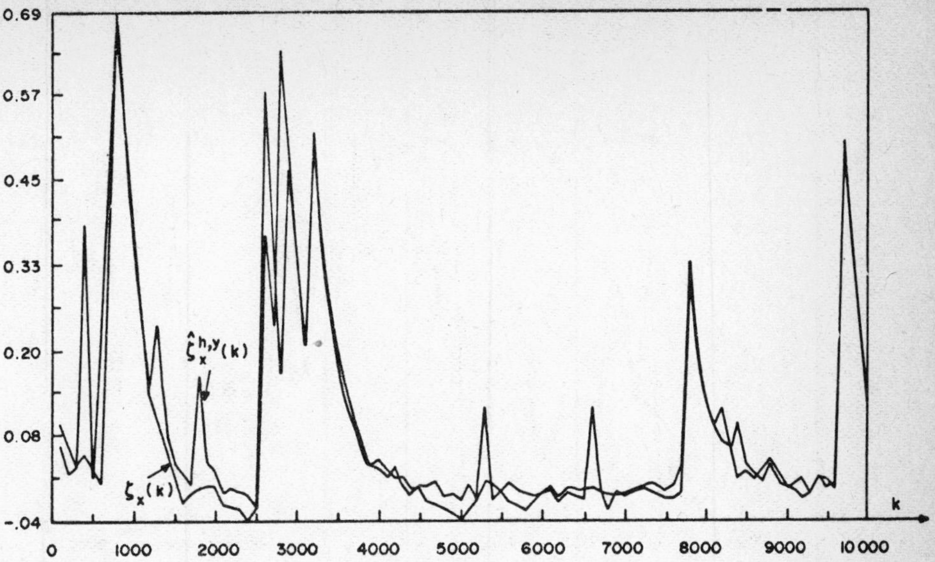

Fig.8.1a: $\zeta_X(k)$ and $\hat{\zeta}_X^{h,y}(k)$ as functions of k for the system given by equations (8.39), where $q_1 = 1$.

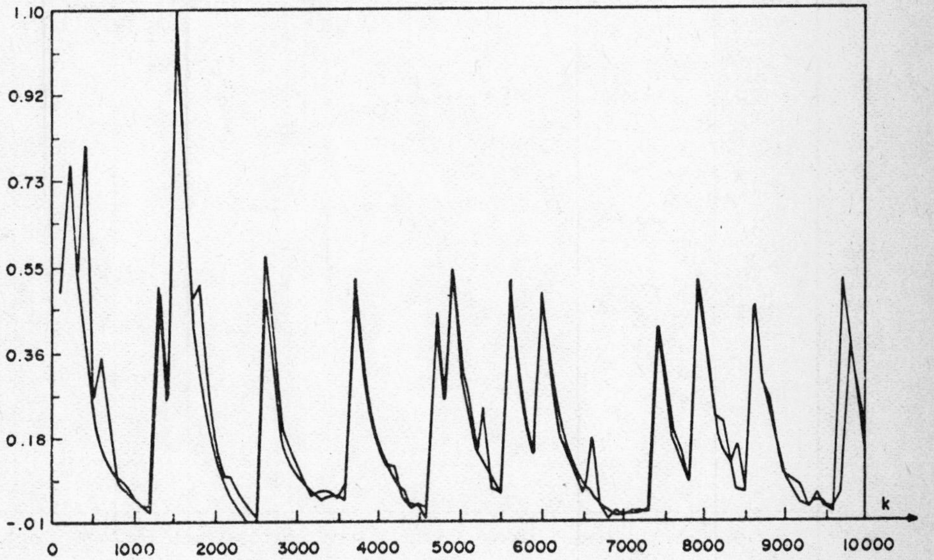

Fig.8.1b: $\zeta_X(k)$ and $\hat{\zeta}_X^{h,y}(k)$ as functions of k for the system given by equations (8.39), where $q_1 = 2$.

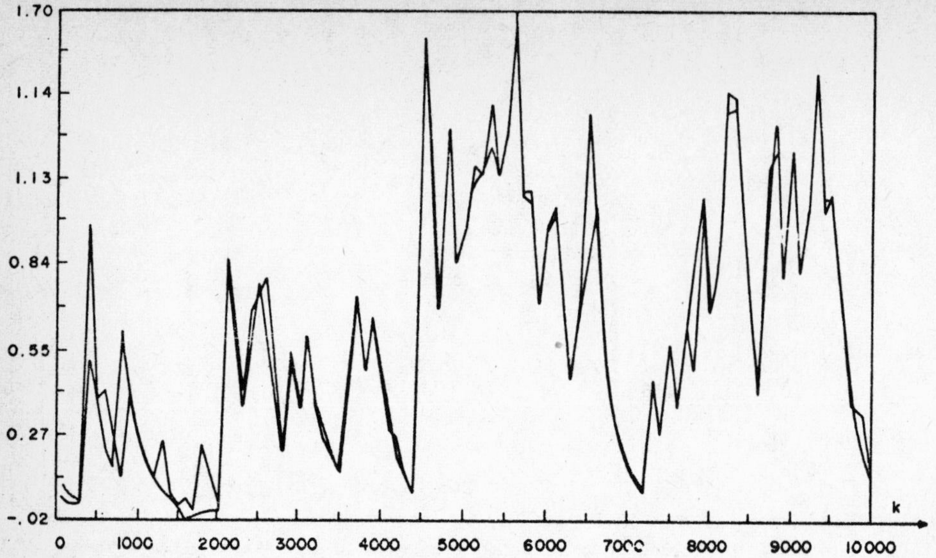

Fig.8.1c: $\zeta_x(k)$ and $\hat{\zeta}_x^{h,y}(k)$ as functions of k for the system given by equations (8.39), where $q_1 = 4$.

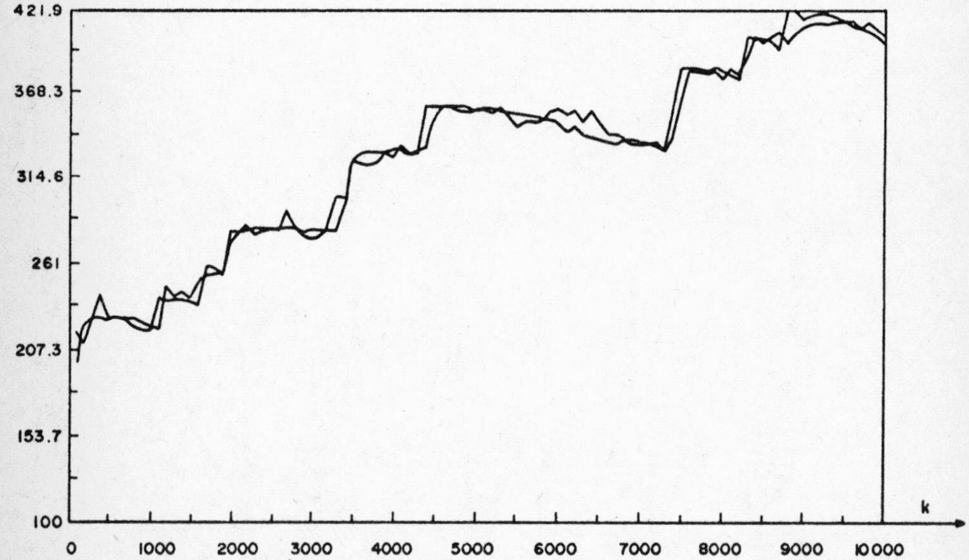

Fig.8.2: $\zeta_x(k)$ and $\hat{\zeta}_x^{h,y}(k)$ as functions of k for the system given by equations (8.41), where $q_1 = 1$.

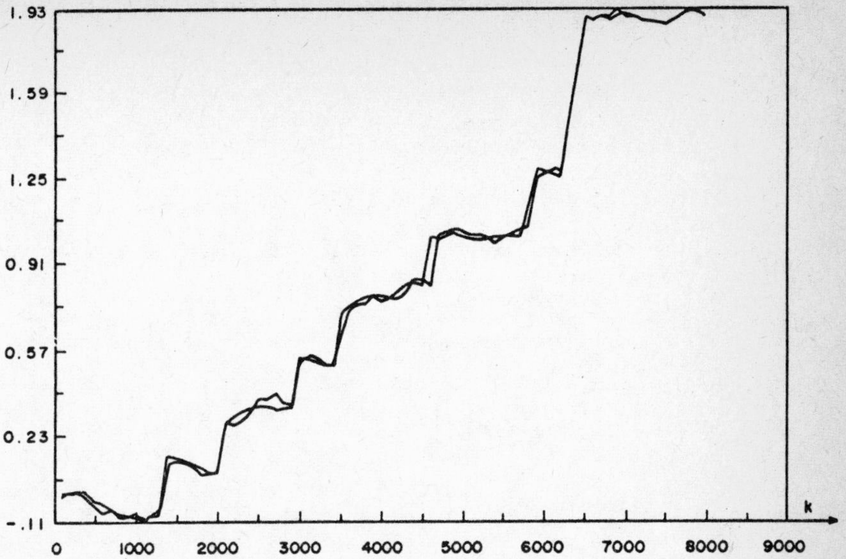

Fig.8.3a: $\zeta_x(k)$ and $\hat{\zeta}_x^{h,y}(k)$ as functions of k for the system given by equations (8.43), where $\sigma = 0.05$ and $q_1 = 1$.

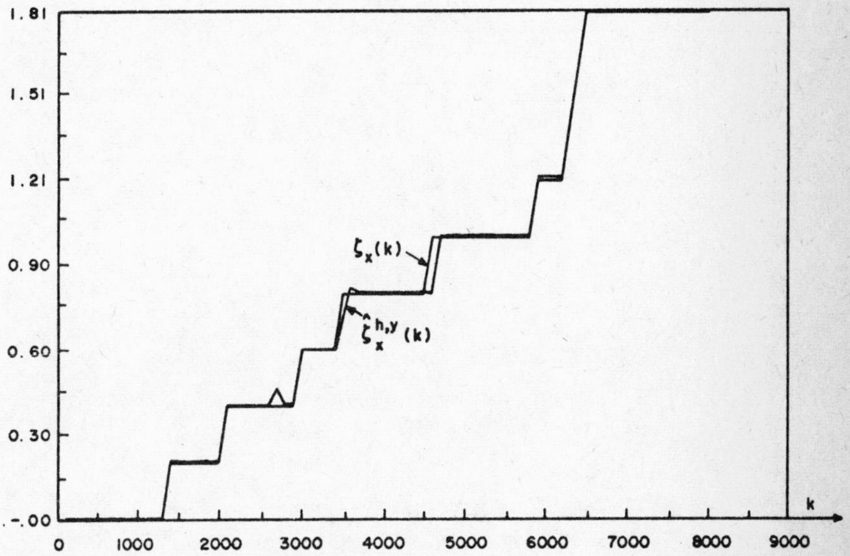

Fig.8.3b: $\zeta_x(k)$ and $\hat{\zeta}_x^{h,y}(k)$ as functions of k for the system given by equations (8.43), where $\sigma = 0.0$ and $q_1 = 1$.

8.5 EXAMPLES: THE CASE m=2

In this section cases are considered where $\zeta_x = \{\zeta_x(t),\ t \geq 0\}$ (equation (8.1)) is an $\mathbb{R}^2$-valued Markov process. In these cases the set D_h is given by equation (3.71) and the equations for computing $\hat{\zeta}_x^{h,y}$,(i.e. equations (8.28)-(8.29) and (8.31) where dy replaces dy^h in (8.28) reduce to

$$\begin{aligned} dP^y_{i,j}(t) = [& \sum_{\ell=1}^{r} q_\ell P^y_{i-\nu_{1\ell},j-\nu_{2\ell}}(t) + \lambda(i-1,j;i,j)P^y_{i-1,j}(t) \\ & + \lambda(i+1,j;i,j)P^y_{i+1,j}(t) + \lambda(i,j;i,j)P^y_{i,j}(t) \\ & + \lambda(i,j+1;i,j)P^y_{i,j+1}(t) + \lambda(i,j-1;i,j)P^y_{i,j-1}(t)]dt \\ & + P^y_{i,j}(t) \sum_{\ell=1}^{p} \gamma_\ell^{-2}(g_\ell(ih,jh) - \hat{g}_\ell(t))(dy_\ell(t) - \hat{g}_\ell(t)dt) \end{aligned} \tag{8.45}$$

$$-L_1 \leq i \leq L_1 \quad , \quad -L_2 \leq j \leq L_2 \ , \quad t \in (0,T)$$

$$\hat{g}_\ell(t) = \sum_{i=-L_1}^{L_1} \sum_{j=-L_2}^{L_2} g_\ell(ih,jh)P^y_{i,j}(t) \quad , \quad \ell=1,\dots,p \quad , \ t \in [0,T] \tag{8.46}$$

and

$$\hat{\zeta}_x^{h,y}(t) = \sum_{i=-L_1}^{L_1} \sum_{j=-L_2}^{L_2} (ih,jh)P^y_{i,j}(t) \quad , \quad t \in [0,T] \ , \tag{8.47}$$

where

$$\begin{cases} \lambda(i,j;i,j) \triangleq -[\sum_{\ell=1}^{2} (\sigma_\ell^2 + h|f_\ell(ih,jh)|) + h^2 \sum_{\ell=1}^{r} q_\ell)/h^2 \\ -L_1 \leq i \leq L_1 \quad , \quad -L_2 \leq j \leq L_2 \end{cases} \tag{8.48}$$

$$\begin{cases} \lambda(i-1,j;i,j) \triangleq (\sigma_1^2/2 + h\ f_1^+(ih-h,jh))/h^2 \\ -L_1 + 1 \leq i \leq L_1 \quad , \quad -L_2 \leq j \leq L_2 \end{cases} \tag{8.49}$$

$$\begin{cases} \lambda(i+1,j;i,j) \triangleq (\sigma_1^2/2 + h\, f_1^-(ih+h,jh))/h^2 \\ -L_1 \le i \le L_1-1 \quad , \quad -L_2 \le j \le L_2 \end{cases} \tag{8.50}$$

$$\begin{cases} \lambda(i,j+1;i,j) \triangleq (\sigma_2^2/2 + h\, f_2^-(ih,jh+h))/h^2 \\ -L_1 \le i \le L_1 \quad , \quad -L_2 \le j \le L_2-1 \end{cases} \tag{8.51}$$

$$\begin{cases} \lambda(i,j-1;i,j) \triangleq (\sigma_2^2/2 + h\, f_2^+(ih,jh-h))/h^2 \\ -L_1 \le i \le L_1 \quad , \quad -L_2+1 \le j \le L_2 \end{cases} \tag{8.52}$$

$$\begin{cases} \lambda(-L_1-1,j;-L_1,j) = \lambda(L_1+1,j;L_1,j) = 0, \quad -L_2 \le j \le L_2 \\ \lambda(i,L_2+1;i,L_2) = \lambda(i,-L_2-1;i,-L_2) = 0 \ , \ -L_1 \le i \le L_1. \end{cases} \tag{8.53}$$

The following set of systems was considered:

(a)
$$\begin{cases} dx_1 = [\text{sign}(x_2) + 10x_1(0.5 - |x_1| - |x_2|)]dt + 0.01dW_1 + c_{11}dN_1 \\ \qquad\qquad\qquad\qquad\qquad\qquad\qquad\qquad\qquad t > 0 \\ dx_2 = [-\text{sign}(x_1) + 10x_2(0.5 - |x_1| - |x_2|)]dt + 0.01dW_2 + c_{21}dN_1 \end{cases} \tag{8.54}$$

$$dy_i = x_i dt + 0.005dv_i \quad , \quad t > 0 \quad , \quad i=1,2 \ . \tag{8.55}$$

In this case the set D was taken to be

$$D \triangleq \{x \in \mathbb{R}^2 : |x_i| < 1.2 + \delta, \ \ i=1,2\} \ , \ \delta < 0.1 \tag{8.56}$$

and the following set of parameters was used: $L_1 = L_2 = 12$, $h = 0.1$, $c_{11} = c_{21} = 4h, q_1 = 1,2$; $\Delta = 10^{-3}$, $N = 10^4$.

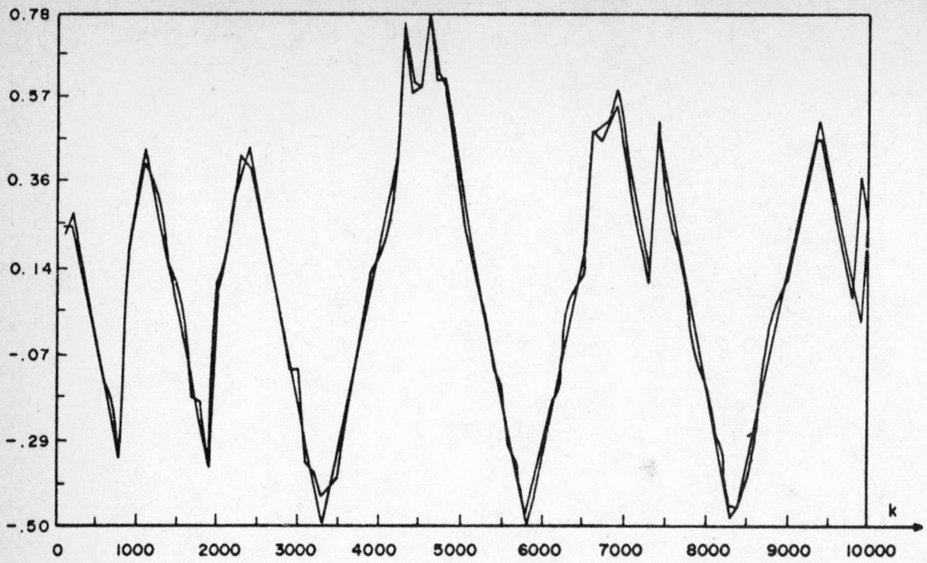

Fig.8.4a: $\zeta_{x1}(k)$ and $\hat{\zeta}_{x1}^{h,y}(k)$ as functions of k for the system given by equations (8.54)-(8.55), where $q_1 = 1$.

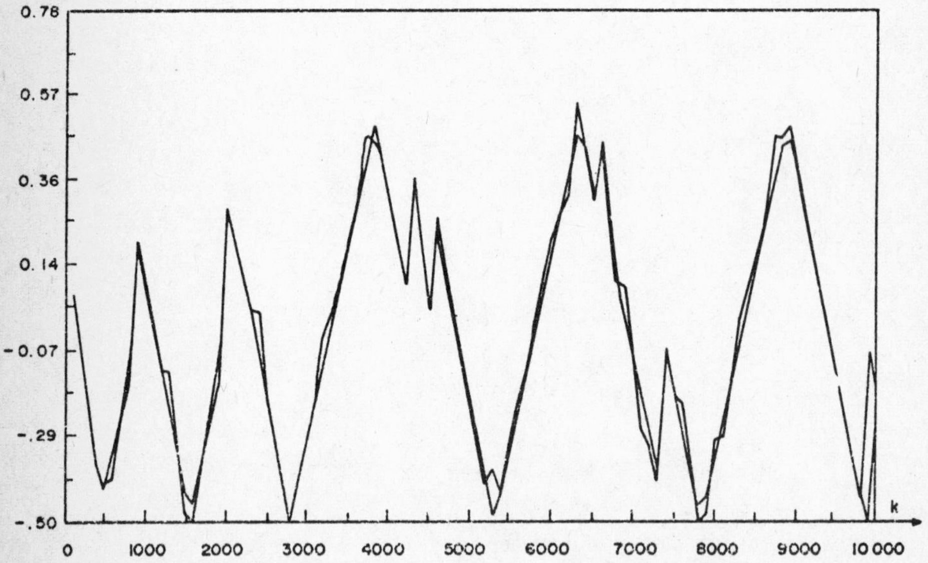

Fig.8.4b: $\zeta_{x2}(k)$ and $\hat{\zeta}_{x2}^{h,y}(k)$ as functions of k for the system given by equations (8.54)-(8.55), where $q_1 = 1$.

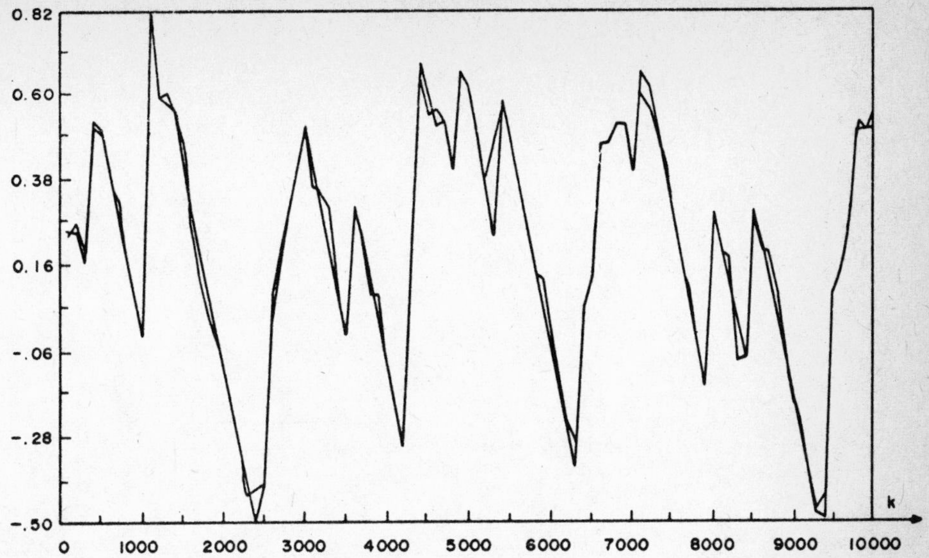

Fig.8.4c: $\zeta_{x1}(k)$ and $\hat{\zeta}^{h,y}_{x1}(k)$ as functions of k for the system given by equations (8.54)-(8.55), where $q_1 = 2$.

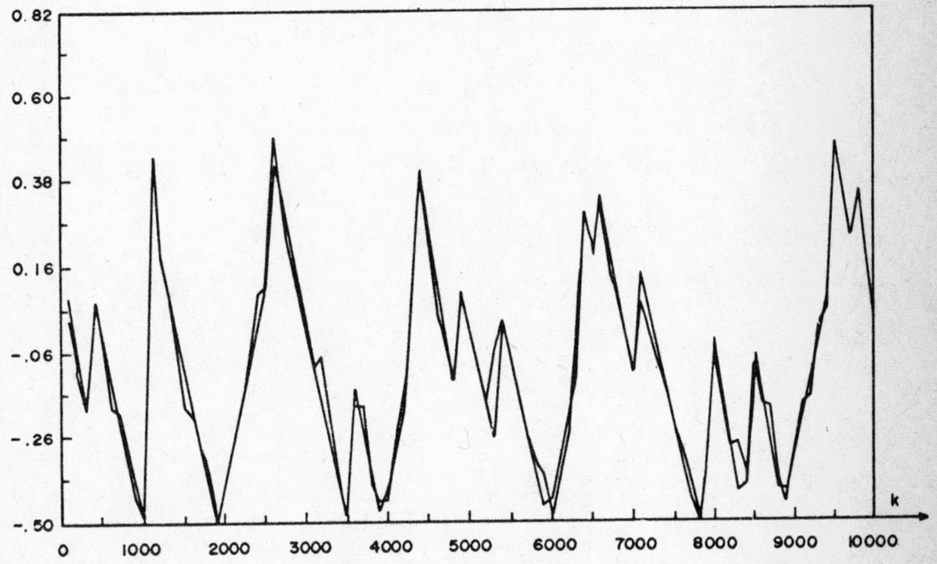

Fig.8.4d: $\zeta_{x2}(k)$ and $\hat{\zeta}^{h,y}_{x2}(k)$ as functions of k for the system given by equations (8.54)-(8.55), where $q_1 = 2$.

(b)
$$\begin{cases} dx_1 = x_2 dt \\ dx_2 = [10\,\mathrm{sat}(1-x_1) - 3x_2]dt + 10(0.2dW + 0.1c\,dN) \end{cases} \qquad t > 0 \qquad (8.57)$$

$$dy_i = x_i dt + 0.1 dv_i \quad , \quad i=1,2, \quad t > 0 \qquad (8.58)$$

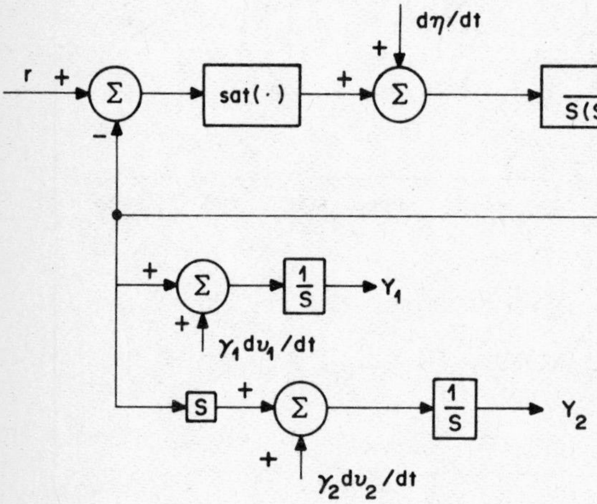

Fig.8.5: Block-diagram representation of the system given by (8.57)-(8.58)

In Fig.8.5 $d\eta/dt$ stands for $0.2dW/dt + 0.1c\,dN/dt$; the function sat(•) is given by (5.67), and $r(t) = 1$, $t \geq 0$. Set D was taken to be

$$D \triangleq \{x \in \mathbb{R}^2 \ : \ |x_i| < 4 + \delta, \ \ i=1,2\} \ , \ \delta < 0.1 \qquad (8.59)$$

and the following set of parameters was used: $L_1 = L_2 = 40$, $h = 0.1$, $c = 5h$, $q_1 = 1$, $\Delta = 10^{-3}$, $N = 10^4$.

(c)
$$\begin{cases} dx_1 = x_2 dt \\ dx_2 = [10\,\mathrm{sign}(1-x_1) - 3x_2]dt + 10(0.2dW + 0.1c\,dN) \end{cases} \qquad t > 0 \qquad (8.60)$$

$$dy_i = x_i dt + 0.1 dv_i \quad , \quad i=1,2 \quad , \quad t > 0 \tag{8.61}$$

(see Fig. 8.5). Set D and the parameters were here taken to be the same as in the previous example (case (b)), but with $q_1 = 1,2$.

Typical extracts from the results are given in the following figures. All the graphs in this section were plotted using the set of points $\{t'_k = 99\Delta + 100k\Delta \; : \; k=0,1,\dots,99\}$.

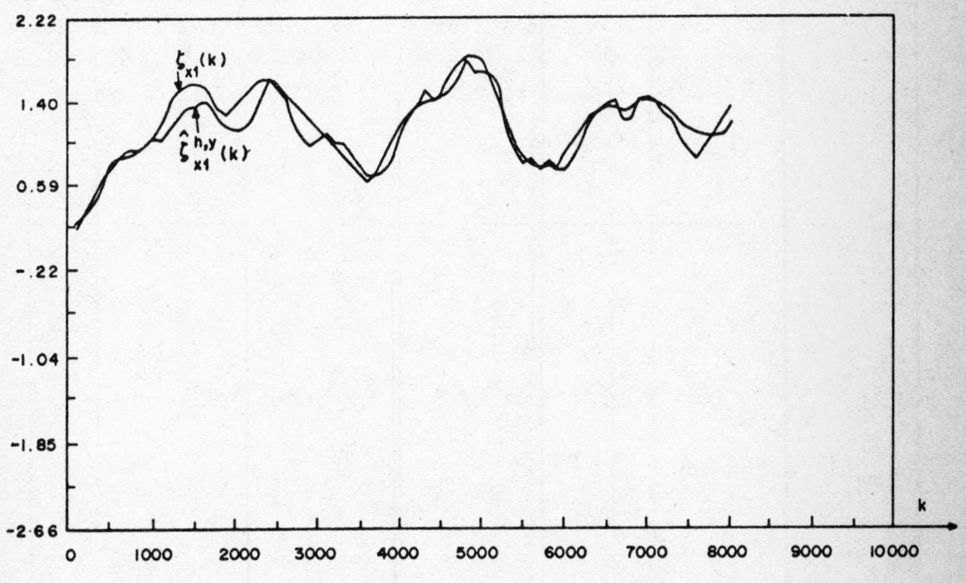

Fig.8.6a: $\zeta_{X1}(k)$ and $\hat{\zeta}^{h,y}_{X1}(k)$ as functions of k for the system given by equations (8.57)-(8.58).

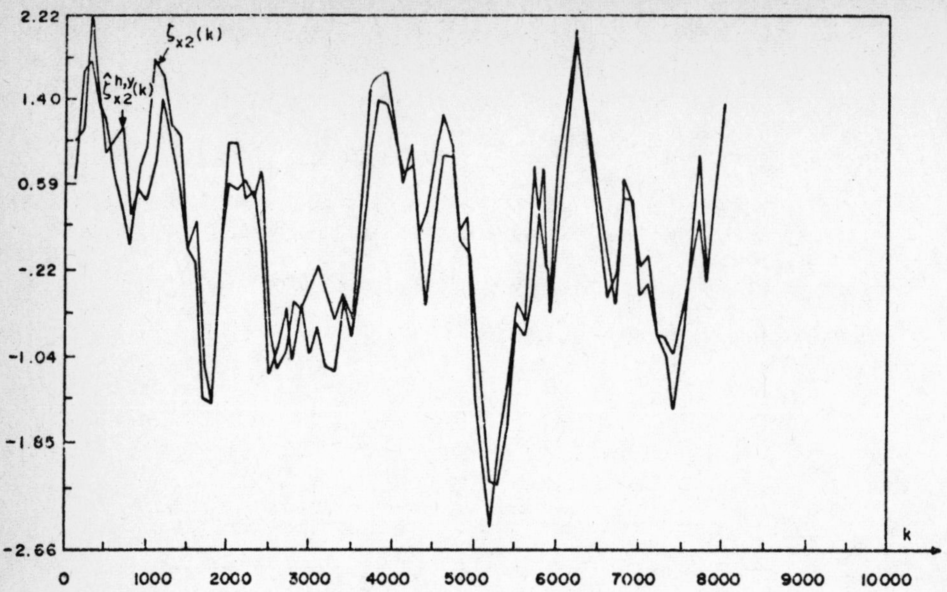

Fig.8.6b: $\zeta_{x2}(k)$ and $\hat{\zeta}_{x2}^{h,y}(k)$ as functions of k for the system given by equations (8.57)-(8.58).

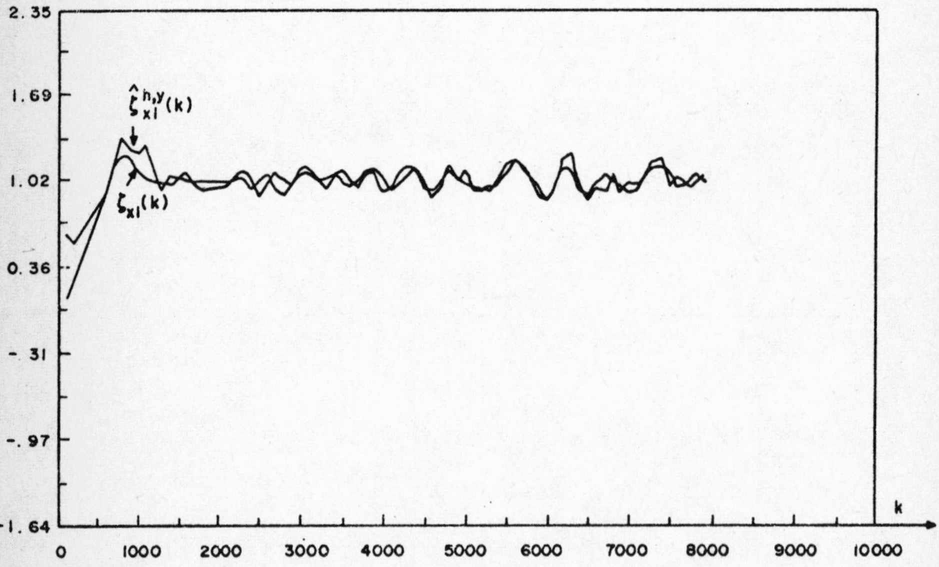

Fig.8.7a: $\zeta_{x1}(k)$ and $\hat{\zeta}_{x1}^{h,y}(k)$ as functions of k for the system given by equations (8.60)-(8.61), where $q_1 = 1$.

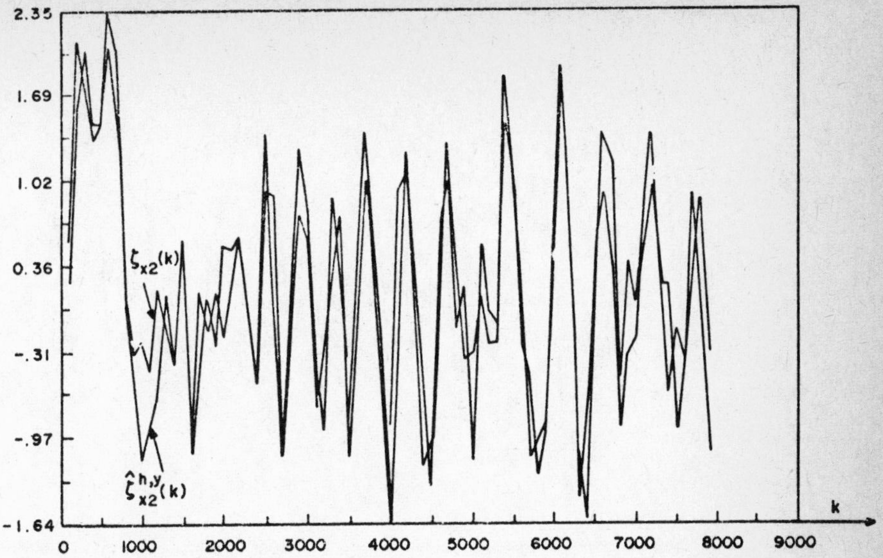

Fig.8.7b: $\zeta_{X2}(k)$ and $\hat{\zeta}^{h,y}_{X2}(k)$ as functions of k for the system given by equations (8.60)-(8.61), where $q_1 = 1$.

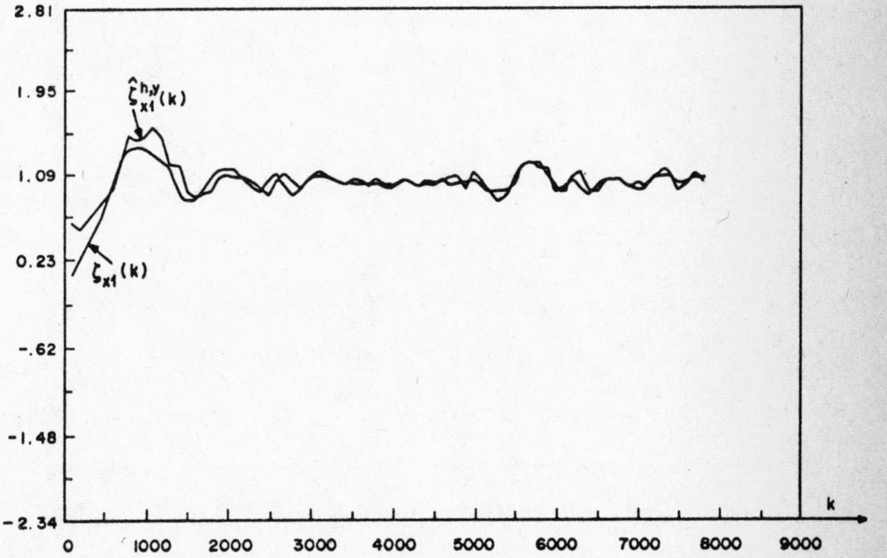

Fig.8.7c: $\zeta_{X1}(k)$ and $\hat{\zeta}^{h,y}_{X1}(k)$ as functions of k for the system given by equations (8.60)-(8.61), where $q_1 = 2$.

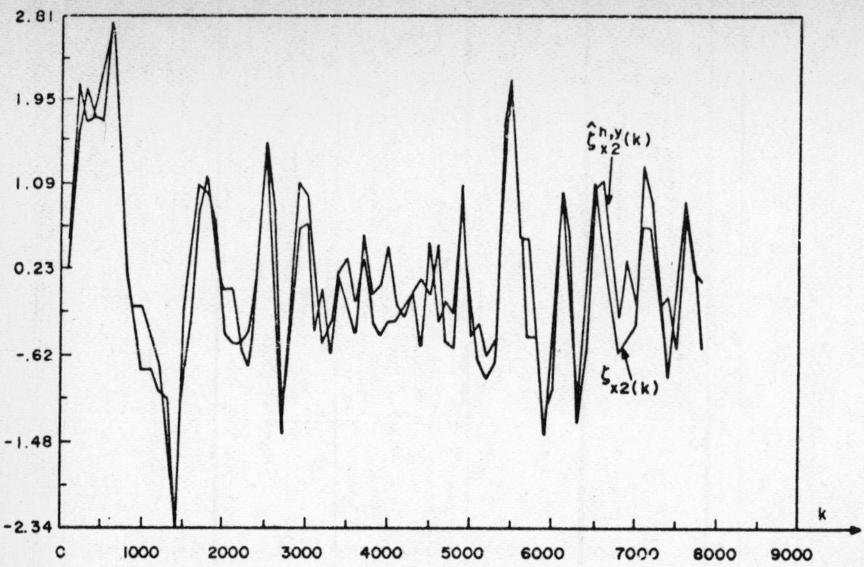

Fig.8.7d: $\zeta_{x2}(k)$ and $\hat{\zeta}^{h,y}_{x2}(k)$ as functions of k for the system given by equations (8.60)-(8.61), where $q_1 = 2$.

8.6 AN EXTENSION OF EQUATION (8.1)

Consider the $\mathbb{R}^m$-valued Markov process $\zeta_x = \{\zeta_x(t),\ t \geq 0\}$ satisfying the following equation

$$\zeta_x(t) = x + \int_0^t f(\zeta_x(s-))ds + BW(t) + \int_0^t C(\zeta_x(s-))dN(s), t \geq 0, x \in \mathbb{R}^m \tag{8.62}$$

and let the observation process Y be given by equation (8.2), where $f : \mathbb{R}^m \to \mathbb{R}^m$, $g : \mathbb{R}^m \to \mathbb{R}^p$, $B \in \mathbb{R}^{m\times m}$, $\Gamma \in \mathbb{R}^{p\times p}$. **W**, **V** and **N** are as described in Section 8.1. Let $C(x) = \{c_{ij}(x)\}$. It is assumed that $c_{ij} : \mathbb{R}^m \to \mathbb{R}$, i=1,...,m, j=1,...,r are continuous and bounded on any bounded domain in $\mathbb{R}^m$. Given h > 0, define the following set of m×r matrices, $\Lambda^{(k)}(x)$, $x \in \mathbb{R}^m_h$, k=1,...,r by

$$h\, \Lambda^{(k)}(x)\varepsilon^k = \arg \min_{a \in \mathbb{R}^m_h} |C(x)\varepsilon^k - a| \tag{8.63}$$

(where $|x| = [\sum_{i=1}^m x_i^2]^{\frac{1}{2}}$ and ε^k, k=1,...,r are defined in Section 8.2).

By following the development given in Sections 8.2-8.3 it can easily be shown that the equations for computing $\{\hat{\zeta}_x^{h,y}(t),\ t \in [0,T]\}$ are in this case given by

$$\begin{cases} dP_a^y(t) = \sum_{b \in D_h} \lambda(b,a) P_b^y(t)dt \\ \quad + P_a^y(t) \sum_{\ell=1}^{p} \gamma_\ell^{-2}(g_\ell(a) - \hat{g}_\ell(t))(dy_\ell(t) - \hat{g}_\ell(t)dt) \\ a \in D_h \quad , \quad t \in (0,T) \end{cases} \tag{8.64}$$

$$\hat{g}_\ell(t) = \sum_{a \in D_h} g_\ell(a) P_a^y(t) \quad , \quad \ell=1,\dots,p \quad , \ t \in [0,T] \tag{8.65}$$

$$\hat{\zeta}_x^{h,y}(t) = \sum_{a \in D_h} a P_a^y(t) \quad , \quad t \in [0,T] \tag{8.66}$$

where the function $\lambda : \mathbb{R}_h^m \times R_h^m \to \mathbb{R}$ is defined by

$$\lambda(x,x) \triangleq -[\sum_{i=1}^{m} (\sigma_i^2 + h|f_i(x)|) + h^2 \sum_{i=1}^{r} q_i]/h^2 \quad , \ x \in D_h . \tag{8.67}$$

$$\lambda(x,x \pm e^i h) \triangleq (\sigma_i^2/2 + h\, f_i^{\pm}(x))/h^2, \ i=1,\dots,m \quad , \quad x \in D_h \tag{8.68}$$

$$\lambda(x,x + h\Lambda^{(j)}(x)\varepsilon^j) \triangleq q_j \quad , \quad j=1,\dots,r \quad , \quad x \in D_h \tag{8.69}$$

$$\lambda(x,y) \triangleq 0 \quad , \quad x \in D_h \quad , \quad y \in U_x \tag{8.70}$$

$$\lambda(x,y) \triangleq 0 \quad , \quad x \in \mathbb{R}_h^m - D_h \quad , \quad y \in \mathbb{R}_h^m \tag{8.71}$$

where

$$U_x \triangleq \{y \in \mathbb{R}_h^m : y \neq x,\ y \neq x \pm e^i h \ , \ i=1,\dots,m, \text{ and } y \neq x + h\Lambda^{(j)}(x)\varepsilon^j \ \ j=1,\dots,r\} \tag{8.72}$$

It is assumed here that $\Lambda^{(j)}(x)\varepsilon^j \neq e^i$, $i=1,\dots,m$, $j=1,\dots,r$.

8.7 REFERENCES

8.1 J.R.Fisher, Optimal nonlinear filtering, in *Advances in Control Systems*, Ed. C.T. Leondes, pp. 197-300, Academic Press, New York, 1967.

8.2 T.P. McGarty, *Stochastic Systems and State Estimation*, John Wiley & Sons, New York, 1974.

8.3 H. Kwakernaak, Filtering for systems excited by Poisson white noise, in *Control Theory, Numerical Methods and Computer Systems Modelling*, Ed. A. Bensoussan and J.L. Lions, Lecture Notes in Economics and Mathematical Systems, 107, pp. 468-492, Berlin, 1975.

8.4 S.P. Au, State estimation for linear systems driven simultaneously by Wiener and Poisson processes, Ph.D. Thesis, Coordinated Science Laboratory and Dept. of Elect. Eng., University of Illinois at Urbana-Champaign, 1979.

8.5 S.I. Marcus, Modeling of nonlinear systems driven by semimartingales with applications to nonlinear filtering, Proceedings CISS, 1981.

8.6 S.I. Marcus, Low dimensional filters for a class of finite state estimation problems with Poisson observations, *Systems & Control Letters*, 1, pp. 237-241, 1982.

8.7 S.P. Au, A.H. Haddad and H.V. Poor, A state estimation algorithm for linear systems driven simultaneously by Wiener and Poisson processes, *IEEE Trans. on Automatic Control*, AC-27, pp. 617-626, 1982.

8.8 H.J. Kushner and C.F. Yu, Probability methods for the convergence of finite difference approximations to partial differential-integral equations II, *J. of Math. Anal. and App.*, 45, pp. 54-72, 1974.

8.9 H.J. Kushner and G. DiMasi, Approximations for functionals and op= timal control problems on jump diffusion processes, *J. of Math. Anal. and App.*, 63, pp. 772-800, 1978.

CHAPTER 9

PREDICTION VIA MARKOV CHAINS APPROXIMATION

9.1 INTRODUCTION

Let an $\mathbb{R}^m$-valued Markov process $\zeta_x = \{\zeta_x(t) = (\zeta_{x1}(t),\ldots,\zeta_{xm}(t)),\ t \geq 0\}$ satisfy the following set of equations

$$\zeta_{xi}(t) = x_i + \int_0^t f_i(\zeta_x(s))ds + \sigma_i W_i(t),\ t \geq 0,\ i=1,\ldots,m, \quad (9.1)$$

with the noisy observation of ζ_x given by

$$y_i(t) = \int_0^t g_i(\zeta_x(s))ds + \gamma_i v_i(t),\ t \geq 0,\ i=1,\ldots,p\ , \quad (9.2)$$

where $x = (x_1,\ldots,x_m)$, and $y(t) = (y_1(t),\ldots,y_p(t))$; $f : \mathbb{R}^m \to \mathbb{R}^m$ and $g : \mathbb{R}^m \to \mathbb{R}^p$ are given functions satisfying the conditions stated in Sec= tion 3.1; σ_i, $i=1,\ldots,m$ and γ_j, $j=1,\ldots,p$ are given positive numbers. $\mathbf{W} = \{W(t) = (W_1(t),\ldots,W_m(t)),\ t \geq 0\}$ and $\mathbf{V} = \{v(t) = (v_1(t),\ldots,v_p(t)),\ t \geq 0\}$ are $\mathbb{R}^m$-valued and $\mathbb{R}^p$-valued standard Wiener processes respectively. It is assumed that $\mathbf{W}$ and $\mathbf{V}$ are mutually independent.

Denote by F_t^y, $t \in [0,T]$, the smallest σ-field generated by the family of random elements $Y^t = \{y(s)\ ;\ 0 \leq s \leq t\}$. The problem dealt with in this chapter is to find an approximation $\hat{\zeta}_x^{h,y}(k,s)$ to

$$\hat{\zeta}_x(t,s) = E[\zeta_x(t \wedge \tau_T -) | F^y_{s \wedge \tau_T -}],\quad t \in [s,T]\ ,\quad s \in [0,T] \quad (9.3)$$

at the instants $t_k = k\Delta$, $k=1,2,\ldots$, where $\tau_T = \tau_T(x)$ is the first exit time of ζ_x from an open and bounded domain $D \subset \mathbb{R}^m$. The minimum least-squares estimate of $\zeta_x(t)$ given Y^s, $s < t$, is $E[\zeta_x(t) | F_s^y]$. The problem of finding $\{E[\zeta_x(t)|F_s^y], 0 \le s < t \le T\}$ is called the *(nonlinear) prediction problem*. This problem is closely related to the filtering problem and its treatment usually follows the treatment of the filtering problem.

In this chapter we follow the same procedures as those applied in Chapter 3. First, a continuous-time Markov chain $\{\zeta_x^h(t\wedge\tau_T^h), t \in [0,T]\}$, which constitutes an approximation to $\{\zeta_x(t\wedge\tau_T), t \in [0,T]\}$, is constructed on D_h (see Sections 3.2-3.3). Second, an optimal least-squares filter is derived for the on-line computation of $\hat{\zeta}_x^h(t,s) = E[\zeta_x^h(t\wedge\tau_T^h-)|F_{s\wedge\tau_T^h-}^{y,h}]$. Third, an estimator $\{\hat{\zeta}_x^{h,y}(k,s), k\Delta \in [s,T]\}$ is constructed as an approximation to $\{\hat{\zeta}_x(k\Delta,s), k\Delta \in [s,T]\}$ (equation (9.3)) and this estimator is simulated for a variety of examples.

9.2 THE EQUATIONS OF OPTIMAL PREDICTION

Let $\{\zeta_x^h(t), t \in [0,T]\}$, D, D_h, $\tau_T(x)$, $\tau_T^h(x)$, $\{y^h(t), t \in [0,T]\}$, $\{F_t^{y,h}, t \in [0,T]\}$, $\{\tilde{P}_\alpha(t), \alpha \in \mathbb{R}_h^m, t \in [0,T]\}$ and $\{\zeta_x^h(t\wedge\tau_T^h), t \in [0,T]\}$ be defined by equations (3.6)-(3.14),(3.19), $D_h = \mathbb{R}_h^m \cap D$, (3.20), (3.21), (3.22); $F_t^{y,h} = \sigma(y^h(s);\ 0 \le s \le t)$, $t \in [0,T]$; (3.26) and (3.6)-(3.10), (3.31), (3.35) respectively. Assume that $\sup_{t \in [0,T]} E|\zeta_x^h(t)|^2 < \infty$ and $\int_0^T E|h_t|^2 dt < \infty$, where h_t is given by equation (3.23). Define

$$\tilde{P}_\alpha(t,s) \triangleq P(\zeta_x^h(t) = \alpha|F_s^{y,h}),\ 0 \le s < t \le T,\ x \in D_h,\ \alpha \in \mathbb{R}_h^m . \tag{9.4}$$

Then,

$$\tilde{\zeta}_x^h(t,s) \triangleq E[\zeta_x^h(t)|F_s^{y,h}] = \sum_{\alpha \in \mathbb{R}_h^m} \alpha\, \tilde{P}_\alpha(t,s),\quad 0 \le s < t \le T \tag{9.5}$$

i.e., $\tilde{\zeta}_x^h(t,s)$ is the optimal least-squares estimate of $\zeta_x^h(t)$, given $\{y^h(u), 0 \le u \le s\}$.

Let

$$\chi_\alpha(t) \triangleq \begin{cases} 1 & \text{if } \zeta_x^h(t) = \alpha \\ & \\ 0 & \text{otherwise.} \end{cases} \tag{9.6}$$

Then, since $F_s^{y,h} \subset F_t^{y,h}$, $s < t$, it follows that

$$\tilde{P}_\alpha(t,s) = E[\chi_\alpha(t)|F_s^{y,h}] = E[E[\chi_\alpha(t)|F_t^{y,h}]|F_s^{y,h}] = E[\tilde{P}_\alpha(t)|F_s^{y,h}] . \tag{9.7}$$

By using equations (3.28) we obtain

$$\begin{cases} E[\tilde{P}_\alpha(t)|F_s^{y,h}] = \tilde{P}_\alpha(s) + E[\int_s^t \sum_{\gamma \in \mathbb{R}_h^m} \lambda(\gamma,\alpha)\tilde{P}_\gamma(u)du|F_s^{y,h}] \\ + E[\int_s^t \tilde{P}_\alpha(u) \sum_{\ell=1}^p \gamma_\ell^{-2}(g_\ell(\alpha) - \hat{\hat{g}}_\ell(u))(dy_\ell^h(u) - \hat{\hat{g}}_\ell(u)du)|F_s^{y,h}] \\ \alpha \in \mathbb{R}_h^m \quad , \quad 0 \le s < t \le T, \end{cases} \tag{9.8}$$

where $\hat{\hat{g}}$ is given by equations (3.29).

Denote

$$m_\ell(t) \triangleq y_\ell^h(t) - \int_0^t \hat{\hat{g}}_\ell(u)du, \quad t \in [0,T], \quad \ell=1,\ldots,p \tag{9.9}$$

Then, by using equations (9.9), (3.22) and (3.29), it follows that

$$\begin{aligned} E[m_\ell(t) - m_\ell(s)|F_s^{y,h}] &= E[\int_s^t g_\ell(\zeta_x^h(u))du + \gamma_\ell(\quad v_\ell(t) - v_\ell(s)) \\ &\quad - \int_s^t \sum_{\alpha \in \mathbb{R}_h^m} g_\ell(\alpha)\tilde{P}_\alpha(u)du|F_s^{y,h}], \quad \ell=1,\ldots,p \end{aligned} \tag{9.10}$$

Assuming that $\sup_{t \in [0,T]} E|g(\zeta_x^h(t))|^2 < \infty$, and using the property $F_s^{y,h} \subset F_u^{y,h}$, $s < u$, equations (9.10) yield

$$\begin{aligned} E[m_\ell(t) - m_\ell(s)|F_s^{y,h}] &= E[\int_s^t E[g_\ell(\zeta_x^h(u))|F_u^{y,h}]du - \int_s^t \sum_{\alpha \in \mathbb{R}_h^m} g_\ell(\alpha)\tilde{P}_\alpha(u)du|F_s^{y,h}] \\ &= E[\int_s^t \sum_{\alpha \in \mathbb{R}_h^m} g_\ell(\alpha)\tilde{P}_\alpha(u)du - \int_s^t \sum_{\alpha \in \mathbb{R}_h^m} g_\ell(\alpha)\tilde{P}_\alpha(u)du|F_s^{y,h}] = 0, \quad 0 \le s < t \le T, \\ &\qquad \ell=1,\ldots,p. \end{aligned} \tag{9.11}$$

Hence $\{m(t) = (m_1(t),\ldots,m_p(t)),\ t \in [0,T]\}$ is a $(F_t^{y,h},P)$-square integrable martingale. Consequently, see [9.1], the process

$$\begin{cases} M_\alpha(t) \overset{\Delta}{=} \int_0^t \tilde{P}_\alpha(u) \sum_{\ell=1}^p \gamma_\ell^{-2}(g_\ell(\alpha) - \hat{\hat{g}}_\ell(u))(dy_\ell^h(u) - \hat{\hat{g}}_\ell(u)du) \\ t \in [0,T], \quad \alpha \in \mathbb{R}_h^m \end{cases} \tag{9.12}$$

is a $(F_t^{y,h},P)$-square integrable martingale from which it follows that

$$E[M_\alpha(t) - M_\alpha(s) | F_s^{y,h}] = 0,\ 0 \le s < t \le T,\ \alpha \in \mathbb{R}_h^m . \tag{9.13}$$

Thus, by using equations (9.12)-(9.13), equations (9.8) reduce to

$$E[\tilde{P}_\alpha(t)|F_s^{y,h}] = \tilde{P}_\alpha(s) + E[\int_s^t \sum_{\gamma \in \mathbb{R}_h^m} \lambda(\gamma,\alpha)\tilde{P}_\gamma(u)du | F_s^{y,h}] . \tag{9.14}$$

Or, by using equation (9.7) equation (9.14) yields

$$\tilde{P}_\alpha(t,s) = \tilde{P}_\alpha(s) + \int_s^t \sum_{\gamma \in \mathbb{R}_h^m} \lambda(\gamma,\alpha)\ E[\tilde{P}_\gamma(u)|F_s^{y,h}]du . \tag{9.15}$$

Thus, equations (9.7) and (9.15) imply

$$\begin{aligned} &\tilde{P}_\alpha(t,s) = \tilde{P}_\alpha(s) + \int_s^t \sum_{\gamma \in \mathbb{R}_h^m} \lambda(\gamma,\alpha)\tilde{P}_\gamma(u,s)du \\ &0 \le s < t \le T \quad , \quad \alpha \in \mathbb{R}_h^m, \end{aligned} \tag{9.16}$$

where $\{\tilde{P}_\alpha(s),\ \alpha \in \mathbb{R}_h^m\}$ are determined by equations (3.28)-(3.29), and $\{\lambda(\gamma,\alpha),\ \gamma,\alpha \in \mathbb{R}_h^m\}$ by equations (3.6)-(3.10).

Define

$$P_\alpha(t,s) \overset{\Delta}{=} P(\zeta_x^h(t \wedge \tau_T^h-) = \alpha | F_{s\wedge\tau_T^h-}^{y,h}),\ 0 \le s < t \le T,\ \alpha \in D_h . \tag{9.17}$$

Then, the equations for $\hat{\zeta}_x^h(t,s) = E[\zeta_x^h(t\wedge\tau_T^h-)|F_{s\wedge\tau_T^h-}^{y,h}]$ are given by

$$\begin{aligned} &P_\alpha(t,s) = P_\alpha(s) + \int_s^t \sum_{\gamma \in D_h} \lambda(\gamma,\alpha)P_\gamma(u,s)du \\ &0 \le s < t \le T \quad , \quad \alpha \in D_h, \end{aligned} \tag{9.18}$$

where $\{\lambda(\gamma,\alpha),\ \gamma,\alpha \in D_h\}$ are determined by equations (3.6)-(3.10) and (3.31) and $\{P_\alpha(s),\ \alpha \in D_h\}$ by equations (3.36)-(3.37), and

$$\hat{\zeta}_x^h(t,s) = \sum_{\alpha \in D_h} \alpha\, P_\alpha(t,s) \quad , \quad 0 \le s < t \le T. \tag{9.19}$$

Given $Y^s = \{y(u),\ 0 \le u \le s\}$, $s \in [0,T]$. Then, in order to compute an approximation to $\hat{\zeta}_x(t,s)$ (equation (9.3)), first equations (3.36)-(3.37) are solved on the time interval $[0,s]$, where in equations (3.36) the increment dy^h is replaced by dy. Let $\{P_\alpha^y(u),\ \alpha \in D_h,\ u \in [0,s]\}$ denote the solution to equations (3.36)-(3.37), where dy replaces dy^h in (3.36). Second, equations (9.18) are solved on the time interval $[s,T]$, where in (9.18) $P_\alpha^y(s)$ replaces $P_\alpha(s)$, $\alpha \in D_h$. Denote the solution to equations (9.18), where $P_\alpha^y(s)$ replaces $P_\alpha(s)$, $\alpha \in D_h$, by $\{P_\alpha^y(t,s),\ \alpha \in D_h,\ t \in [s,T]\}$. Third, a process $\hat{\zeta}_x^{h,y} = \{\hat{\zeta}_x^{h,y}(t,s),\ t \in [0,T]\}$ is defined by

$$\hat{\zeta}_x^{h,y}(t,s) \triangleq \begin{cases} \sum_{\alpha \in D_h} \alpha\, P_\alpha^y(t) & \text{if } t \in [0,s] \\ \\ \sum_{\alpha \in D_h} \alpha\, P_\alpha^y(t,s) & \text{if } t \in [s,T]. \end{cases} \tag{9.20}$$

The process $\{\hat{\zeta}_x^{h,y}(t,s)\}$ serves here as an approximation to $\{\hat{\zeta}_x(t,s)\}$ (equation (9.3)). In the next section a procedure for computing $\{\hat{\zeta}_x^{h,y}(k\Delta,s),\ k\Delta \in [0,T]\}$ is suggested. We assume that $\Pi_\alpha = P_\alpha^y(0)$, $\alpha \in D_h$ are unknown.

9.3 AN ALGORITHM FOR COMPUTING $\hat{\zeta}_x^{h,y}(t,s)$

Let $\varepsilon > 0$, $\Delta > 0$, $N_1\Delta = s$, $N\Delta = T$, $P_\alpha^y(0) = \prod_{i=1}^m (2L_i+1)^{-1}$, $\alpha \in D_h$, $\hat{g}_\ell(0) = \prod_{i=1}^m (2L_i+1)^{-1} \sum_{\alpha \in D_h} g_\ell(\alpha)$ and $\hat{\zeta}_x^{h,y}(k,s) \triangleq \hat{\zeta}_x^{h,y}(k\Delta,N_1\Delta)$. Note that $hL_i = a_i$, $i=1,\ldots,m$ (see equation (3.19)).

Algorithm 9.3

1. $k:=0$

2. If $k \le N_1$, then $\gamma(k)=1$; Otherwise $\gamma(k)=0$.

3. For $\alpha \in D_h$ calculate

$$\begin{cases} P^y_\alpha(k+1):=P^y_\alpha(k) + \sum_{\gamma \in D_h} \lambda(\gamma,\alpha)P^y_\gamma(k)\Delta \\ \quad + \gamma(k)P^y_\alpha(k)\sum_{\ell=1}^{p} \gamma_\ell^{-2}(g_\ell(\alpha) - \hat{g}_\ell(k))(y_\ell(k+1)-y_\ell(k) - \hat{g}_\ell(k)\Delta) \end{cases} \tag{9.21}$$

$$P^y_\alpha(k+1):=\max(0,P^y_\alpha(k+1)) \tag{9.22}$$

4. $$Z(k+1):= \sum_{\alpha \in D_h} P^y_\alpha(k+1) \tag{9.23}$$

5. If $Z(k+1) \ge \varepsilon$, then for $\alpha \in D_h$, $P^y_\alpha(k+1):=P^y_\alpha(k+1)/Z(k+1)$; (9.24)

   Otherwise stop.

6. For $\ell=1,\dots,p$ calculate

$$\hat{g}_\ell(k+1):= \sum_{\alpha \in D_h} g_\ell(\alpha)P^y_\alpha(k+1) \tag{9.25}$$

$$\hat{\zeta}^{h,y}_x(k+1,s):= \sum_{\alpha \in D_h} \alpha\, P^y_\alpha(k+1). \tag{9.26}$$

7. If $k=N$ or if $\hat{\zeta}^{h,y}_x(k+1,s) \notin D$, then stop. Otherwise, $k:= k+1$ and go to 2.

The problem of establishing conditions for the weak convergence of $\{\hat{\zeta}^{h,y}_x(t,s),\ t \in [0,T]\}$ to $\{\hat{\zeta}_x(t,s),\ t \in [0,T]\}$ (equation (9.3)) as $h \downarrow 0$ is outside the scope of this work. Instead, the role of $\{\hat{\zeta}^{h,y}_x(t,s), t \in [0,T]\}$, as an approximation to $\{\hat{\zeta}_x(t,s),\ t \in [0,T]\}$, is demonstrated in the sequel by means of numerical experimentation with several examples.

9.4 EXAMPLES

In this section the notations used are the same as those used in Section 3.6.

(a)
$$\begin{cases} dx_1 = x_2 dt \\ dx_2 = [10\ \text{sat}(1-x_1) - 3x_2]dt + 2dW_2 \end{cases} \qquad t > 0 \tag{9.27}$$

$$dy_i = x_i dt + \gamma_i dv_i \ , \ t > 0 \ , \ i=1,2 \tag{9.28}$$

(see Fig.8.5). In this case D is given by

$$D \overset{\Delta}{=} \{x \in \mathbb{R}^2 : |x_i| < 4 + \delta, \ i=1,2\} \ , \ \delta < 0.1 \tag{9.29}$$

and the following set of parameters was used: $\gamma_1 = \gamma_2 = 0.1, 0.2$; $h=0.1$, $L_1 = L_2 = 40$, $\Delta = 10^{-3}$, $N_1 = 2000, 5000$; $N = 6 \cdot 10^3$.

(b)
$$\begin{cases} dx_1 = 0.1\ x_1 dt + 0.01 dW_1 \\ dx_2 = 0.2 x_2 dt + 0.01 dW_2 \end{cases} \qquad t > 0 \tag{9.30}$$

$$dy_i = x_i^3 dt + 0.001 dv_i \ , \quad t > 0 \quad , \quad i=1,2. \tag{9.31}$$

In this case D is given by

$$D \overset{\Delta}{=} \{x \in \mathbb{R}^2 : |x_i| < 1.2 + \delta, \ i=1,2\} \ \delta < 0.1 \tag{9.32}$$

and the following set of parameters was used: $h=0.1$, $L_1 = L_2 = 12$, $\Delta = 10^{-3}$, $N_1 = 3000, 7500$; $N = 9500$.

(c)
$$\begin{cases} dx_1 = [\alpha\,\text{sign}(x_2) + 50x_1(0.6 - |x_1| - |x_2|)]dt + 0.001dW_1 \\ dx_2 = [-\alpha\,\text{sign}(x_1) + 50x_2(0.6 - |x_1| - |x_2|)]dt + 0.001dW_2 \end{cases} \quad t > 0 \tag{9.33}$$

$$dy_i = \text{sign}(x_i)dt + 0.005 dv_i, \quad t > 0, \ i=1,2. \tag{9.34}$$

In this case D is given by

$$D \triangleq \{x \in \mathbb{R}^2 : |x_i| < 1.2 + \delta,\ i=1,2\}, \quad \delta < 0.1 \tag{9.35}$$

and the following set of parameters was used: h=0.1, α = 0.2,1; $L_1 = L_2 = 12$, $\Delta = 10^{-3}$, N_1 = 5000, 7500; $N = 10^4$.

(d)
$$dx_i = 0.05dW_i, \quad t > 0, \quad i=1,2 \tag{9.36}$$

$$dy_i = x_i^3 dt + 0.005dv_i, \quad t > 0, \quad i=1,2. \tag{9.37}$$

In this case D is given by

$$D \triangleq \{x \in \mathbb{R}^2 : |x_i| < 1 + \delta,\ i=1,2\}, \ \delta < 0.05 \tag{9.38}$$

and the following set of parameters was used: h = 0.05, $L_1 = L_2 = 20$, $\Delta = 10^{-3}$, N_1 = 5000, $N = 10^4$.

All the graphs in this section were plotted using the set of points $\{t'_k = 100k\Delta : k=0,1,\ldots,100\}$.

REFERENCES

9.1 H. Kunita and S. Watanabe, On square integrable martingales, *Nagoya Math. J.*, 30, pp 209-245, 1967.

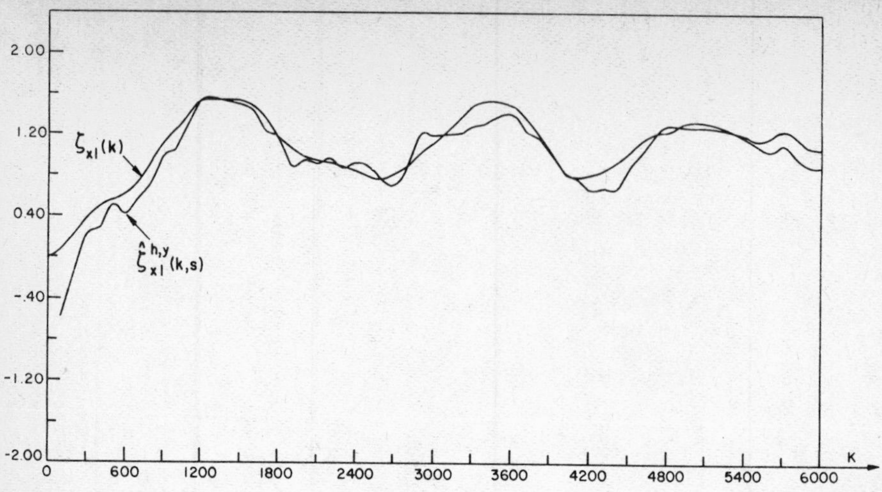

Fig.9.1a: $\zeta_{X1}(k)$ and $\hat{\zeta}^{h,y}_{X1}(k,s)$ as functions of k for the system given by equations (9.27)-(9.28), where $\gamma_1 = \gamma_2 = 0.1$ and $N_1 = 5000$.

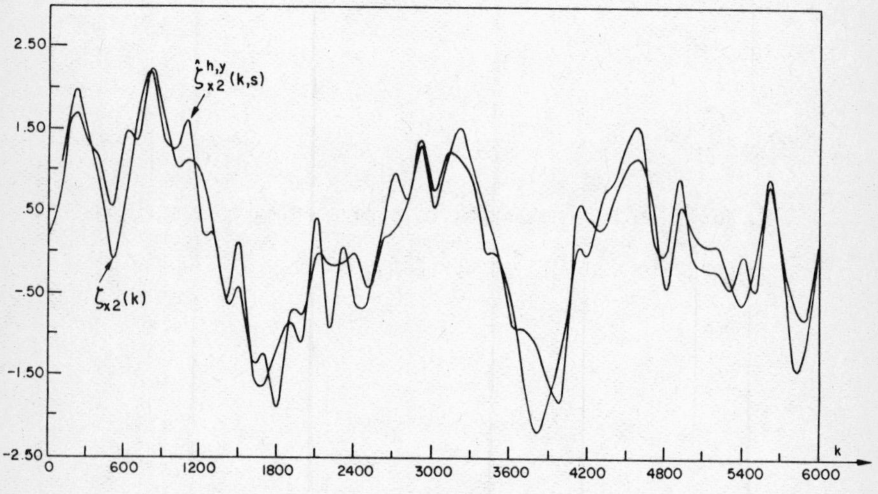

Fig.9.1b: $\zeta_{X2}(k)$ and $\hat{\zeta}^{h,y}_{X2}(k,s)$ as functions of k for the system given by equations (9.27)-(9.28), where $\gamma_1 = \gamma_2 = 0.1$ and $N_1 = 5000$.

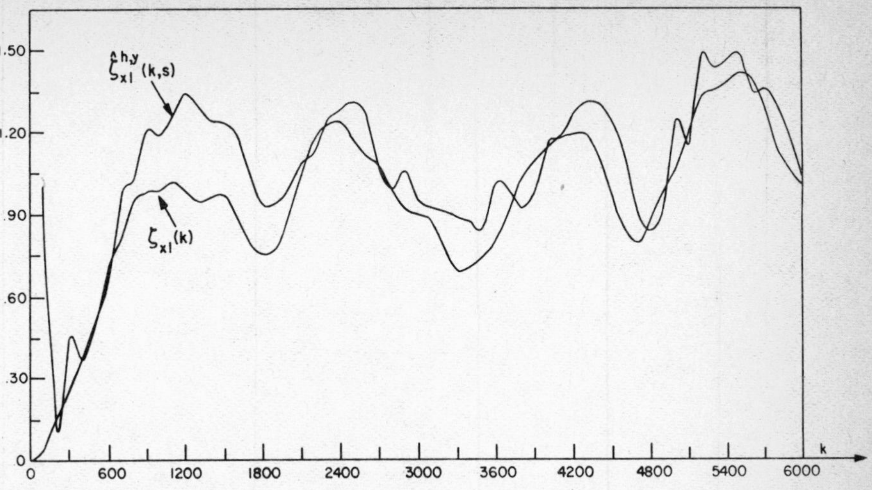

Fig.9.1c: $\zeta_{x1}(k)$ and $\hat{\zeta}_{x1}^{h,y}(k,s)$ as functions of k for the system given by equations (9.27)-(9.28), where $\gamma_1 = \gamma_2 = 0.2$ and $N_1 = 2000$.

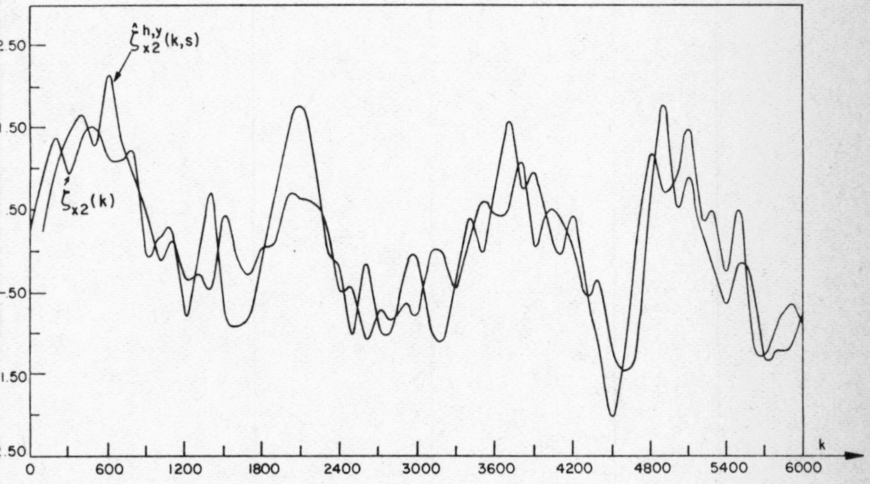

Fig.9.1d: $\zeta_{x2}(k)$ and $\hat{\zeta}_{x2}^{h,y}(k,s)$ as functions of k for the system given by equations (9.27)-(9.28), where $\gamma_1 = \gamma_2 = 0.2$ and $N_1 = 2000$.

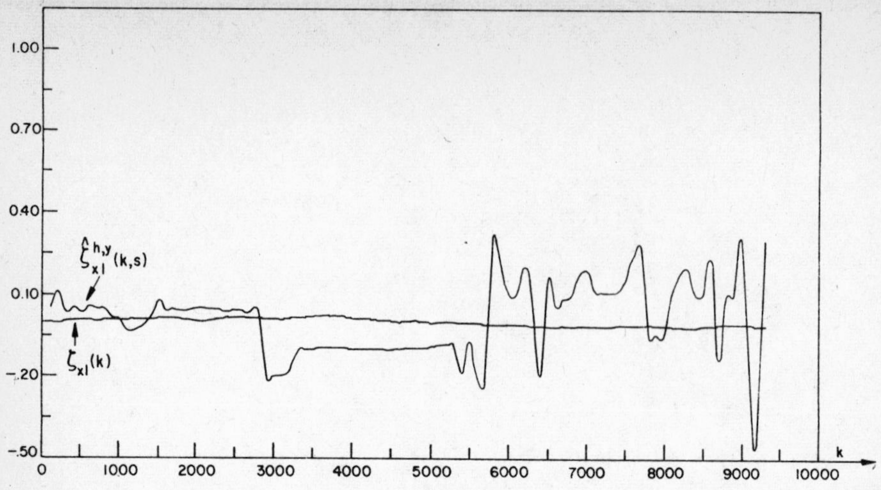

<u>Fig.9.2a</u>: $\zeta_{X1}(k)$ and $\hat{\zeta}^{h,y}_{X1}(k,s)$ as functions of k for the system given by equations (9.30)-(9.31), where N_1 = 7500.

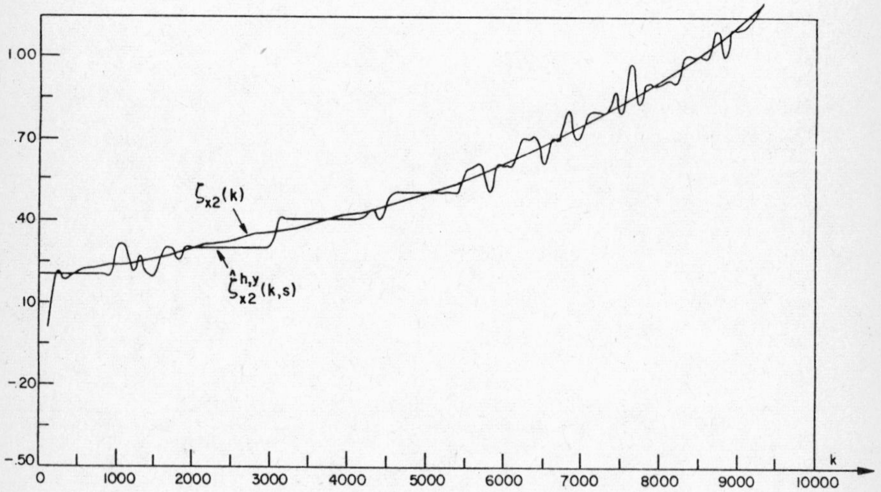

<u>Fig.9.2b</u>: $\zeta_{X2}(k)$ and $\hat{\zeta}^{h,y}_{X2}(k,s)$ as functions of k for the system given by equations (9.30)-(9.31), where N_1 = 7500.

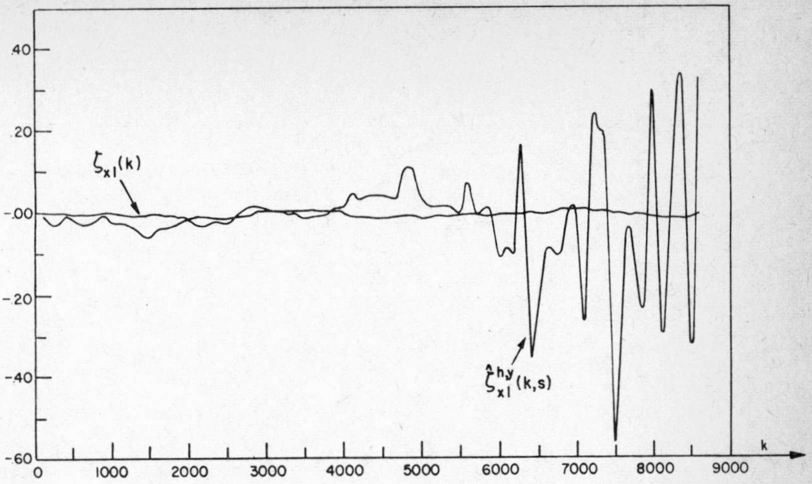

Fig.9.2c: $\zeta_{x1}(k)$ and $\hat{\zeta}^{h,y}_{x1}(k,s)$ as functions of k for the system given by equations (9.30)-(9.31), where N_1 = 3000.

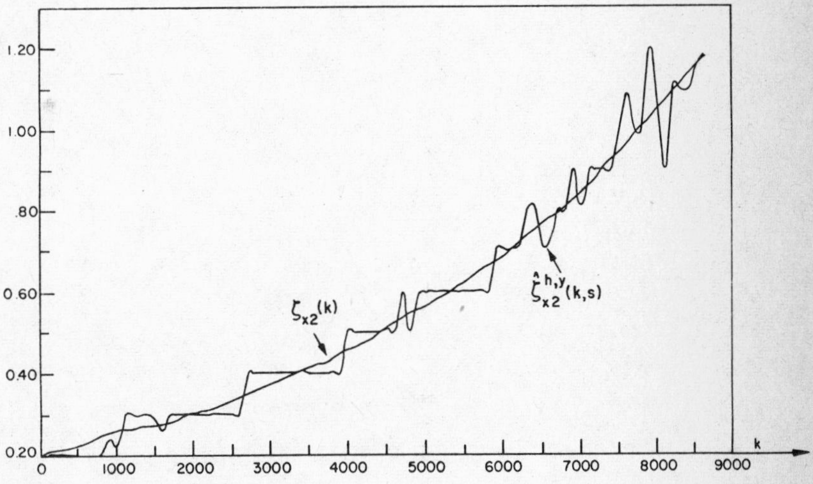

Fig.9.2d: $\zeta_{x2}(k)$ and $\hat{\zeta}^{h,y}_{x2}(k,s)$ as functions of k for the system given by equations (9.30)-(9.31), where N_1 = 3000.

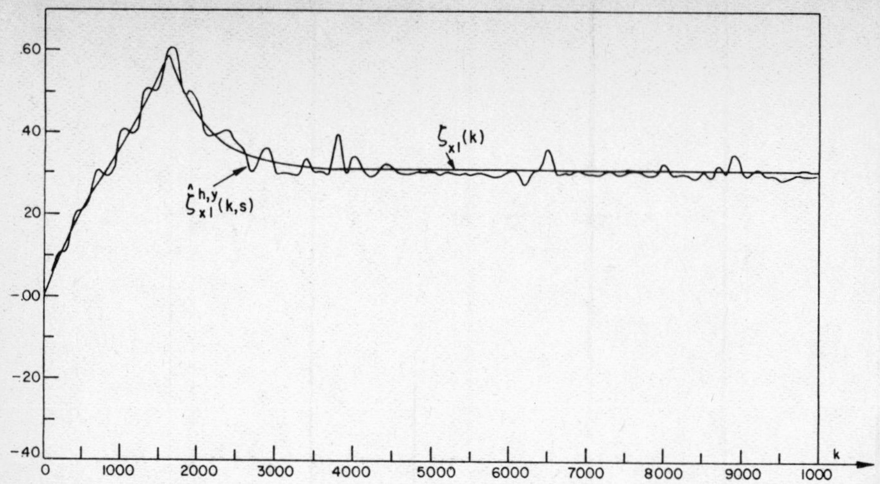

Fig.9.3a: $\zeta_{X1}(k)$ and $\hat{\zeta}^{h,y}_{X1}(k,s)$ as functions of k for the system given by equations (9.33)-(9.34), where $\alpha = 0.2$ and $N_1 = 5000$.

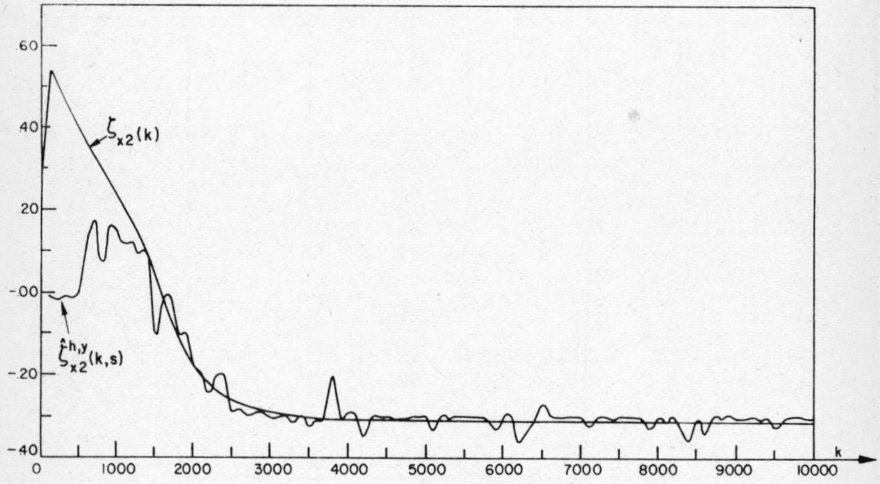

Fig.9.3b: $\zeta_{X2}(k)$ and $\hat{\zeta}^{h,y}_{X2}(k,s)$ as functions of k for the system given by equations (9.33)-(9.34), where $\alpha = 0.2$ and $N_1 = 5000$.

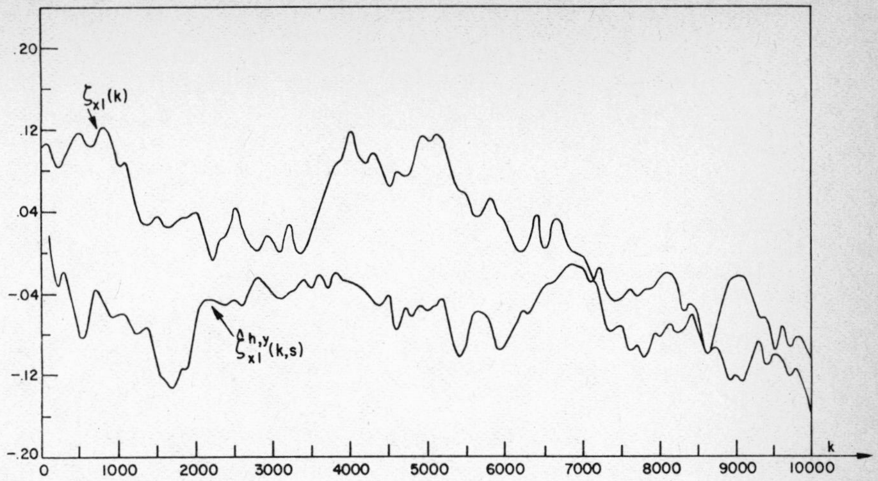

Fig.9.4a: $\zeta_{x1}(k)$ and $\hat{\zeta}^{h,y}_{x1}(k,s)$ as functions of k for the system given by equations (9.36)-(9.37).

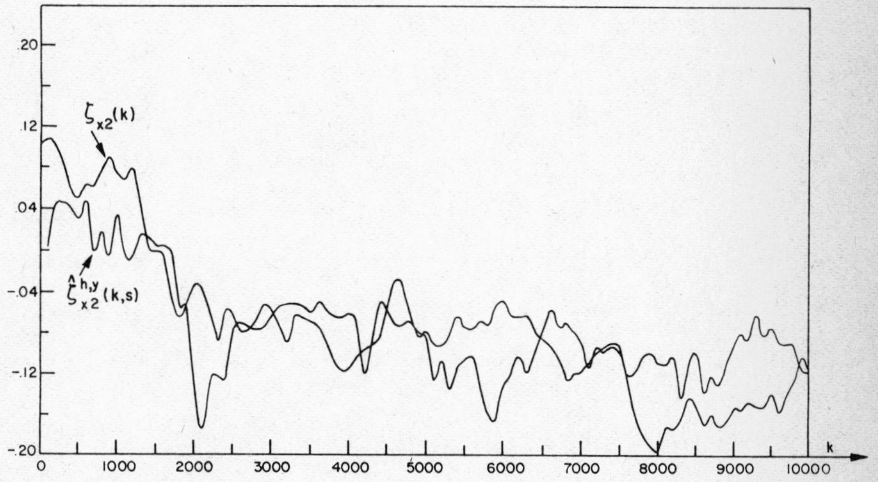

Fig.9.4b: $\zeta_{x2}(k)$ and $\hat{\zeta}^{h,y}_{x2}(k,s)$ as functions of k for the system given by equations (9.36)-(9.37).

CHAPTER 10

SOME EXTENSIONS OF LINEAR FILTERING

10.1 LINEAR FILTERING WITH NON-GAUSSIAN INITIAL CONDITIONS

10.1.1 Introduction

Consider the linear system given by

$$dx(t) = A(t)x(t)dt + G(t)dW(t),\ t \in (0,T);\ x(0) = \theta \tag{10.1}$$

$$dy(t) = H(t)x(t)dt + \Gamma(t)dv(t),\ t \in (0,T);\ y(0) = 0, \tag{10.2}$$

where $x(t) = (x_1(t),\dots,x_m(t))'$ is the state vector and $y(t) = (y_1(t),\dots, y_p(t))'$ is the output measurement vector. $A(t)$, $G(t)$, $H(t)$ and $\Gamma(t)$, $t \in [0,T]$, are given $m\times m$, $m\times m$, $p\times m$ and $p\times p$ matrices satisfying the con= ditions stated in Section 1.3. $W = \{W(t),\ t \geq 0\}$ and $V = \{v(t),\ t \geq 0\}$ are an $\mathbb{R}^m$-valued and an $\mathbb{R}^p$-valued standard Wiener process respectively, and θ is an $\mathbb{R}^m$-valued random element such that $E|\theta|^2 < \infty$. It is assumed that θ, W and V are mutually independent.

Denote by $\zeta = \{\zeta(t) = (\zeta_1(t),\dots,\zeta_m(t)),\ t \in [0,T]\}$ the solution to equations (10.1). If the distribution of θ is Gaussian, then $\hat{\zeta} = \{\hat{\zeta}(t) = E[\zeta(t)|\mathcal{F}_t^y],\ t \in [0,T]\}$, where $\mathcal{F}_t^y = \sigma(y(u),\ 0 \leq u \leq t)$, is determined by the Kalman-Bucy filter, i.e., equations (1.75)-(1.78). But (see Section 1.3) if the distribution of θ is non-Gaussian, then $\hat{\zeta}$ is determined by

$$\begin{aligned} d\hat{\zeta}(t) &= A(t)\hat{\zeta}(t)dt + P(t)H'(t)R^{-1}(t)[dy(t) - H(t)\hat{\zeta}(t)dt],\ t \in (0,T) \\ \hat{\zeta}(0) &= E\theta, \end{aligned} \tag{1.75}$$

where $R^{-1}(t)$ is given by equation (1.76). However, the matrix $P(t)$

$$P(t) = E[(\zeta(t)-\hat{\zeta}(t))(\zeta(t)-\hat{\zeta}(t))'|\mathcal{F}_t^y],\ t \in [0,T], \tag{1.61}$$

which appears in equation (1.75), is a random matrix.

The system given by equations (10.1)-(10.2), with $G(t) = I_m$ and $\Gamma(t) = I_p$ (where I_m and I_p denote the unit m×m and p×p matrices respectively), and where θ is an $\mathbb{R}^m$-valued random element such that

$$P(\theta \in A) = \int_A p(x)dx \quad , \quad A \in \mathbb{R}^m \tag{10.3}$$

and $p(x)$, $x \in \mathbb{R}^m$ is a given function, is dealt with in Ref [10.1]. It is shown there that $\hat{\zeta}$ is determined by

$$\begin{aligned} d\hat{\zeta}(t) &= [A(t)-G(t,\eta(t))H'(t)H(t)]\hat{\zeta}(t)dt + G(t,\eta(t))H'(t)dy(t) \\ & \qquad t \in (0,T) \quad (10.4) \\ d\eta(t) &= (H(t)\Phi(t))'H(t)[\Phi(t)c(t,\eta(t)) - \hat{\zeta}(t)]dt + (H(t)\Phi(t))'dy(t), \end{aligned}$$

where $G(t,x)$ and $c(t,\eta)$ are determined by

$$\begin{cases} dQ(t)/dt = A(t)Q(t) + Q(t)A'(t) - Q(t)H'(t)H(t)Q(t) - I_m, \ t \in (0,T) \\ Q(0) = 0, \end{cases} \tag{10.5}$$

$$\begin{cases} d\Phi(t)/dt = [A(t) - Q(t)H'(t)H(t)]\Phi(t), \ t \in (0,T) \\ \Phi(0) = I_m, \end{cases} \tag{10.6}$$

$$\psi(t) \triangleq \int_o^t \Phi'(u)H'(u)H(u)\Phi(u)du, \tag{10.7}$$

$$q(t,\eta) \triangleq \int_{\mathbb{R}^m} \exp[-\tfrac{1}{2}(x'\psi(t)x - 2x'\eta)]p(x)dx, \tag{10.8}$$

$$c(t,\eta) \triangleq \int_{\mathbb{R}^m} x \exp[-\tfrac{1}{2}(x'\psi(t)x - 2x'\eta)]p(x)dx/q(t,\eta) \tag{10.9}$$

and

$$G(t,\eta) \triangleq Q(t) + \Phi(t)[\int_{\mathbb{R}^m} xx' \exp[-\tfrac{1}{2}(x'\psi(t)x - 2x'\eta)]p(x)dx/q(t,\eta) - c(t,\eta)c'(t,\eta)]\Phi'(t). \tag{10.10}$$

In general, however, equations (10.4)-(10.10) are too complex for im=

plementation. An alternative approach for the computation of $\hat{\zeta}$ is to apply the following procedure: *Adaptive estimation via parallel processing.*

Suppose that $\zeta(0) = \theta \in \{x^{(1)},\ldots,x^{(L)}\}$, where $x^{(i)} \in \mathbb{R}^m$ $i=1,\ldots,L$ are given vectors. (In the case when $\zeta(0) = \theta \in \Theta$, where Θ is a compact set in $\mathbb{R}^m$, choose L and $\{x^{(i)}\}$ such that the set $\{x^{(1)},\ldots,x^{(L)}\}$ will 'approximate' Θ in some sense. However, in this case only an approximation to $\hat{\zeta}$ is computed). Define

$$\chi_i(t) \triangleq \begin{cases} 1 & \text{if } \zeta(0) = \theta = x^{(i)} \\ 0 & \text{otherwise;} \end{cases} \tag{10.11}$$

then,

$$P_i(t) \triangleq E[\chi_i(t)|F_t^y] = P(\zeta(0) = x^{(i)}|F_t^y),\ t \in [0,T],\ i=1,\ldots,L. \tag{10.12}$$

By using the relation

$$E[\phi(\zeta(t))|F_t^y] = \sum_{i=1}^{L} E[\phi(\zeta(t))|F_t^y,\ \zeta(0) = x^{(i)}]P_i(t) \tag{10.13}$$

for any measurable function $\phi : \mathbb{R}^m \to \mathbb{R}$ and inserting $F(\theta,\zeta(t)) = \chi_i(t)$ in equation (1.25), the following equations are obtained:

$$dP_i(t) = P_i(t)(\hat{\zeta}^{(i)}(t)-\hat{\zeta}(t))'(\Gamma^{-1}(t)H(t))'\Gamma^{-1}(t)(dy(t)-H(t)\hat{\zeta}(t)dt) \tag{10.14}$$

$t \in (0,T)$, $i=1,\ldots,L$,

where

$$\hat{\zeta}^{(i)}(t) = E[\zeta(t)|F_t^y,\ \zeta(0) = x^{(i)}],\ t \in [0,T],\ i=1,\ldots,L \tag{10.15}$$

and, by using equation (10.13),

$$\hat{\zeta}(t) = \sum_{i=1}^{L} \hat{\zeta}^{(i)}(t)P_i(t),\ t \in [0,T]. \tag{10.16}$$

From the theory of Kalman-Bucy filtering it follows that $\hat{\zeta}^{(i)}$, $i=1,\ldots,L$, are determined by

$$\begin{cases} d\hat{\zeta}^{(i)}(t)=A(t)\hat{\zeta}^{(i)}(t)dt + P(t)H'(t)R^{-1}(t)[dy(t)-H(t)\hat{\zeta}^{(i)}(t)dt], t \in (0,T) \\ \\ \hat{\zeta}^{(i)}(0) = x^{(i)} \end{cases} \tag{10.17}$$

i=1,...,L

and

$$\begin{cases} dP(t)/dt=A(t)P(t)+P(t)A'(t)+G(t)G'(t)-P(t)H'(t)R^{-1}(t)H(t)P(t), t \in (0,T) \\ \\ P(0) = 0. \end{cases} \tag{10.18}$$

The set of equations (10.14),(10.16),(10.17) and (10.18) constitute the filter equations for computing $\hat{\zeta} = \{\hat{\zeta}(t),\ t \in [0,T]\}$. Equations (10.18) are independent of the observed process Y and, as a result, can be com= puted off-line. Here, instead of solving equations (10.18), the following set of (matrix) equations is solved (see, for example, Ref. [10.2]):

$$\begin{pmatrix} dX(t)/dt \\ \\ dY(t)/dt \end{pmatrix} = \begin{pmatrix} -A'(t) & H'(t)R^{-1}(t)H(t) \\ \\ G(t)G'(t) & A(t) \end{pmatrix} \begin{pmatrix} X(t) \\ \\ Y(t) \end{pmatrix} t \in (0,T) \tag{10.19}$$

$$X(0) = I_m \qquad Y(0) = 0$$

$$P(t) = Y(t)X^{-1}(t) \quad , \quad t \in [0,T]. \tag{10.20}$$

REMARK:

In this Section we are dealing with the augmented Markov process $\{(\theta,\zeta(t)),\ t \in [0,T]\}$. Hence, the filtering formula (1.25) remains valid if we there replace the expression $F(\zeta(t))$ by $F(\theta,\zeta(t))$.

10.1.2 Example

Consider the following system of equations

$$\begin{aligned} d^2x/dt^2 &= -\alpha dx/dt + W_x \\ dW_x/dt &= -aW_x + \sigma\eta_x, \end{aligned} \tag{10.21}$$

where α, a and σ are given positive numbers and $\{\eta_x(t), t \geq 0\}$ is a Gaussian white noise. For more details on this system see Ref [10.3]. Denote $x_1 \triangleq x$, $x_2 \triangleq dx/dt$ and $x_3 \triangleq W_x$. Then equations (10.21) can be written as

$$\begin{pmatrix} dx_1 \\ dx_2 \\ dx_3 \end{pmatrix} = \begin{pmatrix} 0 & 1 & 0 \\ 0 & -\alpha & 1 \\ 0 & 0 & -a \end{pmatrix} \begin{pmatrix} x_1 \\ x_2 \\ x_3 \end{pmatrix} dt + \begin{pmatrix} 0 & 0 & 0 \\ 0 & 0 & 0 \\ 0 & 0 & \sigma \end{pmatrix} \begin{pmatrix} dW_1 \\ dW_2 \\ dW_3 \end{pmatrix} \qquad (10.22)$$

where $W = \{W(t) = (W_1(t), W_2(t), W_3(t)), t \geq 0\}$ is an $\mathbb{R}^3$-valued standard Wiener process. The measurement process is given here by

$$\begin{pmatrix} dy_1 \\ dy_2 \end{pmatrix} = \begin{pmatrix} 1 & 0 & 0 \\ 0 & 1 & 0 \end{pmatrix} \begin{pmatrix} x_1 \\ x_2 \\ x_3 \end{pmatrix} dt + \begin{pmatrix} \gamma_1 & 0 \\ 0 & \gamma_2 \end{pmatrix} \begin{pmatrix} dv_1 \\ dv_2 \end{pmatrix} \qquad (10.23)$$

where γ_1 and γ_2 are given positive numbers and $V = \{v(t) = (v_1(t), v_2(t)), t \geq 0\}$ is an $\mathbb{R}^2$-valued standard Wiener process. Denote by $\zeta = \{\zeta(t) = (\zeta_1(t), \zeta_2(t), \zeta_3(t)), t \geq 0\}$ the solution to the equation (10.22). It is assumed that $\zeta(0)$, W and V are mutually independent.

In this example, the matrix $\Lambda(t)$ (equation (10.19)) is given by

$$\Lambda(t) = \Lambda = \begin{bmatrix} 0 & 0 & 0 & \gamma_1^{-2} & 0 & 0 \\ -1 & \alpha & 0 & 0 & \gamma_2^{-2} & 0 \\ 0 & -1 & a & 0 & 0 & 0 \\ 0 & 0 & 0 & 0 & 1 & 0 \\ 0 & 0 & 0 & 0 & -\alpha & 1 \\ 0 & 0 & \sigma^2 & 0 & 0 & -a \end{bmatrix} . \qquad (10.24)$$

Numerical experimentation was carried out for the following set of parameters: $\Delta = 0.005$; $N = 2000, 4000$; $L = 6$; $\alpha = 0.4$; $a = 0.1$;

$$x^{(1)} = \begin{pmatrix} 700 \\ -300 \\ 0.00 \end{pmatrix}, \quad x^{(2)} = \begin{pmatrix} 700 \\ -200 \\ 0.00 \end{pmatrix}, \quad x^{(3)} = \begin{pmatrix} 700 \\ -100 \\ 0.00 \end{pmatrix}, \quad x^{(4)} = \begin{pmatrix} 700 \\ 100 \\ 0.00 \end{pmatrix},$$

$$x^{(5)} = \begin{pmatrix} 700 \\ 200 \\ 0.00 \end{pmatrix} \quad \text{and } x^{(6)} = \begin{pmatrix} 700 \\ 300 \\ 0.00 \end{pmatrix} .$$

Typical extracts from the numerical results are presented in the follow= ing figures:

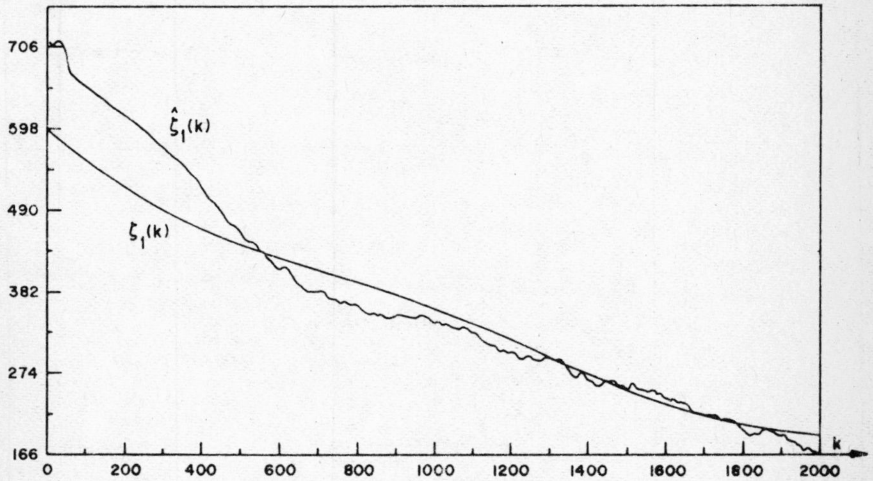

Fig.10.1-a: $\zeta_1(k)$ and $\hat{\zeta}_1(k)$ as functions of k for the case where $\zeta(0) = (600,-100,0.00)'$, $\sigma = 10$, $\gamma_1 = 10$, $\gamma_2 = 40$ and $N = 2000$.

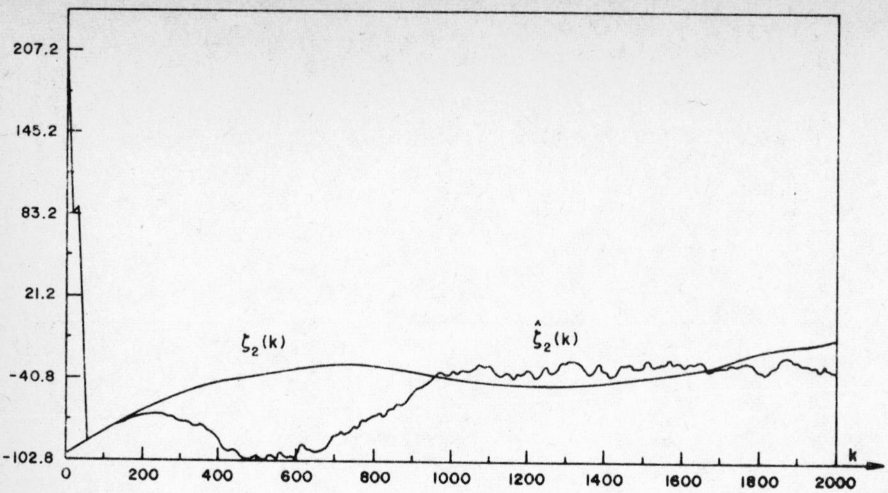

Fig.10.1-b: $\zeta_2(k)$ and $\hat{\zeta}_2(k)$ as functions of k for the case where $\zeta(0) = (600,-100,0.00)'$, $\sigma = 10$, $\gamma_1 = 10$, $\gamma_2 = 40$ and $N = 2000$.

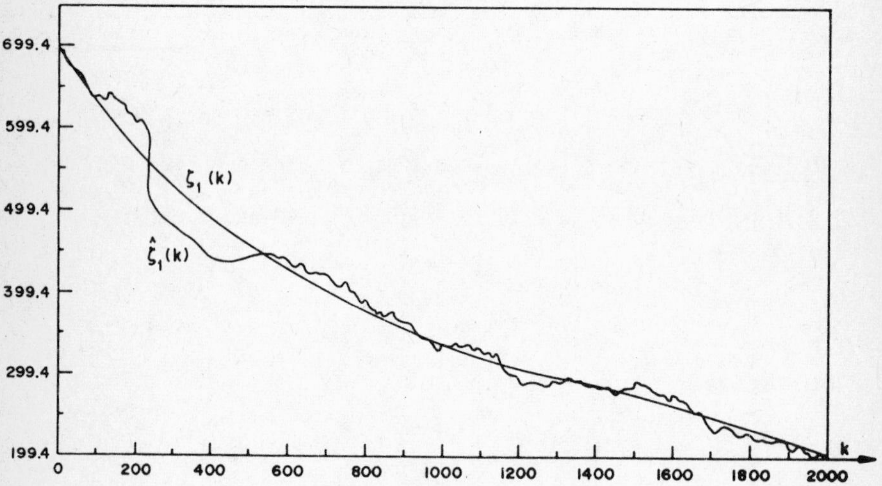

Fig.10.2-a: $\zeta_1(k)$ and $\hat{\zeta}_1(k)$ as functions of k for the case where $\zeta(0) = (700,-150,0.00)'$, $\sigma = 10$, $\gamma_1 = 10$, $\gamma_2 = 40$ and $N = 2000$.

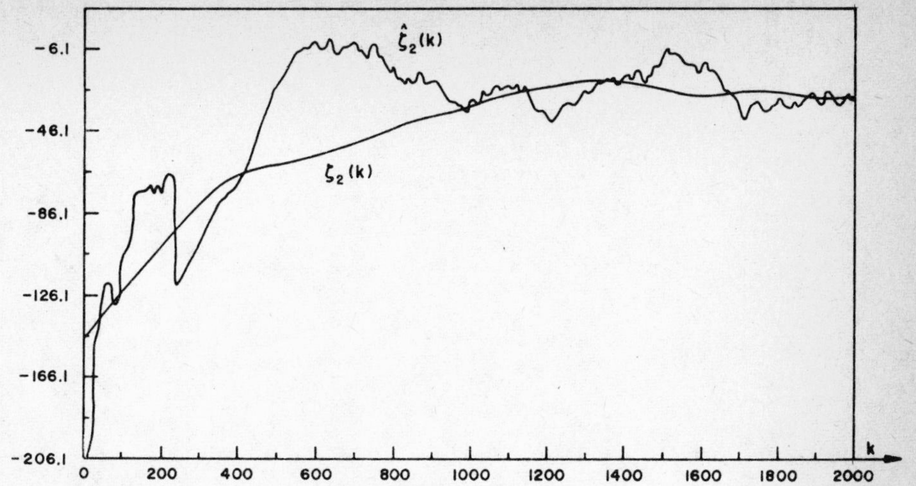

Fig.10.2-b: $\zeta_2(k)$ and $\hat{\zeta}_2(k)$ as functions of k for the case where $\zeta(0)$ = (700,-150,0.00)', σ = 10, γ_1 = 10, γ_2 = 40 and N = 2000.

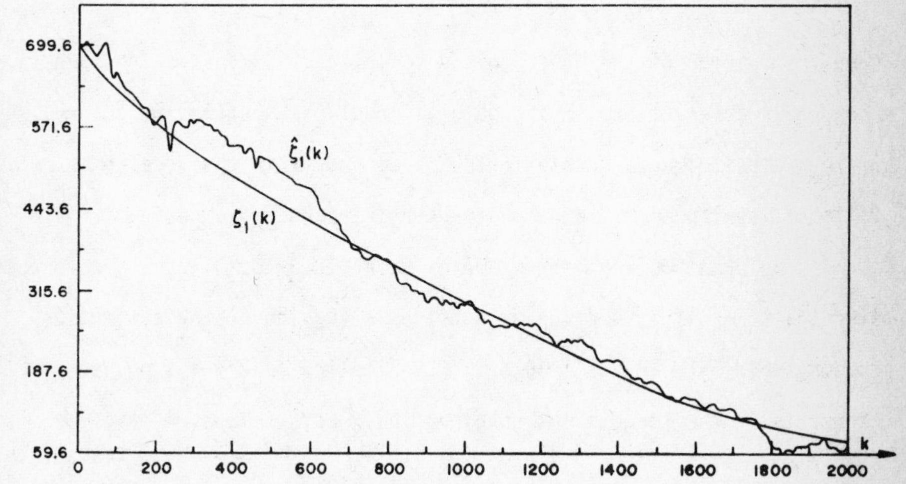

Fig.10.3-a: $\zeta_1(k)$ and $\hat{\zeta}_1(k)$ as functions of k for the case where $\zeta(0)$ = (700,-150,0.00)', σ = 10, γ_1 = 20, γ_2 = 80 and N = 2000.

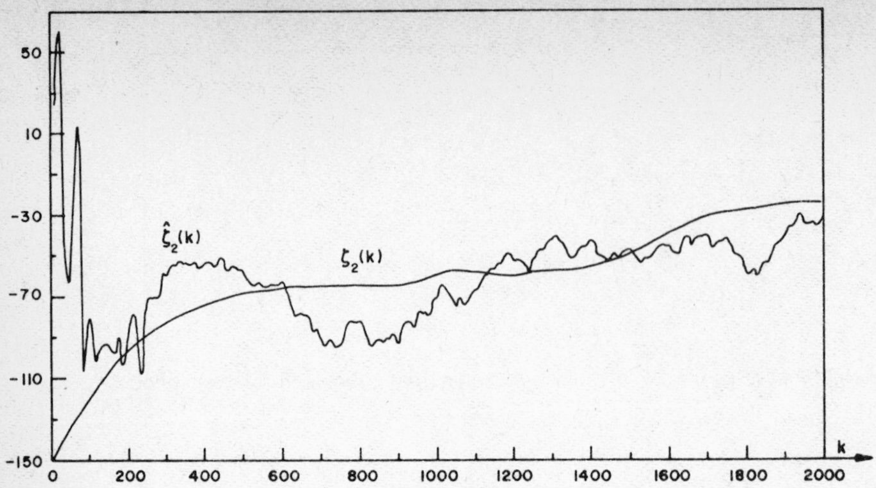

Fig.10.3-b: $\zeta_2(k)$ and $\hat{\zeta}_2(k)$ as functions of k for the case where $\zeta(0)$ = (700,-150,0.00)', σ = 10, γ_1 = 20, γ_2 = 80 and N = 2000.

10.2 ESTIMATION OF MANEUVERING TARGETS

10.2.1 Introduction

The determination of the position and velocity of a maneuvering target using radar observations is a problem in nonlinear estimation theory. During the past decade much effort has been spent in the development of digital filtering algorithms for tracking airborne maneuvering targets. For more information see, for example, Ref. [10.3-10.4] and the references cited there. In this section we consider a version of the estimation problem dealt with in Ref. [10.3]. For simplicity we only consider the components of the target's motion along the x-axis. Thus we have

$$\begin{aligned} \ddot{x} &= -\alpha\dot{x} + u_x + W_x \\ & \qquad\qquad\qquad\qquad t > 0, \qquad\qquad (10.25) \\ \dot{W}_x &= -aW_x + \sigma\eta_x \end{aligned}$$

where α is the drag coefficient, u_x is the command input in the x direc= tion, W_x is the acceleration process acting in the x direction, $\{\eta_x(t),$

$t \geq 0\}$ is a Gaussian white noise and σ is a given positive number.

Denote $x_1 \triangleq x$, $x_2 \triangleq dx/dt$ and $x_3 \triangleq W_x$; then equations (10.25) can be written as

$$\begin{pmatrix} dx_1 \\ dx_2 \\ dx_3 \end{pmatrix} = \begin{pmatrix} 0 & 1 & 0 \\ 0 & -\alpha & 1 \\ 0 & 0 & -a \end{pmatrix} \begin{pmatrix} x_1 \\ x_2 \\ x_3 \end{pmatrix} dt + \begin{pmatrix} 0 \\ u_x \\ 0 \end{pmatrix} dt + \begin{pmatrix} 0 \\ 0 \\ \sigma \end{pmatrix} dW(t), t > 0 \tag{10.26}$$

where $W = \{W(t), t \geq 0\}$ is an $\mathbb{R}$-valued standard Wiener process.

The tracking measurements along the x-axis are given by

$$dy = (1,0,0) \begin{pmatrix} x_1 \\ x_2 \\ x_3 \end{pmatrix} dt + \gamma dv \quad , \quad t > 0, \ y(0) = 0, \tag{10.27}$$

where γ is a given positive number and $V = \{v(t), t \geq 0\}$ is an $\mathbb{R}$-valued standard Wiener process. u_x, the maneuvering target's input is here assumed to be a random variable with range $\{u_1,\ldots,u_L\}$, where u_i, $i=1,\ldots,L$ are given numbers. Hence, the maneuverability of the target is here re= presented by the random input u_x and the random acceleration process $x_3 = W_x$, both expressions appearing in equation (10.26). Denote by $\zeta = \{\zeta(t) = (\zeta_1(t),\zeta_2(t),\zeta_3(t)), t \geq 0\}$ the solution to equation (10.26). It is assumed that $\zeta(0)$, u_x, W and V are mutually independent and that $E|\zeta(0)|^2 < \infty$.

In the next section equations are derived for the computation of $\hat{\zeta}(t) = E[\zeta(t)|\mathcal{F}_t^y]$, where $\mathcal{F}_t^y = \sigma(y(s), 0 \leq s \leq t)$, $t \in [0,T]$.

10.2.2 The Minimum Variance Filter Equations

Denote

$$A \triangleq \begin{pmatrix} 0 & 1 & 0 \\ 0 & -\alpha & 1 \\ 0 & 0 & -a \end{pmatrix} \quad H \triangleq (1\ 0\ 0) \quad B \triangleq (0\ 1\ 0)' \quad G \triangleq (0\ 0\ \sigma)' \tag{10.28}$$

$$\chi_i(t) \triangleq \begin{cases} 1 & u_x = u_i \\ & \qquad\qquad i=1,\dots,L. \\ 0 & \text{otherwise} \end{cases} \tag{10.29}$$

Then

$$P_i(t) = E[\chi_i(t)|F_t^y] = P(u_x = u_i|F_t^y),\ t \in [0,T],\ \ i=1,\dots,L. \tag{10.30}$$

By using the relation

$$E[\phi(\zeta(t))|F_t^y] = \sum_{i=1}^{L} E[\phi(\zeta(t))|F_t^y,\ u_x = u_i]P_i(t),\ t \in [0,T] \tag{10.31}$$

for any measurable function $\phi : \mathbb{R}^m \to \mathbb{R}$, and inserting $F(\zeta(t),u_x) = \chi_i(t)$ in equation (1.25), the following equations are obtained:

$$dP_i(t) = \gamma^{-2}P_i(t)(\hat{\zeta}^{(i)}(t) - \hat{\zeta}(t))'H'(dy(t) - H\hat{\zeta}(t)dt)$$

$$t \in (0,T),\ i=1,\dots,L, \tag{10.32}$$

where

$$\hat{\zeta}^{(i)}(t) = E[\zeta(t)|F_t^y\ ,\ u_x = u_i],\ t \in [0,T],\ i=1,\dots,L, \tag{10.33}$$

and, by using equation (10.31)

$$\hat{\zeta}(t) = \sum_{i=1}^{L} \hat{\zeta}^{(i)}(t)P_i(t),\ t \in [0,T]. \tag{10.34}$$

From the theory of Kalman-Bucy filtering it follows that $\hat{\zeta}^{(i)}$, $i=1,\dots,L$, are determined by

$$d\hat{\zeta}^{(i)}(t) = A\hat{\zeta}^{(i)}(t)dt + Bu_i dt + \gamma^{-2}P(t)H'(dy(t) - H\hat{\zeta}^{(i)}(t)dt)$$

$$t \in (0,T),\ \ i=1,\dots,L \tag{10.35}$$

$$\hat{\zeta}^{(i)}(0) = E\zeta(0)\ ,\ \ i=1,\dots,L$$

$$dP(t)/dt = AP(t) + P(t)A' + GG' - \gamma^{-2}P(t)H'HP(t),\ \ t \in (0,T)$$

$$P(0) = E[(\zeta(0) - E\zeta(0))(\zeta(0) - E\zeta(0))']. \tag{10.36}$$

The set of equations (10.32),(10.34) and (10.35)-(10.36) constitute the filter equations for computing $\hat{\zeta} = \{\hat{\zeta}(t),\ t \in [0,T]\}$. Note that $P_i(t)$, $\hat{\zeta}^{(i)}(t)$, $i=1,\ldots,L$ and $\hat{\zeta}(t)$ (equations (10.32),(10.34)-(10.35)) can be computed in parallel, while equations (10.36) can be solved off-line.

In this chapter (as in Section 10.1), instead of solving equations (10.36), the following set of equations is solved:

$$\begin{pmatrix} dX(t)/dt \\ \\ dY(t)/dt \end{pmatrix} = \begin{pmatrix} -A' & \gamma^{-2}H'H \\ \\ GG' & A \end{pmatrix} \begin{pmatrix} X(t) \\ \\ Y(t) \end{pmatrix} \quad t \in (0,T) \tag{10.37}$$

$$X(0) = I, \quad Y(0) = P(0) \tag{10.38}$$

$$P(t) = Y(t)X^{-1}(t) \quad , \quad t \in [0,T]. \tag{10.39}$$

Also of interest is the following estimator:

$$\hat{u}_x(t) \triangleq E[u_x|\mathcal{F}_t^y] = \sum_{i=1}^{L} u_i\, P_i(t) \;,\; t \in [0,T]. \tag{10.40}$$

Remark

In this Section we deal with the augmented Markov process $\{(\zeta(t),u_x),\ t \in [0,T]\}$. Hence, the filtering formula (1.25) remains valid if we there replace the expression $F(\zeta(t))$ by $F(\zeta(t),u_x)$.

10.2.3 Example

Numerical experimentation was carried out for the following set of para= meters: $\alpha = 0.4$, $a = 0.1$, $\sigma = 10$; $\gamma = 1, 10, 50$; $u_x = 100$, $L = 7$, $\{u_i\} = \{-150,-100,-50,0,50,100,150\}$, $\Delta = 0.005$, $N = 2000$, $\zeta(0) = (500, 160,0)'$, $\hat{\zeta}(0) = (700,150,0)'$ and

$$P(0) = \begin{pmatrix} 40\,000 & -2\,000 & 0 \\ -2\,000 & 100 & 0 \\ 0 & 0 & 0 \end{pmatrix}$$

Some extracts from the numerical results are presented in the following figures:

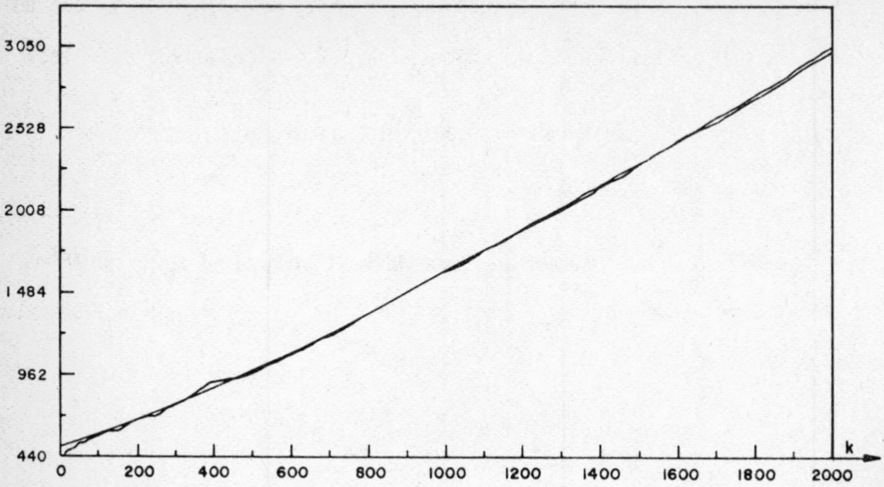

Fig.10.4-a: $\zeta_1(k)$ and $\hat{\zeta}_1(k)$ as functions of k for the case where $\gamma = 10$.

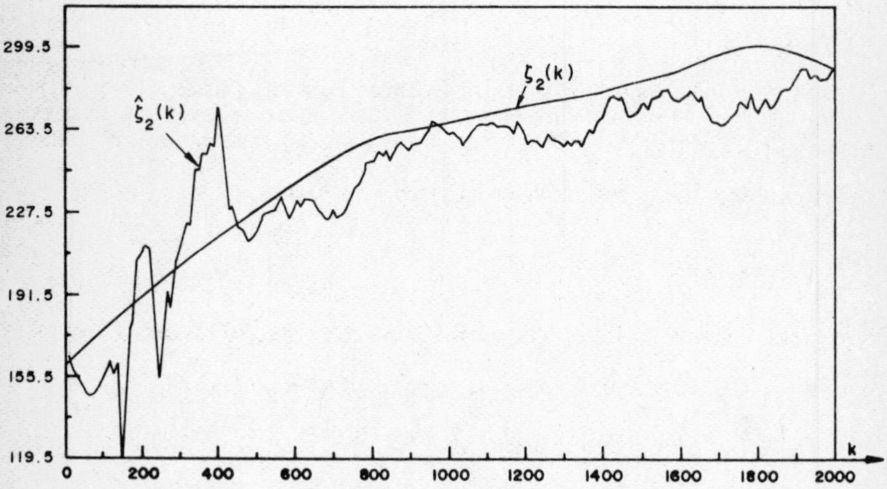

Fig.10.4-b: $\zeta_2(k)$ and $\hat{\zeta}_2(k)$ as functions of k for the case where $\gamma = 10$.

10.3 A DETECTION PROBLEM

Consider the linear system given by

$$dx(t) = A(t)x(t)dt + G(t)dW(t) \quad , \quad t \in (0,T);\ x(0) = x_o \tag{10.41}$$

$$dy(t) = \gamma H(t)x(t)dt + \Gamma(t)dv(t) \quad , \quad t \in (0,T);\ \ y(0) = 0, \tag{10.42}$$

where $A(t)$, $G(t)$, $H(t)$ and $\Gamma(t)$ are given matrices as described in Section 1.3. $W = \{W(t),\ t \geq 0\}$ and $V = \{v(t),\ t \geq 0\}$ are an $\mathbb{R}^m$-valued and an $\mathbb{R}^p$-valued standard Wiener process respectively, and x_o is an $\mathbb{R}^m$-valued Gaussian random element such that $E|x_o|^2 < \infty$. γ is a random variable such that

$$P(\gamma=1) = q \text{ and } P(\gamma=0) = 1-q, \tag{10.43}$$

where $q > 0$ is given. It is assumed that x_o,γ, W and V are mutually independent. Denote by $\zeta = \{\zeta(t) = (\zeta_1(t),\ldots,\zeta_m(t)),\ t \in\ [0,T]\}$ the solution to (10.41).

The minimum variance estimate of $(\zeta(t),\gamma)$, based on the measurements $Y^t = \{y(u),\ 0 \leq u \leq t\}$, is given by $\hat{\zeta}(t) = E[\zeta(t)|\mathcal{F}_t^y]$ and $\hat{\gamma}(t) = E[\gamma|\mathcal{F}_t^y]$. The problem of finding $\{\hat{\gamma}(t),\ t \in [0,T]\}$ is actually a detection problem. Here, one has to decide between the two possible observation processes

$$\text{(i)} \qquad y(t) = \int_0^t \Gamma(u)dv(u) \quad , \quad t \in [0,T] \tag{10.44}$$

and

$$\text{(ii)} \qquad y(t) = \int_0^t H(u)\zeta(u)du + \int_0^t \Gamma(u)dv(u) \quad , \quad t \in [0,T]. \tag{10.45}$$

A discrete in time version of the detection-estimation problem considered here is dealt with in Ref. [10.5], and a minimum mean-square error linear filter is derived. In this section, the minimum mean-square error filter (or, the minimum variance estimate) is derived for the system given by equations (10.41)-(10.42). Using this result, the equations for the minimum mean-square error filter for the discrete in time system, are easy to obtain. Define

$$P_i(t) = P(\gamma=i|\mathcal{F}_t^y) \quad , \quad t \in [0,T], \quad i=0,1. \tag{10.46}$$

By using the relation

$$E[\phi(\zeta(t))|\mathcal{F}_t^y] = \sum_{i=0}^{1} E[\phi(\zeta(t))|\mathcal{F}_t^y , \quad \gamma=i]P_i(t) \tag{10.47}$$

for any measurable function $\phi : \mathbb{R}^m \to \mathbb{R}$, and inserting $F(\zeta(t),\gamma) = \gamma$ in equation (1.25)(in this case $A_t\gamma \equiv 0$), we obtain the following equation

$$dP_1(t) = P_1(t)(1-P_1(t))(\hat{\zeta}^{(1)}(t))'H'(t)R^{-1}(t)(dy(t)-P_1(t)H(t)\hat{\zeta}^{(1)}(t)dt)$$
$$t \in (0,T) \tag{10.48}$$
$$P_1(0) = q,$$

where

$$\hat{\zeta}^{(i)}(t) \triangleq E[\zeta(t)|\mathcal{F}_t^y,\gamma=i], \quad t \in [0,T] \quad , \quad i=0,1 \tag{10.49}$$

$$R^{-1}(t) = (\Gamma^{-1}(t))'\Gamma^{-1}(t) \quad , \quad t \in [0,T] \quad , \tag{10.50}$$

and (as follows from equation (10.47))

$$\hat{\zeta}(t) = \sum_{i=0}^{1} \hat{\zeta}^{(i)}(t)P_i(t) = \hat{\zeta}^{(o)}(t)(1-P_1(t)) + \hat{\zeta}^{(1)}(t)P_1(t), \ t \in [0,T]. \tag{10.47'}$$

Note that $P_1(t) = \hat{\gamma}(t) \quad , \quad t \in [0,T]$.

$\hat{\zeta}^{(o)}(t)$ is the minimum variance estimate of $\zeta(t)$ based on Y^t in the case where

$$dy(t) = \Gamma(t)dv(t) \quad , \quad t \in (0,T) \tag{10.44'}$$

Hence, $\hat{\zeta}^{(o)}(t)$ is determined by

$$d\hat{\zeta}^{(o)}(t) = A(t)\hat{\zeta}^{(o)}(t)dt \quad , \quad t \in (0,T)$$
$$\tag{10.51}$$
$$\hat{\zeta}^{(o)}(0) = Ex_o.$$

$\hat{\zeta}^{(1)}(t)$ is the minimum variance estimate of $\zeta(t)$ based on Y^t in the case where

$$dy(t) = H(t)\zeta(t)dt + \Gamma(t)dv(t) \quad , \quad t \in (0,T). \tag{10.45'}$$

Hence, $\hat{\zeta}^{(1)}(t)$ is determined by the following Kalman-Bucy filter

$$d\hat{\zeta}^{(1)}(t) = A(t)\hat{\zeta}^{(1)}(t)dt + P(t)H'(t)R^{-1}(t)(dy(t)-H(t)\hat{\zeta}^{(1)}(t)dt),\ t \in (0,T) \tag{10.52}$$

$$\hat{\zeta}^{(1)}(0) = Ex_0$$

$$dP(t)/dt = A(t)P(t) + P(t)A'(t) + G(t)G'(t) - P(t)H'(t)R^{-1}(t)H(t)P(t),\ t \in (0,T) \tag{10.53}$$

$$P(0) = E[(x_0 - Ex_0)(x_0 - Ex_0)'].$$

In conclusion, Fig. 10.5 shows the block diagram of the minimum mean-square filter for the system given by equations (10.41)-(10.42).

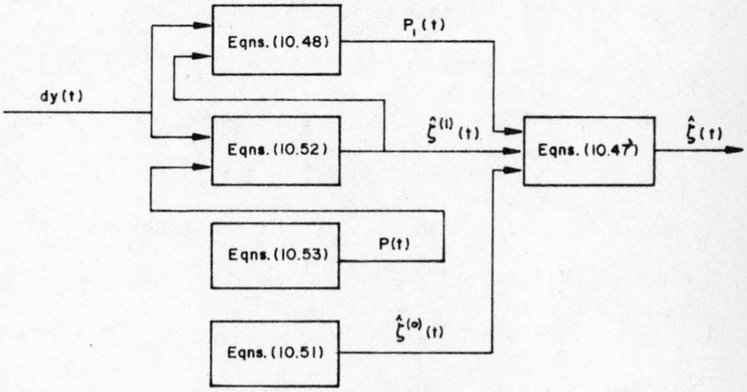

Fig.10.5: The block diagram of the optimal filter.

Remark:

In this section we deal with the augmented Markov process $\{(\zeta(t),\gamma),t \in [0,T]\}$. Hence, the filtering formula (1.25) remains valid if we there replace the expression $F(\zeta(t))$ by $F(\zeta(t),\gamma)$.

10.4 STATE AND COVARIANCE ESTIMATION

Consider the linear system given by

$$dx(t) = Ax(t)dt + GdW(t) \quad , \quad t \in (0,T);\ x(0) = x_o \tag{10.54}$$

$$dy(t) = Hx(t)dt + \Gamma dv(t) \quad , \quad t \in (0,T);\ y(0) = 0, \tag{10.55}$$

where A, H and Γ are given $m \times m$, $p \times m$ and $p \times p$ matrices respectively; $W = \{W(t),\ t \geq 0\}$ and $V = \{v(t),\ t \geq 0\}$ are an $\mathbb{R}^m$-valued and an $\mathbb{R}^p$-valued standard Wiener process respectively; x_o is an $\mathbb{R}^m$-valued Gaussian random element such that $E|x_o|^2 < \infty$, and G is an unknown $m \times m$ matrix. It is assumed that Γ^{-1} exists and that x_o, W and V are mutually independent.

Note that the covariance of the noise term in (10.54) is GG'dt. Denote by $\zeta = \{\zeta(t) = (\zeta_1(t),\ldots,\zeta_m(t)),\ t \in [0,T]\}$ the solution to (10.54). The minimum variance estimate of $(\zeta(t),G)$, based on the measurements $Y^t = \{y(u),\ 0 \leq u \leq t\}$, is given by $\hat{\zeta}(t) = E[\zeta(t)|\mathcal{F}_t^y]$ and $\hat{G}(t) = E[G|\mathcal{F}_t^y]$, where $\mathcal{F}_t^y = \sigma(y(u),\ 0 \leq u \leq t)$. The problem of constructing a filter for the computation of $\{(\hat{\zeta}(t),\ \hat{G}(t)),\ t \in [0,T]\}$ is a special case of more general problems, namely, adaptive filtering and identification of GG' and $\Gamma\Gamma'$ based on Y^t. In order to acquaint himself with these problems the reader is advised to consult Refs. [10.6 - 10.9] and the references cited there. Here, a different approach for an adaptive filter is suggested.

Suppose that $G \in \{G^{(1)},\ldots,G^{(L)}\}$, where $G^{(i)}$, $i = 1,\ldots,L$ are given $m \times m$ matrices. Define

$$\chi_i(t) \triangleq \begin{cases} 1 & \text{if } G = G^{(i)} \\ 0 & \text{otherwise;} \end{cases} \tag{10.56}$$

then

$$P_i(t) \triangleq E[\chi_i(t)|F_t^y] = P(G = G^{(i)}|F_t^y), \ t \in [0,T], \ i = 1,\dots,L. \tag{10.57}$$

By using the relation

$$E[\phi(\zeta(t))|F_t^y] = \sum_{i=1}^{L} E[\phi(\zeta(t))|F_t^y, \ G = G^{(i)}]P_i(t), \tag{10.58}$$

for any measurable function $\phi : \mathbb{R}^m \to \mathbb{R}$, and considering the augmented Markov process $\{(\zeta(t),G), \ t \in [0,T]\}$, and inserting $F(\zeta(t),G) = \chi_i(t)$ in equation (1.25), the following equations are obtained:

$$dP_i(t) = P_i(t)(\hat{\zeta}^{(i)}(t) - \hat{\zeta}(t))'H'R^{-1}(dy(t) - H\hat{\zeta}(t)dt) \tag{10.59}$$

$$t \in (0,T), \quad i = 1,\dots,L, \quad R^{-1} \triangleq (\Gamma^{-1})'\Gamma^{-1},$$

where

$$\hat{\zeta}^{(i)}(t) \triangleq E[\zeta(t)|F_t^y, \ G = G^{(i)}], \ t \in [0,T], \quad i = 1,\dots,L \tag{10.60}$$

and, by using (10.58),

$$\hat{\zeta}(t) = \sum_{i=1}^{L} \hat{\zeta}^{(i)}(t)P_i(t), \quad t \in [0,T]. \tag{10.61}$$

From the theory of Kalman-Bucy filtering it follows that $\hat{\zeta}^{(i)}$, $i = 1,\dots,L$ are determined by

$$\begin{cases} d\hat{\zeta}^{(i)}(t) = A\hat{\zeta}^{(i)}(t)dt + Q_i(t)H'R^{-1}(dy(t) - H\hat{\zeta}^{(i)}(t)dt), \ t \in (0,T) \\ \hat{\zeta}^{(i)}(0) = Ex_o, \end{cases} \tag{10.62}$$

$$i = 1,\dots,L$$

and

$$\begin{cases} dQ_i(t)/dt = AQ_i(t) + Q_i(t)A' + G^{(i)}(G^{(i)})' - Q_i(t)H'R^{-1}HQ_i(t), \ t \in (0,T) \\ Q_i(0) = E[(x_o - Ex_o)(x_o - Ex_o)'], \end{cases} \tag{10.63}$$

$$i = 1,\dots,L.$$

Also,

$$\hat{G}(t) = \sum_{i=1}^{L} G^{(i)}P_i(t), \quad t \in [0,T]. \tag{10.64}$$

The set of equations (10.59), (10.61) and (10.62)-(10.64) constitute the filter equations for computing $\{(\hat{\zeta}(t),\hat{G}(t)),\ t \in [0,T]\}$.

10.5 CONCLUSIONS

For each of the problems posed in Sections 10.1 - 10.4 equations for optimal (minimum variance) filters have been derived. In each of the cases these equations constitute adaptive, parallel processing filters. The results obtained in this chapter can serve as a starting point for constructing approximate adaptive, parallel processing filters for correspondingly more complicated problems.

10.6 REFERENCES

10.1 V.E. Benes and I. Karatzas, Estimation and control for linear, par= tially observable systems with non-Gaussian initial distribution, *Stochastic Processes and their Applications*, 14, pp 233-248, 1983.

10.2 A.H. Jazwinski, *Stochastic Processes and Filtering Theory*, Academic Press, New York, 1970.

10.3 R.L. Moose, H.F. Vanlandingham and D.H. McCabe, Modelling and esti= mation for tracking maneuvering targets, *IEEE Trans. on Aerospace and Electronic Systems*, 15, pp 448-456, 1979.

10.4 R.A. Singer, Estimating optimal tracking filter performance for manned maneuvering targets, *IEEE Trans. on Aerospace and Electronic Systems*, pp. 473-483, 1970.

10.5 N.E. Nahi, Optimal recursive estimation with uncertain observation, *IEEE Trans. on Information Theory*, IT-15, pp 457-462, 1969.

10.6 H.W. Brewer, Identification of the noise characteristics in a Kalman filter, *Control and Dynamic Systems*, Edited by C.T. Leondes, Vol. 12, pp. 491-581, Academic Press, New York, 1976.

10.7 R.F. Ohap and A.R. Stubberud, Adaptive minimum variance estimation in discrete-time linear systems, *Control and Dynamic Systems*, Edited by C.T. Leondes, Vol. 12, pp. 583-624, Academic Press, New York, 1976.

10.8 L. Chin, Advances in adaptive filtering, *Control and Dynamic Systems*, Edited by C.T. Leondes, Vol. 15, pp. 277-356, Academic Press, New York, 1979.

10.9 K. Ohnishi, Direct recursive estimation of noise statistics., *Control and Dynamic Systems*, Edited by C.T. Leondes, Vol. 16, pp. 249-297, Academic Press, New York, 1980.

Lecture Notes in Control and Information Sciences

Edited by A. V. Balakrishnan and M. Thoma

Vol. 43: Stochastic Differential Systems
Proceedings of the 2nd Bad Honnef Conference
of the SFB 72 of the DFG at the University of Bonn
June 28 – July 2, 1982
Edited by M. Kohlmann and N. Christopeit
XII, 377 pages. 1982.

Vol. 44: Analysis and Optimization of Systems
Proceedings of the Fifth International
Conference on Analysis and Optimization of Systems
Versailles. December 14–17, 1982
Edited by A. Bensoussan and J. L. Lions
XV, 987 pages, 1982

Vol. 45: M. Arató
Linear Stochastic Systems
with Constant Coefficients
A Statistical Approach
IX, 309 pages. 1982

Vol. 46: Time-Scale Modeling of Dynamic Networks
with Applications to Power Systems
Edited by J. H. Chow
X, 218 pages. 1982

Vol. 47: P. A. Ioannou, P. V. Kokotovic
Adaptive Systems with Reduced Models
V, 162 pages. 1983

Vol. 48: Yaakov Yavin
Feedback Strategies for Partially
Observable Stochastic Systems
VI, 233 pages, 1983

Vol. 49: Theory and Application of Random Fields
Proceedings of the IFIP-WG 7/1
Working Conference
held under the joint auspices of the
Indian Statistical Institute
Bangalore, India, January 1982
Edited by G. Kallianpur
VI. 290 pages. 1983

Vol. 50: M. Papageorgiou
Applications of Automatic Control Concepts
to Traffic Flow Modeling and Control
IX, 186 pages. 1983

Vol. 51: Z. Nahorski, H.F. Ravn, R.V.V. Vidal
Optimization of Discrete Time Systems
The Upper Boundary Approach
V, 137 pages 1983

Vol. 52: A. L. Dontchev
Perturbations, Approximations and Sensitivity Analysis
of Optimal Control Systems
IV, 158 pages. 1983

Vol. 53: Liu Chen Hui
General Decoupling Theory of Multi
Process Control Systems
XI, 474 pages. 1983

Vol. 54: Control Theory for Distribut
Parameter Systems and Application
Edited by F. Kappel, K. Kunisch,
W. Schappacher
VII, 245 pages. 1983.

Vol. 55: Ganti Prasada Rao
Piecewise Constant Orthogonal Fun
and Their Application to Systems an
VII, 254 pages. 1983.

Vol. 56: Dines Chandra Saha, Ganti
Identification of Continuous
Dynamical Systems
The Poisson Moment Functional
(PMF) Approach
IX, 158 pages. 1983.

Vol. 57: T. Söderström, P. G. Stoica
Instrumental Variable Methods
for System Identification
VII, 243 pages. 1983.

Vol. 58: Mathematical Theory of
Networks and Systems
Proceedings of the MTNS-83 Interna
Symposium
Beer Sheva, Israel, June 20–24, 198
Edited by P. A. Fuhrmann
X, 906 pages. 1984

Vol. 59: System Modelling and Optim
Proceedings of the 11th IFIP Confere
Copenhagen, Denmark, July 25-29, 1
Edited by P. Thoft-Christensen
IX, 892 pages. 1984

Vol. 60: Modelling and Performance
Evaluation Methodology
Proceedings of the International Sem
Paris, France, January 24–26, 1983
Edited by F. Bacelli and G. Fayolle
VII, 655 pages. 1984

Vol. 61: Filtering and Control of Rando
Processes
Proceedings of the E.N.S.T.-C.N.E.T. C
Paris, France, February 23–24, 1983
Edited by H. Korezlioglu, G. Mazziotto
J. Szpirglas
V, 325 pages. 1984